THE SECRET UNIVERSE

THE
SECRET UNIVERSE

AN EXPLORATION OF THE EXTRATERRESTRIAL DEITY HYPOTHESIS AND THE MYSTERIES OF THE UNIVERSE

The Esoteric Edition of The Genesis of Revelation

AERIK VONDENBURG

Twin Pillars Publishing

All rights reserved. Without limiting the rights under copyright reserved, no part of this publication may be reproduced, stored in or introduced into a retrieval system, or transmitted, in any form, or by any means (electronic, mechanical, photocopying, recording, or otherwise) without the prior written permission of both the copyright owner and the above publisher of this book.

Printed in the United States of America.

Twin Pillars Publishing

ISBN-13: 978-0-692-05455-0
ISBN-10: 0692054553

Copyright 2013
Published 2017

Scripture taken from:

THE EXPANDED BIBLE. Copyright© 2011 by Thomas Nelson, Inc. Used by permission. All rights reserved.

GOD'S WORD®,©1995 God's Word to the Nations ("Names of God"). Used by Permission of Baker Publishing Group.

THE HOLY BIBLE, NEW INTERNATIONAL VERSION®, NIV® Copyright© 1973, 1978, 1984, 2011 by Biblica, Inc.® Used by permission. All rights reserved worldwide.

WORLD ENGLISH BIBLE. Public Domain.

CONTENTS

Preface, 1
I. The Quest for Truth, 6
II. God and the Godhead, 11
III. The Inter-Worldly Reality, 47
IV. The Creation of Evolution, 127
V. The Tree of Knowledge, 168
VI. The Tree of Life, 175
VII. The Extra-Sensory Reality, 263
VIII. The Metaphysics of Hyperspace, 279
IX. The Serpent, 295
X. The Theocracy of Yahweh, 315
XI. The Theocracy of Jehovah and the Mission of the Savior, 405
XII. Epilogue, 501

This book is the Esoteric Edition of an earlier work by the author titled: *The Genesis of Revelation: Secrets of the Bible Revealed and a Case for Reformation.* The following pages can be passed over by those who have already read that edition: 324-351; 405-425; 435-515.

Dedicated to the memory of Giordano Bruno.

A NOTE FROM THE AUTHOR

Despite the controversial nature of the findings that are presented in this work, it is not my intention to denounce the primary principles of organized religion, nor any particular ethnic group. The information presented herein that does not conform to traditional Judeo-Christian doctrine pertains only to the mistaken interpretations of man.
　—Aerik Vondenburg

PREFACE

The long journey that lead me into the research and writing of this book began in childhood. It was during those impressionable years that I was first told that the world, and all living things in it, had been created by God. I remember being perplexed by the concept of such an eternal being. If God exists, and if "he" created the universe, where then did God come from? And, if God once literally appeared in person in ancient times, why then does he not do so now? I eventually realized that I would never receive a credible answer to these childhood questions.

It was in my teenage years that I began to disassociate myself from my religious upbringing as I became more aware of the logical deductions of science. However, by the time that I was in my twenties I came to realize that mainstream scientists harbored their own dogmatic views. For example, I knew that the secular physicalist[1] bias that they held against subjects pertaining to the paranormal was unjustifiable after I personally experienced paranormal phenomena. The most stunning of which was the sighting of a luminescent anthropomorphic energy form. These experiences were, at least for myself, empirical evidence of what countless witnesses have been reporting for millennia. I therefore came to realize that just because mainstream scientists have not yet discovered a way to apprehend the anomalous aspects of nature does not mean that such aspects do not exist.

It was also during that same period in my life that I witnessed light phenomena in the sky. The flat-shaped luminescent object that I saw moving outside of my bedroom window appeared one afternoon. The object was completely enveloped in an ethereal—as opposed to electrical—orange light. It moved silently in a steady horizontal

---

[1] Physicalism is a monistic philosophy that asserts that all reality is based on physical substance. It is a modern-age form of materialism. In this book, materialism refers to the interest in material items.

course, as if on some invisible track, before stopping. After holding completely still for a moment, it recommenced its course until it disappeared into a nearby cloud—which indicated to me that it was intelligently controlled. I am unsure if the object was cylinder or disk shaped. From my vantage point, I could only see that is was flat.

Although the experience was peaceful, indeed even sublime, when the object stopped I did become unnerved. This is because I had heard the stories of unwilling people who had been taken aboard such craft and supposedly examined by black-eyed alien beings. Indeed, the feeling that I got at the time is that the occupant or perhaps occupants of the object were seeking to make contact—possibly even with myself. I do not know for sure if it was a coincidence that the object departed after I began to be concerned, but that is what happened. I have since learned that the visitors who have no intention of abducting anyone against their will are not only extrasensory but sensitive to our fears and prejudices, and will usually retreat if they sense that a person is not receptive to them. I suspect that this is what happened.

Something else happened during the sighting that was so unusual that I usually omit it when recounting what happened because of how admittedly unbelievable it sounds. Nevertheless, I feel it necessary to submit a completely honest account. Before the object departed it moved into a nearby cloud. Approximately one minute later, after what appeared to be faint sun spots were seen moving around in the cloud in a spiraling motion, the entire cloud suddenly began to move, as if it had been activated into motion. It appeared to me as if whatever had moved into the cloud was, at that moment, using the thick cumulus to conceal itself as it moved over the populated suburban region that I was living in at the time. I later tried to rationalize what I had witnessed by assuming that a gust of wind could have suddenly pushed the cloud; however, the other clouds that were not far away did not seem to be effected by such a gust. It also seems unlikely that it could have been coincidence that the cloud behaved unusually after an Unidentified Flying Object happened to fly into it. I have therefore come to dismiss the gust of wind explanation.

At that time, I had never heard of anyone else reporting such UFO-cloud phenomena, and I had no intention of being the first. However, I later discovered during the research of this book that others have had similar experiences. One such sighting was recounted in Terence McKenna's[2] book *True Hallucinations*, in which he too reported seeing a flying saucer exerting some type of elemental influence on a cloud. Another more recent instance was captured on a cell-phone video by a resident of Philadelphia in 2015 (Speigal 2015). What appears in the video is a rapidly moving and direction-changing flying cloud that appears to be intelligently controlled. The man who shot the video (Hector Garcia) claimed that he could also see lights inside the cloud[3]. I have been unable to uncover any information that this incident, and others like it that have been posted on video-sharing websites, are all nothing but hoaxes or misidentifications. Moreover, in the case of my own encounter, I am certain that the object that I witnessed was not a misidentification, and that it clearly operated independently of my own cognition.

It therefore seems evident that somebody or perhaps something is using clouds (that may perhaps be artificially generated?) to conceal themselves as they traverse over heavily populated regions. The reason for this may either be to reduce the chances of inciting mass hysteria, or perhaps to avoid having to deal with the fighter jets that are sometimes dispatched to intercept them—which has indeed occurred in the past[4].

---

[2] Terence McKenna was an ethnobotonist, philosopher, shaman, and author of the books: *The Invisible Landscape, The Archaic Revival, Food of the Gods, and True Hallucinations.*

[3] www.huffingtonpost.com/2015/07/08/shape-changing-ufo-philadelphia_n_7749204.html

[4] One exceptional example is the 1976 Tehran UFO incident. The presence of a UFO in Iranian airspace was reported by several witnesses and confirmed by radar. The two fighter jets that were dispatched to engage the UFO both lost power as they approached the object. Power was restored only after the F-4s disengaged. U.S. Defense Intelligence Agency documents related to this incident were released under the

After that experience, I debated whether or not I should report what had happened. However, in those days I did not feel that I knew enough about the UFO/extraterrestrial situation, and was unsure of what kind of drama that I would be getting myself into by going public. Therefore, I kept what had happened to myself and only a few friends who I thought might be open to it (I suspect that many others are doing the same). However, I eventually did contact a UFO investigation organization years later—if my memory is correct it was MUFON[5]—not to report my sighting, but to see if anyone else had seen it. What I learned is that if anyone else did see it, they too did not report it; or, perhaps, no one had seen it because the craft was concealed inside the cloud.

The impetus that compelled me to research and write these books was a need for understanding. After the years of research that I devoted myself to after that experience, I now understand that these types of events have been occurring for centuries. Although, the ancient people did not attribute such experiences to extraterrestrials, but rather to the gods. Indeed, reports of otherworldly beings in the skies who once presided over the affairs of human-kind can be found all throughout the ancient texts—including the Bible.

I eventually realized that the only way to truly understand what had happened, and what was continuing to happen, was to not only read the ancient records for myself but to re-read the Bible without any preconceived interpretations. However, actually doing so proved to be more difficult than expected. This is because breaking through common traditional perspectives that have been so successfully ingrained into the collective mentality of the world over the course of thousands of years is no casual task. No doubt the reader of this book will face a similar challenge.

What I also discovered in the Bible were mysterious references to "secret" matters. I learned that some of these secrets were known to

---

Freedom of Information Act and is posted (along with many other previously classified documents) at the website: www.theblackvault.com/casefiles/the-1976-iran-incident/

[5] MUFON stands for the Mutual UFO Network. It is the largest and oldest civilian UFO investigation organization in the USA.

some of the earliest Christians but have since been lost. Of course, I felt compelled to inquire if the hidden matters could have anything to do with the extraterrestrial question; and if not, what other explanations might be possible.

Over the course of my investigation, it became apparent that some of the previous findings that were submitted by others in this field (e.g., Erich von Daniken and Zecheria Sitchin) were not accurate and in need of reform. What also happened is that the discoveries eventually extended past the extraterrestrial question; and yet everything was undeniably connected as well.

CHAPTER I

THE QUEST FOR TRUTH

THE technological achievements that have been conceived by the engineers of our modern world has led humanity into an Age of Information. In this novel period in human history, just about everything that has ever been written or spoken has become available at the touch of a button. However, when looking back through the annals of the past it is possible to find cases where the free flow of information has not only been unavailable but even purposely suppressed. Despite the progress that has been made in recent times, the legacy of this problem continues to affect our world to this day.

There are essentially three types of information controllers. The first is the despot. The despot intentionally withholds information in order to prevent others from becoming informed, and thus empowered. The second is the protector. The protector withholds information that might disturb the peace of the status quo (i.e., the "vital lie"). The third type of controller is the sage. The sage withholds information in order to protect its sacred integrity from the immature. The sage only disseminates esoteric information to those who he or she deems worthy and capable of comprehending and accepting it. I propose that it is mostly because of these three factors, along with both innocent misinterpretations and information that is simply unavailable to anyone, that the truth has not been made more available to the public.

Of course, without access to the facts misunderstandings inevitably arise. This axiom is no more true than in matters that pertain to the Bible. The thesis that is presented in this book not only explains how these misunderstandings are related to the "hidden" "secret" matters that are referred to in the Bible but will also present a case for full disclosure and reform.

* * *

A passage in the New Testament reports that the truth shall set us free (John 8.32); which raises a deeper question that has long challenged philosophers throughout the ages; namely, what is the truth?

When evaluating the validity of information using an epistemological method, an investigator must consider how the information was obtained in the first place. For example, was it acquired directly from personal—i.e., empirical—experience (*a posteriori*/"knowledge by acquaintance")—or indirectly (*a priori*/ "knowledge by description")? This is a significant factor, especially in the case of an indirect situation, since the disposition of an individual may effect how the information is interpreted. An individual's personal bias may influence the interpretation of the data toward a direction that is in accordance with that individual's pre-established belief system.

The theory of social constructivism suggests that what most people recognize as the truth is actually a fabrication (i.e., a hyperreal simulation)[6] that has been formulated by an authority figure or institution. We will see in parts to come how this practice has led to the greatest degree of misconception, since it leaves open the possibility of the institution, and/or the individual, of effecting the information with either mistaken or in some cases even intentionally misleading conceptions, before passing along that interpretation to the audience which it administers to. The challenge of the investigator is to search beyond such extraneous prescriptions, in order to locate the

---

[6] For or more on this subject see: *Simulacra and Simulation* by Jean Baudrillard. (University of Michigan Press, 1994).

original root source from which the information was initially derived (i.e., *advaita*).[7]

The investigator must begin this inquiry by first clearing his or her consciousness from all previously established interpretations—most of which were formulated in less favorable times—and approach the data with a clear state of mind (i.e., *tabula rasa*). The foundation of this examination should begin with the premise that it is best not to be too certain of anything. It is said, for instance, that what made Socrates the wisest man in ancient Greece is that he was the only man who *knew* that he knew nothing. Likewise, it is only when we have freed ourselves from all affected influences that we can be more open to that which simply *is*;[8] and that our state of mind should not be confined to any one particular scientific or religious interpretation—and yet, this investigation will be taking the perspectives of both of these two rival establishments into consideration (which is in accord with the thesis-antithesis-synthesis dialectic).[9]

Another means by which a proposed truth can be evaluated is by comparing it with other information that is related to it. When put under a greater contextual lens, the information may take on an entirely new meaning. For example, we will see that when biblical information is compared to corresponding texts that have been uncovered in the region that used to be Mesopotamia, a surprising new and revolutionary picture of that ancient world, and the religious institutions that developed around it, arises. In this case, information receives greater meaning, clarity, and credibility when it is congruous with other data that is related to it. This type of inferential contextualism is referred to as coherence theory (i.e., coherentism).

When attempting to understand information that is related to the mythological and religious texts of the world, it is also necessary to

---

[7] In the Hindu tradition, *advaita* (i.e., *uttara mimamsa*) refers to the distinction between reality and appearance.

[8] This is also the type of process employed by the revolutionary philosopher, Rene Descartes.

[9] This is the type of Socratic method that was espoused by the influential eighteenth- and nineteenth-century German philosopher, G.W.F. Hegel.

consider the influence of subjective preconception in the translation process. This issue is especially significant in the case of translators who are in some way influenced, either consciously or subconsciously, by preexisting and possibly mistaken interpretations.

When it comes to the books of the Bible, it will be shown that the common picture that the public has been provided that pertains to the characters, events, and doctrines that are reported in these books, has been influenced by numerous fallacious factors that have obscured the original truth. We will find that once the outer layers of devotional veneer are stripped away, what is left is a picture of a world that is significantly different than the one that has been provided to us by the institutions of man.

To begin with, we must free our consciousness from the artificial projections that have been cast onto the wall of Plato's cave—as described in the "Allegory of the Cave" in Plato's work: The Republic.[10] This classic analogy represents the archetypal struggle between delusion and enlightenment. It is an experience that is also reminiscent of the ancient story of the young man Siddhartha Gautama, who set out on a quest for greater understanding, and who eventually achieved both inner peace and a higher mental perspective. Indeed, it seems that the enlightened perspective is revealed in the unattached and meditative state; a state that is free from the distracting noise of every-day life and the afflicting misconceptions of man and the institutions of his construction. The world of delusion that the Buddha triumphed over is referred to in the Eastern religious tradition as the "veil" (*maya*) of the devil (*mara*).

---

[10] Although it is common practice to italicize the titles of books, this rule does not apply to specific classic and lengthy religious works, such as the Torah and the Bible. One of the objectives of this work is to encourage the impartial evaluation of information in order to formulate more accurate deductions. This aim is made more difficult if certain rules, and thus perceptions, apply to some texts but not to others. Therefore, for the purpose of fostering a more objective perspective, all works will be treated equally. Accordingly, only works outside of the ancient era will be italicized.

Few cross over the river. Most are stranded on this side. On the riverbank they run up and down.
—The Buddha, The Dhammapada

But small is the gate and narrow the road that leads to life, and only a few find it.
—Jesus, Matthew 7.14

* * *

Throughout this book, both inductive hypothesis and deductive conclusions will be presented. Moreover, two distinct dialectical theories will also be considered. The first is what can be referred to as maximalist theory, which posits that although the ancient records (especially the Bible) have obviously been affected by devotional embellishment, the reports are nevertheless based on actual people and events. The second is the minimalist approach, which holds that the ancient records are more fictional rather than factual accounts. I believe that both perspectives deserve to be considered; although, unlike mainstream scholarly minimalism, the minimalist exegesis that is considered in this work maintains the premise that the ancient people were not simply fabricating stories, philosophies, and religions entirely from out of nothing. I make the case that at least primary segments of the accounts were based on genuine source experiences that were expanded and embellished throughout the years.

In regard to the maximalist interpretation: even if the historicity of the antiquarian accounts were to one day be conclusively discredited, I maintain that this does not diminish the value that the following findings that directly relate to the ancient texts provide to the fields of mythology and religion.

CHAPTER II

GOD AND THE GODHEAD

According to current cosmological calculations, the universe was created by an explosion of primordial energy that erupted over 13 billion years ago. Exactly where this energy came from, and what catalyst sparked the so-called "big-bang,"[11] remains a mystery—or at least, unproven.

Of course, the explanation that is offered by many of the religions of the world attributes the formation of the universe to a supreme being; although, where such an entity itself could have originated remains just as much of a mystery. The fundamental issue of this matter reduces down to the differences between the religious doctrines of creationism (i.e., "intelligent design") and the scientific theory of evolution.

Dealing with the question of the existence of God has been a kind of thorn-in-the-side of science for centuries. There would seem to be no evidence whatsoever that would support the belief in the existence of a supernatural creator being; especially the deity of the biblical

---

[11] The big-bang theory is the most prominent and accepted of all theories at this time. Examples of evidence for this event exists in the form of an expanding red-shifting universe, and remnant background microwave radiation. The framework for the big-bang theory is also consistent with Einstein's general relativity, which has proven to be true. Furthermore, the discovery of the Higgs-boson particle also indicates that the universe is unstable and finite, which also concurs with the big-bang paradigm.

tradition who allegedly appeared in person in the ancient Near East.

In some instances, the deity was said to have appeared in physical humanoid form. (More on this subject later.) However, it seems that the being that the preservers of this tradition speak of in the present era is not so much of a physical manifestation, but rather more of an omniscient and invisible energy system. For the rational observer, such an ineffable abstraction would seem to be implausible, since it posits that God, who once existed in a literal anthropomorphic form in ancient times has inexplicably withdrawn himself from this world, never to be known in such personal terms ever again. How could this be? If such a supernatural creator being exists, why then would "he" now conceal himself? A rational person would have every reason to be more than just a little skeptical of this supernatural and inconsistent scenario. If there is any truth to this claim, then the applications of science should be able to verify it.

In this case, does such evidence exist?

* * *

We must begin this investigation by turning our attention back to the primary traditional source of popular conceptions of God and the birth of the universe: the Hebrew Bible.

Atheist claim that the Hebrew deity Yahweh/Jehovah was just one of the many fictitious gods of the primitive ancient world, and that the Bible was written by priest-scribes who were motivated to construct a mythology that would help create and support their own careers.

It is true that the Old Testament was written by priest-scribes from both the northern kingdom of Israel and especially the southern kingdom of Judah. Some of their accounts were based on Semitic and Mesopotamian oral folktales and texts (Blenkinsopp 1992: 93-94; Coogan 1978: 20-21, 77-80; Finkelstein and Silberman 2001: 281; Greenstein 2007: 56; Kramer 1963: 148-149, 290-299; MacDonald 1988: 17; Van Seters 1992: 3, 22; 1999: 114, 117), as well as their own written records (e.g., the Book of Generations), which they then modified to make adhere to Yahwehist[12] creed.

It is also true that at least some of the stories of the Bible do not agree with the archaeological evidence. For example, archaeological data indicates that the destruction of Jericho that is described in the Bible never happened (Finkelstein and Silberman 2001: 81-82)—or at least did not occur within the time-frame of traditional biblical chronology (Bruins and Van Der Plicht 1995: 218). Furthermore, there is also no evidence of a four-decade long exodus of approximately two million Israelites from out of Egypt; nor is there an Egyptian record of plagues of frogs or locusts, or a pharaoh or his men drowning in the Red Sea—or even the Reed Sea. It must also be considered that the absence of approximately two million workers—which is what the biblical report indicates—would have been devastating to the Egyptian economy. Such an event should have turned up in the archaeological record as well; and yet it does not. There is also no report in the recovered Egyptian records of two million people moving past the Egyptian garrison check-points that were stationed all around the region. Biblical scholar Richard Elliott Friedman presents a strong case in his book *Who Wrote the Bible?* that posits that *if* there was an exodus it would have involved only *one* of the tribes of Israel—most likely the Levites. This not only explains why this tribe is reported being especially involved in the exodus story and the lack of archaeological evidence but it also explains why some of the Levites had Egyptian names (e.g., Hophni, Phinehas, and Moses). (More on this subject later.)

However, it must also be acknowledged that other evidence exists that does corroborate the biblical account. For example, not only do the biblical enemies of the Judean Israelites (e.g., Merodach, Ashurbanipal, Cyrus, Darius, and Nebuchadnezzar) appear in the Assyrian and Babylonian records but the Judean Israelite kings, such as Ahab (Rainey 2001: 146-147), Ahaz (Torrey 1940: 27-28),

---

[12] The term Yahwehist will be used in this work to denote any follower of the deity Yahweh. This term should not be conflated with the term "Jahwist," which is used by scholars to denote the biblical author who is referred to as J source.

Hezekiah (Pritchard 1969: 287-288), Hoshea, Josiah (Pritchard 1969: 284, 568-569), and King David (Finkelstein and Mazar 2007: 14), also appear in non-biblical ancient records. Furthermore, in the case of the Mesha stele, we have a Moabite report that parallels the account that is recorded in 2 Kings 3. What is also significant about the Mesha stele is that Yahweh is also specifically referred to (Cross 1997: 61). Likewise, many of the towns and cities that are reported in the Bible have also been discovered. These are only a few of the many examples. It must therefore be concluded that the biblical record cannot be entirely disregarded.

Eventually, over time, the truth of what really happened all those centuries ago became obscured for two primary reasons: (1) The stories were initially passed on orally before being written down; in many cases, centuries after the time that they allegedly occurred. Of course, over time the accounts naturally lost veracity. (2) The accounts also became less accurate due to the influence of devotional embellishment. Indeed, it must be understood that it was the priest-scribe's motivation to aggrandize the record for two primary reasons: (1) To supposedly avert disaster for Israel and Judah by convincing the people that Yahweh was worthy to be both feared and venerated above all others. (2) Because the personal status and livelihoods of the priests depended on people believing their message. The books of the Torah, and eventually the entire Tanakh (i.e., the Old Testament/Hebrew Bible), were the priest's way of impelling an impression upon the laity, and thereby instilling a tradition of devotion, not only to Yahweh but to themselves as Yahweh's agents on Earth. It is therefore likely that behind all of the inaccurate legends there lies only a remnant core of unadulterated truth, and that it is from this minute authentic source that entire world institutions of historic significance have emerged.

What the findings reveal is that the ancient people were not simply fabricating stories and gods purely from their imaginations, but rather they were basing their accounts on something.

* * *

According to the Old Testament/Tanakh, Yahweh vehemently denied the relevance, although not necessarily the existence, of the other gods. Indeed, the scriptures also report that he was not alone. For example, in passages that are traditionally ascribed to Yahweh, we find the following, in which the deity (i.e., "Elohim") refers to himself in plural terms:

> Then Elohim said, "Let us make humans in our image, in our likeness [. . .]"
> —Genesis 1.26
>
> Then Yahweh Elohim[13] said, "The man has become like one of us [. . .]"
> —Genesis 3.22

The Hebrew word *Elohim* is usually translated as "God"; however, this is not the original definition. In many cases, the singular form for the word God does not appear in the text—which would have been El, or Eloah. Instead, the plural term Elohim is used.[14] Indeed, this definition explains the plural pronouns that exist alongside it.[15]

---

[13] Here we find an example of a Yahwehist editor/scribe (i.e., the "Redactor") combining YHWH (J source) with the Elohim (P source).

[14] Or *Elim* (Exodus 15.11, Psalm 895-7, 29.1).

[15] In the Amarna letters (i.e., El-Amarna tablets), the plural term for gods also appears to be used in a singular context. The Amarna letters are a correspondence between Egypt and its representatives in Canaan and the Amorite kingdom of Amurru (in north Lebanon and Syria) that date back to around the fourteenth century BCE. The plural form of God that is found in this text would seem to indicate that mixing singular and plural terms was a common practice during that era; however, the correspondences were actually written by the governors of the Pharaoh's empire in Syria-Palestine, which was a society that was in the process of evolving into the Yahwehist Hebrew culture and language during that time, and may be influenced by that emerging religious doctrine and linguistics. Indeed, the biblical sites of "Megiddo" and "Jerusalem" are specifically referred to in the letters (Matthews and Benjamin 2006: 147-149). Indeed, it is a Canaanite who purportedly addresses the Pharaoh as

One of the interpretations that is commonly used by devotees of the Judeo-Christian tradition to explain this contradiction to monotheism is that the plural terms were used in order to convey supernatural omnipresence, which is also known as the so-called "majestic," "royal we," or "triune" *pluralis majestatis*. This is sometimes also related to the concept of the Christian trinity of God, which is the Father, the Son, and the Holy Spirit. However, according to the Bible itself, this interpretation cannot always be justified.

The subject of the Elohim is complex. This is because the word can refer to different types of entities. For example, the word may not only refer to Yahweh (Genesis 24.12, etc.) but to any supernatural being, including spirits. In 1 Samuel 28.13, for example, it is reported that an "Elohim" was seen coming up out of the Earth. However, this otherworldly being is not Yahweh, nor is it the Son or the Holy Spirit, but rather the spirit of the deceased prophet, Samuel. The word Elohim may not only refer to members of the council assembly of supernatural beings (Psalm 82.1) but to foreign gods as well (1 Kings 11.33). Indeed, in Psalm 82 Yahweh reprimands the other Elohim as he pronounces "judgment among the gods":

> I said, "You are gods [Elohim]. You are all sons of the Most

---

"my gods, my sun-god," not the Pharaoh (Pope 1955: 21). Therefore, the plural form for God that is presented in a singular context—*if* it is not a mistake committed by the scribe—could be influenced by the same practice that is found in the Hebrew Bible, and therefore does not represent the common beliefs of others. Furthermore, it cannot be ruled out that the Canaanite who uttered those words was actually calling out to the plural gods alongside the singular "sun god." Moreover, such instances are uncertain aberrations and do not constitute evidence of "the plural majesty" in every case (Pope 1955: 6, 21). We must also consider other Levantine records, such as the Ugaritic texts (as well as the Mesopotamian records) in which the plural *ilm* was used in the context of a pantheon of gods (Pope 1955: 6). Indeed, elsewhere in the Amarna letters itself, the word "gods" is accompanied by plural definite articles and pronouns (Moran 1992: 64). Moreover, the plural pronouns that exist alongside the word Elohim, as well as the passages that describe the Elohim as foreign gods in the Bible (1 Kings 11:33) clearly negates a singular only interpretation.

High [Elyon]."[16]
—Yahweh, Psalm 82.6

It is therefore obvious that Yahweh was not addressing the Son and the Holy Spirit. Furthermore, these other gods seem to be more than mere inanimate idols. Indeed, according to the book of Jeremiah, the other foreign Elohim, such as Chemosh, Milcom, and Bel, were all capable of being "exiled," "confounded," and "punished." These examples not only negate the majesty theory but also indicate the plurality of the Elohim. What will also be made apparent further ahead in this exposition, is that the Elohim of the older Genesis account cannot be compared to the heavenly Father, Son, and Holy Spirit of the New Testament. This is because they were not the same. Moreover, it must be remembered that the Hebrew Yahwehists did not even believe in the Trinity. Therefore, this explanation does not account for the appearance of the singular term before the Christian era.

Another interpretation that is used to explain the existence of not only the plural noun but the plural pronouns that coincide alongside it, is that Yahweh was talking among an assembly of angels; however, there are reasons to believe that this is also not accurate. The word angel is derived from the Greek word *angelos*, which itself derives from the original Hebrew word *mal'ak*. It is a word that simply means messenger (Meier 1999: 45). Although angels are commonly thought of as beautiful white-robed and haloed humanoid beings with wings, this is not the original characterization.

It is first necessary to differentiate between the mal'ak and the attendants of the Elohim who are sometimes more specifically referred to in the Bible as the seraphim and the cherubim. What is not commonly understood is that these were different types of beings (Meier 1999: 47). It is commonly believed by the laity that cherubs

---

[16] Note: Four different translations of the Bible are referred to in this work. In some versions, where the original Hebrew word (e.g., Elohim, Yahweh, etc.) is not transcribed into English, the original Hebrew (and Greek) appellation will be cited in brackets.

are baby angels with wings; however, these characterizations are actually influenced by the later Hellenistic Eros (i.e., Cupid) myths; as well as the subsequent Renaissance era *putti/putto* motif. The word *cherub/cherubim* derives from the Assyrian word *karabu*, the Babylonian *karubu*, which stem from the earlier Akkadian *kuribu* (Mettinger: 1999: 190). These words denote an entity that is mighty and propitious.[17] The cherubim were not angels (i.e., messengers), they were guardian genii. Visual representations of these mythic beings appear in Mesopotamian sculptures—which relate to creatures that were referred to as the shedu and the lamassu. These creatures were originally only sculptures that were stationed at the gateways of the gods and kings. Indeed, these are the same type of imposing figures that allegedly stood guard at the entrance-way of the garden of Eden (Genesis 3.24). However, this account of a cherub in the book of Genesis cannot be submitted as evidence of a literal biological or spiritual entity. This is especially due to the fact that this particular biblical book, that attempts to describe prehistoric events that no scribe was actually present to witness, contains other symbolic, as opposed to literal references, as well (which will be examined later). What must be understood about the cherubs is that they were originally only artistic images; just like the half eagle, half lion image of the Egyptian, Persian, and Grecian griffin. It was at a later time that these creatures were brought to life by the priest-scribes in order to impress upon the reader the supernatural nature of Yahweh— which, for example, is precisely what is happening in the book of Ezekiel. Scholars have discovered many instances in the biblical texts of what is referred to as "pious fraud" (Friedman 1987: 102). Indeed, the book of Ezekiel is especially unreliable and cannot be considered to be an accurate report of what literally occurred. (More on this subject later.)

Likewise, in the book of Isaiah (6.1-8) the *seraphim* are described as six-winged humanoid beasts (A similar type of description is

---

[17] "Jewish Encyclopedia: cherub," Jewishencyclopedia.com, 2002-2011, original publication 1906. Accessed 3-18-14. www.jewishencyclopedia.com/articles /4311-cherub

reported in Revelation 4.6-8). However, what is not commonly known is that the seraphim[18] were modeled after another Mesopotamian half-man half-beast winged mythological creature—the identity of this creature will be revealed further ahead. Moreover, the non-humanoid characteristics of these mythological beings conflicts with biblical information that refer to the humanoid appearance of both the angels and the Elohim. It must also be understood that these chimerical creations never even literally existed.

The Hebrew Bible reports that Lord Yahweh was attended to by a plurality of heterogeneous beings. In Genesis 19, the angels of the Lord are called "men," and are described in anthropomorphic terms. In this case, could these humanoid angels be the Elohim? However, these beings are not called Elohim, but rather they are specifically referred to as angels. However, it is possible that these servant emissaries are of the same genus as beings who are referred to elsewhere in the Old Testament/Tanakh as the *bene (ha) Elohim* (Genesis 6.2; Job 1.6, 38.7, etc.),[19] which literally translates as "sons of gods." Indeed, in Genesis 6, the "sons of the Elohim" are described as physical humanoid beings. These were most likely the same individuals who were either the fathers of the hybrid demi-gods—who were referred to in the book of Genesis as the "Nephilim" (Genesis 6.4)—or were themselves genetic members of that hybrid race. (More on this subject later.) This group might also be related to the Rephaim/Rephaites[20] (Deuteronomy 2.10-11; 3.11), the "sons of

---

[18] In many biblical translations, the word seraphim is translated as "burning serpent," which is how it appears in Numbers 21.6 and Deuteronomy 8.15. However, the word seraph can only be translated as such when it is accompanied by the word for serpent (*nachash*). In this case, the burning adjective may be a reference to the burning sensation of the serpent's poison.

[19] Or the sons of Elyon (*bene elyon*) (Psalm 82.6).

[20] In some instances, the Rephaim were associated with spirit beings in the netherworld (Rouillard 1999: 692, 696, 698). In other instances, the designation refers to the mysterious early era inhabitants of Palestine, southern Syria, Transjordania (Rouillard 1999: 697-698). In these instances, these beings were described as "giants."

Anak" (Numbers 13.33), and a group who are referred to in the Ugaritic texts as the Healers[21] (i.e., "the divine ones") (Coogan 1978: 48-50). However, according to the biblical scriptures, none of these archaic beings were equal with the Elohim, but rather were subordinate descendants. In this case, it is difficult to believe that the sentence uttered by the Elohim: "behold the man has become like one of us" (Genesis 3.22), would have applied to beings who were not equal to the Elohim. The only way that this issue can be partially excused is if we suppose that this was a reference to race, rather than to rank.

Collective texts unearthed from the ancient sites of Sumer, Akkad, Assyria, Babylon and others, tell us that the ancient city-states of Mesopotamia were presided over by a council of divine beings who the people regarded as gods. The so-called "idols" of gods that are referred to in the biblical record were symbolic representations of those living beings. Therefore, the people did not worship the idols themselves, but rather what those idols represented. The presence of these other divine beings explains why Yahweh was "jealous" of the other gods (Exodus 34.14). Indeed, it is difficult to believe that Yahweh would have been envious of inanimate objects.

It is commonly believed that Yahweh denied the existence of the other gods; however, this is not true. Yahweh only commanded that the other gods should not be considered as either superior or equal to him. This is the meaning of the words "before" and "with" in the following statements:

> You shall have no other gods [Elohim] before me.
> —Exodus 20.3
>
> See now that I my self am he. There is no god [Elohim] with me.
> —Deuteronomy 32.39

---

[21] The word "Healers" is translated from the Ugaritic word *rp'um*. This is most likely related to the Hebrew word, *rephaim* (Rouillard 1999: 692, 699).

What Yahweh demanded is that he was to be regarded as the supreme head of the divine council of the Elohim. In fact, this assembly of the gods is even mentioned in the Bible itself!

> God [Elohim] presides in the great assembly; he renders judgment among the "gods" [Elohim].
> —Psalm 82.1

This very same council of gods is also referred to in the texts of the Judean/Israelite neighbors in the region. For example, in the Ugaritic Baal Cycle tablets, this assembly is said to be presided over by the elder god, El.

In this case, could the ancient gods of Mesopotamia, Egypt, and the Indus Valley civilizations—as well as perhaps other civilizations from around the world—have been the very same beings who are described in the Old Testament as the Elohim? According to the following biblical verse, the answer to this question is an unequivocal yes:

> I will do this because they have forsaken me and worshiped Ashtoreth, the goddess [Elohim] of the Sidonians, Chemosh the god [Elohim] of the Moabites, and Molech the god [Elohim] of the Ammonites [...]
> —Yahweh, 1 Kings 11.33

It is therefore most likely not coincidence that the time-span of the biblical Elohim coincides precisely with the time-span of the ancient gods of Mesopotamia and the Levant. Indeed, what will become apparent as we delve deeper into this mystery, is that the reason for the similarities between Yahweh and the other gods of the same era and region is because they derived from the same source.

* * *

The books of the Old Testament/Tanakh were written and compiled by many hands over many years. Each of these scribes and redactors added a point of view that was prevalent during their time, or with their particular group, or by the author/redactor himself. Consequently, the books of the Bible are not entirely free from contradictions. For example, in 1 Samuel 28.5 Saul "prayed to Yahweh, but Yahweh did not answer him." However, in 1 Chronicles 10.13-14 Saul is killed because he did not pray to Yahweh. In Samuel 31.4 and 1 Chronicles 10.4, Saul kills himself; but in 2 Samuel 1.5-10 Saul is killed by an Amalekite. However, in 2 Samuel 21.12 it was the Philistines who killed Saul. In 2 Samuel 24.1 "Yahweh" causes David to number the people (i.e., take a census); but in 1 Chronicles 21.1 it is "Satan" who causes David to number the people. In Numbers 23.19, we are told that God never changes his mind; but in Jeremiah 42.10 God changes his mind.[22] These are only a few examples of some of the many inconsistencies that appear throughout the Bible. Even more contradictions will be examined later.[23]

Most scholars generally agree that the books of the Torah were written and compiled by at least four primary sources, those being: the Elohists, the Jahwehists, the Priestly source, and the Deuteronomists. Although what is known as the "documentary hypothesis" has been called in to question in recent years, the basic premise that the books of the Old Testament/Tanakh were written, compiled, and rewritten by numerous writers and redactors over a period of centuries remains the prevailing explanation. Furthermore, critics of this hypothesis are not only usually associated in some way with the Judeo-Christian institution, and therefore harbor ideological differences with the very concept that it presents, but they have been unable to provide a plausible alternative (Blenkinsopp 1992: viii; Friedman 2003: 1; Smith 2002: xxiii).

---

[22] Also in Genesis 6.6, 1 Samuel 15.11, 2 Samuel 24.16, and 1 Chronicles 21.15.

[23] For more information on this subject see Richard E. Friedman's book, *Who Wrote the Bible?* (Harper Collins, 1987); and Finkelstein and Silberman's book, *The Bible Unearthed: Archaeology's New Vision of Ancient Israel and the Origins of its Sacred Texts* (The Free Press, 2001).

Of course, this information contradicts traditional beliefs that pertain to the Torah (i.e., Pentateuch); namely, that it was authored by Moses himself; hence the customary rubric: "The Five Books of Moses." However, it is now known that this cannot be true because not only does the author refer to Moses in the third person, and not only did the author not even claim to be Moses, and not only did the Edomite kings who are referred to in Genesis 36 exist *after* the time of Moses (Friedman 1987: 18-19), and not only do numerous contradictions in the text indicate that more than one person wrote it (Finkelstein and Silberman 2001: 35, 175-176; Friedman 1987: 17-21), but his death is recorded in the very same record as well! It is also difficult to believe that Moses would have referred to himself as the "humblest man on Earth" (Numbers 12.3). Indeed, scholars acknowledge that the books of the Torah were not compiled and written until centuries after the time of Moses (Blenkinsopp 1992: 2-4; Finkelstein and Silberman 2001: 11-12, 36-38, 68; Friedman 1987: 17-21, 29, 223). However, it is possible that earlier written sources, such as the "Book of the Law," which contained information specifically related to regulations (and should not be confused for the entire Torah itself), could have been written by Moses (according to Deuteronomy 31.24),[24] before it was incorporated into the accounts that were written and compiled by others at a later time.

It is evident that it was the monotheistic Yahwehists (not to be confused with the Jahwist "J Source" scribe), who were the primary group among the original sources, who altered the meaning of the plural Elohim in order to make it adhere to their own emerging monotheistic interpretation. This transition explains the presence of such sentences as "Then Elohim said [. . .]."; as opposed to the additional plural article: Then the Elohim said [. . .]. This gradual monotheizing of the divine council also explains why the Canaanite

---

[24] Richard E. Friedman presents a strong case in his book, *Who Wrote the Bible?* (Harper, 1987) that these laws were written by the Priestly source centuries later. There is also evidence that indicates that the Book of the Law was actually an older version of Deuteronomy, and was not written until the seventh century BCE (Finkelstein and Silberman 2001: 281).

god El was eventually amalgamated with Yahweh (Genesis 35.1, etc.). It can therefore be contended that the common mainstream interpretation, which explains the existence of the plural form for God as denoting the so-called "majestic plural," is not accurate and should be abandoned.

* * *

The plural terms for God also corresponds with the Levantine and Mesopotamian records (which comports with coherence theory), which report the existence of a pantheon council of gods in that same era and region. It is evident that some of these older Mesopotamian accounts, beginning in Sumeria, were integrated into the Judean/Israelite record. Perhaps the Yahwehist scribes believed that this was permissible because the patriarch Abraham was from the Mesopotamian city of Ur (Genesis 11.1, 15.7).

> The extent of the Hebrew debt to Sumer becomes more apparent from day to day as a result of the gradual piecing together and translation of the Sumerian literary works; for as can now be seen, they have quite a number features in common with the books of the Bible. [. . .] The ideas and ideals of the Sumerians—their cosmology, theology, ethics, and system of education—permeated to a greater or lesser extent the thoughts and writings of all the peoples of the ancient Near East. [. . .] And the Hebrews of Palestine, the land where the books of the Bible were composed, redacted, and edited, were no exception.
> —Samuel Noah Kramer,[25] *The Sumerians*

One of the records that the Judeans referred to was the following supplementary text to the *Enuma Elish*:

---

[25] Samuel Noah Kramer, Ph.D., was a twentieth-century American professor of Assyriology at the University of Pennsylvania.

> When the gods in their assembly had made [the world],[26] and had created the heavens, and had formed [the Earth], and had brought living creatures into being [. . .] and [had fashioned] the cattle of the field, and the beasts of the field [. . .]
> —The Babylonian Creation Tablet

> In the beginning God [Elohim] created heaven and earth [. . .] Then God [Elohim] said, "Let the earth produce every type of living creature: every type of domestic animal, crawling animal, and wild animal."
> —Genesis 1.1-24

Indeed, even the Judeo-Christian scholar Joseph Blenkinsopp[27] admits that the Genesis account has drawn upon a well-established tradition that is best represented by the Mesopotamian Atra Hasis text (Blenkinsopp 1992: 93-94).

Other primary Mesopotamian records that were incorporated into the biblical account was the Eridu Flood Story, the Epic of Gilgamesh, and the Atra Hasis text, which not only contains the creation of humans by the gods but the Flood story as well. In the following account, Noah is referred to as Utnapishtim, and the biblical Mount Ararat is referred to as Nisir.

> (Then) I opened a hatchway, and down on my cheek streamed the sunlight, [. . .] Into the distance I gazed, to the furthest bounds of the Ocean. Land was upreared at twelve (points), and the Ark on the Mountain of Nisir Grounded; the Mountain of Nisir held fast, nor gave

---

[26] Words in brackets in this passage represent damaged portions of the original text.

[27] Joseph Blenkinsopp, Ph.D., is the John A. O'Brien Professor Emeritus of Biblical Studies at the University of Notre Dame and former president of the Catholic Biblical Association.

lease to her shifting. [...] (Then), when the seventh day dawned, I put forth a dove, and released (her), (But) to and fro went the dove, and returned (for) a resting-place was not. (Then) I a swallow put forth and released; to and fro went the swallow, She (too) returned, (for) a resting-place was not; I put forth a raven, Her, (too,) releasing; the raven went, too, and the abating of waters Saw; and she ate as she waded (and) splashed, (unto me) not returning.
—Utnapishtim, The Epic of Gilgamesh (The Flood tablet)

After 40 more days Noah opened the window he had made in the ship and sent out a raven. It kept flying back and forth until the water on the land had dried up. Next, he sent out a dove to see if the water was gone from the surface of the ground. The dove couldn't find a place to land because the water was still all over the earth. So it came back to Noah in the ship. He reached out and brought the dove back into the ship. He waited seven more days and again sent the dove out of the ship. The dove came to him in the evening, and in its beak was a freshly plucked olive leaf. Then Noah knew that the water was gone from the earth. He waited seven more days and sent out the dove again, but it never came back to him.
—Genesis 8.6-12

More and more animals disembarked onto the earth. Zi-ud-sura the king prostrated himself before An and Enlil. An and Enlil treated Zi-ud-sura kindly......, they granted him life like a god, they brought down to him eternal life. At that time, because of preserving the animals and the seed of mankind, they settled Zi-ud-sura the king in an overseas country, in the land Dilmun, where the sun rises.

—The Flood Story (Segment E)

Noah built an altar to Yahweh. On it he made a burnt offering of each type of clean animal and clean bird. Yahweh smelled the soothing aroma. Yahweh said to himself, "I will never again curse the ground because of humans, even though from birth their hearts are set on nothing but evil. [. . .]
—Genesis 8.20-21

The Ten Commandments, as well as the laws of Moses, were also significantly similar to other god-given law codes of the same era, such as The Thirty Teachings of Amen-Em-Ope, The Code of Hammurabi, The Code of Ur-Nammu, Code of Shulgi, etc. Accounts of kings who were guided to defeat the enemies of their god (e.g., The Mesha Stele [i.e., the Annals of Mesha]) were also not significantly different than the Yahwehist accounts.[28] Furthermore, the literary form of the covenant between Yahweh and the people of Israel in Deuteronomy is remarkably similar to seventh-century Assyrian vassal treaties (Finkelstein and Silberman 2001: 281). Indeed, the Hebrew version appears to have been documented at a time of Assyrian hegemony in that region.

Although the Judean priest-scribes had been exposed to the records of their neighbors in Mesopotamia, it is also true that a Ugaritic/Canaanite influence is also extant. This explains the similarities in their language, culture, and literature (Cross 1997: 182-183; Day 2002: 16, 232-233; Penchansky 2005: 78; Smith 2002: 2-13, 19-25, 28). For example, both El (KTU 1.6 i 35) and Yahweh (Exodus 40) reside in tents (i.e., tabernacles) when they visit the Earth. Both the servants of El (KTU 1.14 ii 10-20) and the servants of Yahweh (Leviticus 16.4) were required to wash themselves with water before making devotional animal sacrifices. Both El (KTU 1.14

---

[28] For more information about this subject see: Matthews, Victor H., Don C. Benjamin. *Old Testament Parallels: Laws and Stories from the Ancient Near East* (Paulist Press, 2006).

i 35) and Yahweh (1 Kings 3.5) appeared to their followers in dreams. Both El (KTU 1.16 i 10) and Yahweh were called "wise" (Proverbs 3.19) and "holy" (Leviticus 19.2). Likewise, the "sons of El" that appear in the Ugaritic tablets (KTU 1.10 i 1) coincides with the biblical sons of Elohim (Genesis 6.2).[29]

According to the Hebrew Bible, Yahweh himself *was* El (Genesis 35.1, etc.). This is peculiar due to the fact that El was the god of the heathen Canaanites. Also, just like El, Yahweh was also referred to as "Elyon" (Most High) (Day 2002: 16, 21; Smith 2002: 32, 56). It is therefore evident that El and Yahweh were originally separate deities (Coogan 1978: 20; Day 2002: 17; Herrmann 1999: 277-279; Smith 2002: 32-43). This fusion began to occur when Yahwehists—most likely the Levites, who were influenced by the Kenites[30] (more on this subject later)—entered into Canaan, perhaps from an extended sojourn in Egypt via Midian and/or Edom, and began to associate with native Semitic Canaanites who were loyal to El (Friedman 1987: 82). This scenario coincides with the archaeological evidence, which indicates that the violent takeover of Canaan in the time of Joshua that is reported in the Bible, never happened (Finklestein and Silberman 2001: 73, 76-79, 81-83). This is because the Israelites were the Canaanites (Smith 2002: 6-7, 19, 31). Indeed, this explains why their languages were nearly identical (Smith 2002: 20-21).

The similarities are not restricted to El alone, but also apply to Yahweh's adversary, Baal (Coogan 1978: 20-21, 79-80; Cross 1997: 148, 211; Smith 2002: 67, 73-91, 184). For example, both Baal (KTU 1.2 iv 5) and Yahweh (Isaiah 19.1) are said to have rode through the skies in a "cloud." Both revealed themselves on a mountain. Both had a temple built of cedar, and both were gods of the storm (Coogan 1978: 20-21, 77).

In its early history, Judaism was not strictly monotheistic, but rather acknowledged the existence of other gods besides Yahweh and

---

[29] Even more parallels can be found in M. S. Smith's book *The Early History of God* (William B. Eerdmans Publishing, 1990 [2002]): 23-24, 37-41.

[30] The Kenites were related to the Midianites. Midian is believed to be one of the locations where Yahwehism originated (van der Toorn 1999: 912).

El—such Baal, and Asherah (Day 2002: 42, 45, 227; Finkelstein and Silberman 2001: 241-242; Penchansky 2005 77-78; Smith 2002; 2-13, 65-66, 109). The reason why this is not more commonly known is because the Old Testament/Tanakh is mostly written from the perspective of only one specific group, namely, the "Yahweh-only" sect, which did not attain full power until a later period (Day 2002: 228-229; Finkelstein and Silberman 2001: 248; Smith 2002: 195-197; van der Toorn, 1999: 912). Indeed, archaeological evidence indicates that monotheism was actually a late development (Finkelstein and Silberman 2001: 234). The process that moved Canaanite religio-cultural creed away from the polytheism of the early age of the Judges (1200-1000 BCE) (Smith 2002: 54, 57, 206), and to the monolatry/henotheism[31] of the monarchy (1000-587 BCE) (Smith 2002: 3, 8-13, 30-31), and to monotheism of the post-exilic age (beginning in 538 BCE) (Smith 2002: 11, 191, 197), occurred over the course of centuries.[32] The monolatry that existed in Judah and Israel is why Yahweh is referred to as the "God of gods" in Deuteronomy 10.17 and Daniel 11.36 (Day 2002: 228-229; Smith 2002: 3, 13). The truth is that "the God of Abraham," in his original identity as El, was once worshiped together with other regional gods—i.e., the Elohim (Day 2002: 42, 45, 227; Finkelstein and Silberman 2001: 241-242; Penchansky 2005: 77-78; Smith 2002: 2-13, 65-66, 109).

> Just as an earthly king is supported by a body of courtiers, so Yahweh has a heavenly court. Originally, these were gods. But as monotheism became absolute, so these were denoted to the status of angels.

---

[31] Monolatry is the belief in the superiority of a singular god without denying the existence of other gods.

[32] Examples of extant evidence are the Kuntillet Ajrud inscriptions, in which Yahweh is reported together with "his Asherah" (Smith 2002: 118-119, 125).

—John Day, [33] *Yahweh and the Gods and Goddesses of Canaan*

It is true that the word *Elohim* is used in both a plural and a singular context in the Bible (Pope 1955: 10). This is because the Bible was not written by a single author. This contradiction reflects the conversion process that was happening during that era as the monotheistic Yahwehists gradually exerted their influence. It must be understood that the Yahweh-only sect was originally only a minority faction. When the monarchy was dissolved, this event left a power vacuum that those Levitical priest-scribes were able to fill (Finkelstein and Silberman 2001: 248, 310).

It is true that Yahweh was recognized as the national god during the Monarchic Era; however, it should be understood that this recognition does not evince the existence of absolute monotheism during that time (Smith 2002: 197; van der Toorn 1999: 918). Indeed, this is why the Old Testament/Tanakh is filled with indignant tales of rulers who did not submit themselves entirely to Yahwehism before this time. The only exception seems to have occurred during the relatively brief reigns of Hezekiah and Josiah.

Therefore, the plural terms for God were a vestige left over from not only a monolatrous/henotheistic religio-culture but an even earlier polytheistic past. This Canaanite past is both reflected in both the use of the words El and Elohim in the Hebrew Bible. Indeed, mainstream scholarship confirms that Yahwehism developed in the matrix of an earlier cult that was primarily dedicated to the worship of El (Cross 1997: 211). This not only explains why the name El appears as a theophoric element in some biblical names (e.g., Gabri-*el*, Micha-*el*, etc.) but why Yahweh is referred to in the role of El. Therefore, when the scribes altered the meaning of the word Elohim, they were attempting to reconcile a new theology with an older one. However, it should be understood that the use of the plural term in a singular context does not negate the original meaning.

---

[33] John Day, Ph.D., was a professor of Old Testament Studies at the University of Oxford.

What must also be understood is that the version of events that are reported in the Bible were written only from the view of a minority faction. It was this group of mostly priest-scribes from the tribe of Levi who essentially rewrote history when they devised the books that became the Pentateuch/Torah—and eventually the entire Old Testament/Tanakh itself (Finkelstein and Silberman 2001: 249).

* * *

In pictorial representations made by the artists of the time, the gods were portrayed as humanoid in appearance (Figure 1). The Mesopotamian records tell us that they had names like: Anu, Enlil, Enki, and Marduk. Although the texts refer to these beings not only as symbolic mythological figures but rather, in many instances, as actual living beings who were personally involved in the affairs of humankind.

> [The Enlil Ninlil text] illustrates vividly the anthropomorphic character of the Sumerian gods. Even the most powerful and most knowing among them were regarded as human in form, thought, and deed. Like man, they planned and acted, ate and drank, married and raised families, supported large households, and were addicted to human passions and weaknesses. By and large, they preferred truth and justice to falsehood and oppression, but their motives are by no means clear, and man is often at a loss to understand them.
> —Samuel Noah Kramer, *History Begins at Sumer*

(Fig. 1)

Who then were these beings who allegedly lived for inhuman amounts of time and who were said to have created the world, humankind, and the universe itself? Moreover, could it really be possible that these characters were nothing more than the fabrications of primitive minds?

Secular skeptics refute the existence of "gods" by explaining that such supernatural characters were a means by which primitive man could explain the mysteries of the world. When most people think of the ancient gods in the present era, they tend to think of the kind of fanciful fables that were authored by the Greek poets. However, when it comes to the biblical record, it must be acknowledged that the reports, in most cases, are more detailed and historically-based.

One way to explain the incredible feats that were attributed to the gods was that human-beings were ascribing supernatural powers to their own human kings in order to aggrandize the authority figure and compel obedience. Could it be possible, in this case, to conclude that the individuals who were claiming to be "gods" in ancient times were nothing more than human-beings with delusions of grandeur? Was it royal human-beings who commanded the scribe record-keepers to exaggerate their identities and abilities? Or was it the ancestors of those legendary kings and queens who deified them at a later time? It really is not such an impossible notion when we take into account that in the ancient traditions of the Mediterranean region, such as Egypt and Rome, human-beings were occasionally elevated to the status of gods. Indeed, these Mediterranean cultures are related to the earlier Mesopotamian tradition (more on this subject later). In this case, could we also attribute such a rational explanation to the biblical record as well?

Despite the plausibility of such a theory, there are other reasons to believe, which will be explicated in upcoming chapters, that an outright dismissal of all purported extraordinary factors cannot be entirely justified.

* * *

History-changing events were taking place approximately five millennia ago, when Bronze Age civilization developed in Mesopotamia, Egypt, and the Indus Valley. The earliest of the known major civilizations was Sumer (i.e., Shumer)—which is referred to as "Shinar" in the *Bible* (Genesis 10.10, etc.). The records report that the people of that time attributed their existence to the graciousness of the tutelary gods, and that it was those supernatural beings who had bestowed the faculties of civilization unto them.

> Without [the god] Enlil, the great mountain,
> No cities would be built, no settlements founded,
> No stalls would be built, no sheepfolds established,
> No king would be raised, no high priest born, [. . .]
> —Hymn to Enlil

Originally, these mysterious progenitors were not only described in poetic mythological tales but were also sometimes reported to have been involved in the practical everyday affairs of man; from agriculture, to economics, to city planning, to warfare, etc. Indeed, in the Sumerian Enki and the World Order text, it is reported that the god Enki directed the plow and caused grain to grow upon the Earth, and therefore, "multiplied abundance for the people." The scribes not only documented that the original gods were involved in the practical affairs of human-kind but reported their supernatural powers as creators of the world and all living things within it as well.

According to the precepts of a more mystical type of philosophy, there is another explanation for the supernatural aspect. Throughout the ages, the mystics have attributed the creation of nature to a transcendent "Godhead"—which is it not to be equated with the "personal god":

> Godhead awesome as the faraway heavens, as the broad sea.
> —Too the Moon God

> But if someone has a personal god from heaven [. . .]
> The god who has looked upon him will give him great strength.
> —A Hymn to Hendursaga (Segment C)

Unlike the anthropomorphic personal god, the Godhead is defined as more of a supernatural life-force, or transcendent essence (i.e., *ousia*), that pervades the entire active universe. This is the type of pantheistic entity that great thinkers, such as Einstein and Spinoza espoused—that is, a God that is diffused into nature itself.

According to the gnostic Christians, Yahweh (i.e., "Yaldabaoth") was a being who they referred to as the Demiurge. The Demiurge was said to be the creator of the lower material world of man. According to the gnostics, the actual universal creator was something known as the First Principle, or "Bythos." In this case, what exactly is the First Principle of Bythos, and what is its relation to the personal gods of the ancient world? Moreover, do such things even literally exist?

A similar notion can be found in the mystical teachings of the Jewish Kabbalah, which presents the concept of an original and universal *Ein Sof* (i.e., *Ayn-Soph*) provenance. This is commonly interpreted as Yahweh before his manifestation into the lower finite world. In this case, we must inquire if this creative transcendent source be in some way related to the monistic super-force that is postulated by modern-day science? Like the kabbalistic principle of the Ein Sof Godhead, Super-force also breaks down into successive emanations (electromagnetism, gravity, the strong nuclear force, and the weak nuclear force). We find a related hierarchical regime in the kabbalistic model of the Tree of Life, which represents the principle emanations (i.e., *sephirot*) of the Ein Sof Godhead. However, in this metaphysical context, the emanations reveal themselves not as physical forces but rather as intellectual and spiritual attributes. In this model, the manifestations of the source begin to diffuse, diversify, and amalgamate as they move further away from its monad inception point. Physicists find a similar progression in nature as well.

There are different ways that the word *monad* can be interpreted. This is partly because not only does this fundamental unit manifest in different ways but it is part of a larger paradigm that the philosopher Gottfried Leibniz[34] referred to as "the labyrinth of the continuum" (Sleigh 2015: 584). Monads can be defined as individuated energy systems—e.g., atoms; souls; the Godhead, etc.—that serve to form and animate life in the universe. Therefore, if we ourselves are autonomous monads whose energy has individuated apart from the amorphous energy of the universe, it can then be suggested that we ourselves—who have developed sense perception, and who are driven by our "perceptions and appetites"—have evolved into a co-creator role. In this case, it can be hypothesized that the long sought "meaning of life" (i.e., *telos*) may have to do with the animation of life itself. This does not only pertain to bearing progeny but to anything that supports life, such as going to work or helping others.

The other aspect of this experience seems to be imbued with an evolutionary quality. Through millennia of life-times, it seems that we advance further along an evolutionary course that is leading towards some numinous objective. This teleological attractor may be related

---

[34] Gottfried Leibniz was a seventeenth- and eighteenth-century German philosopher, historian, diplomat, and mathematician.

to Origen's[35] concept of the *apocatastasis*, and Pierre Teilhard de Chardin's[36] concept of the Omega Point.

Like the Kabbalists, the mystics of the Sufi movement also developed a cosmology that was based on the idea of the archetypal emanations of the Supreme Spirit (*Ar-Ruh-al-Qudsi*) of God. This henological[37] energy of the universal Godhead source can also be related to Plotinus's concept of the One, Aristotle's Unmoved Mover (i.e., Prime Mover), Pythagoras's Monad, Anaximander's Apeiron, Emerson's Over-soul, as well as the Hindu Aksara and/or Brahman, the Chinese Taiji (i.e., The Great Ultimate), the Native American Great Spirit, and the Bythos of the gnostic Christian Proarche.

However, what will become apparent moving forward is that none of these purported manifestations can be directly identified as the beings that are referred to in the Bible as the "Elohim."

* * *

In order to better understand this esoteric scenario, it is necessary to put this cosmological picture into perspective. What has become apparent is that we inhabit an expanding universe, and that this expansion began billions of years ago at a point called the "singularity." It was from this point that the emanation that generated the universe originated. After what has come to be known as the "big-bang" occurred, elementary particles began to appear and interact with the fundamental forces of nature as atoms condensed into stars, stars into galaxies, and galaxies into clusters of galaxies.

To give the reader an idea of the type of scale we are dealing with, a typical galaxy is approximately 30,000 light years in diameter, and the universe contains over 100 billion galaxies! Although the size of

---

[35] Origen was a third-century Judeo-Christian theologian.

[36] Teilhard de Chardin was a twentieth-century French philosopher, Catholic priest, and paleontologist.

[37] Henology is a term that refers to the unifying metaphysical and theological philosophy, that was originally posited by Plotinus, that pertained to the concept of the transcendent "One."

the universe is not precisely known, it is calculated to have a radius of billions of light years and contains approximately 300 sextillion stars.

But how could all of this have all happened, and what caused it?

One scientific explanation that has been put forth is that this event was initiated by a spontaneous quantum fluctuation in an empty vacuum where all energy was compacted into a point of zero volume. It may have been from within this void that a fluctuation occurred that ignited an atomic (or subatomic) spark that erupted into an expansion of energy. However, such an explanation (which is related to Edward Tryon's zero-energy universe theory—which is also sometimes referred to as the "vacuum genesis") brings up questions pertaining to how energy can be created from a state of zero energy, as well as questions related to the conservation of energy laws, which forbid creation *ex nihilo*. It has been suggested that known quantum anomalies would explain violations in the conservation of energy laws; however, such violations can only occur for short periods of time, and therefore would most likely not be applicable to an entire universe that is billions of years old. Furthermore, as will be elucidated further ahead, what appear to be particles appearing from nothing may actually only be virtual particles that *appear* to be appearing from out of nothing.

It is here where I must submit the theory that the universe is most likely not the result of a vacuum, but rather the result of an energy system that exists outside the boundaries of our own particular space-time continuum—or our so-called "bubble universe." It is in this transcendent modality where time and space either do not exist or more likely exists in an infinite state. This exterior state can be referred to as an Infinite Energy Circuit.

It is first necessary to understand that physicists use mathematics to calculate the activities of nature. Both mathematics (e.g., Cantor's set theory) and physics (e.g., Einstein's gravitational singularity) tell us that purposeful infinite equations are possible and therefore, in specific cases, should not be regarded as meaningless errors.

Such calculations have been troubling for scientists since it is a result that seems to be incompatible with our own limited perceptions. In this case, what we are attempting to comprehend is a factor that exists outside the boundaries of our own particular reality. If this sounds outlandish, then it should be remembered that until only recently—that is, in the context of all of human-history—for many, the idea that we live on a spherical planet, as opposed to a flat Earth, was also considered out of the bounds of reality as well.

The cosmic singularity point seems to be leading us back toward the Infinite Energy Circuit. Indeed, the singularity within a black hole is a point of infinite density and infinite temperature, where gravity and space-time curvature also becomes infinite. This is the same type of singularity that is found at the beginning of the universe. Therefore, the discovery of infinite equations, in some specific cases, are not mistakes, but rather are signs that are leading us back to the original transcendent source. Indeed, astronomers have discovered a link between black-holes and the creation of galaxies (Kormendy and Bender 2009: 142-145). Therefore, it will most likely be from inside a black-hole that we will discover the key to understanding the mysteries of the universe. Indeed, this explains why infinity singularities that cosmologists find at the big-bang and inside black-holes are much more severe than other more mundane singularities (Greene 2011: 98).

According to the special theory of relativity, time and space are interconnected. This most likely indicates that time began when space began (Davies and Gribbin 1992: 120-122, 141). As for the question of time and how it could have come into existence, I submit the following conjecture: Time may have actuated due to the gradual release of steady state energy that occurs in vibratory intervals. We experience these sequential intervals as time.

It can therefore be contended that it is unlikely that the universe is infinite if it had a starting point. This is because true infinity has neither a beginning nor an end. Furthermore, both the Hubble telescope and Einstein's general theory of relativity reveal that the universe is expanding. This most likely indicates that space is being created as the universe inflates. Indeed, the inflation model solves a lot of the problems that physicists deal with, and is therefore, along with other reasons, the most likely scenario (1992: 168-169). Furthermore, matter decays as it ages, and physical processes run down as entropy increases (i.e., the second law of thermodynamics). These are indicators that are pointing toward the conclusion that the universe is not ageless (Barr 2003: 59-60). This finite nature also explains why there are limits, such as the Planck time and length[38] to our universe, which also explains why this universe is filled with so many finite possibilities (Greene 2011: 32-33). Indeed, this also seems to be what the unstable nature of the Higgs Boson particle is also telling us (Klotz 2013). Another example pertains to the limit to how much matter and energy can exist in a region of space before it collapses and punctures a hole in fabric of space and time. Because the universe is most likely not eternal, this could indicate that in some distant eon it may arrive at some ultimate terminus.[39] However, the energy that animates and illuminates the universe will not die out, but rather will be transformed and recycled, either back into the IEC or perhaps into another universe via what physicists refer to as "wormholes." (Perhaps by that time, human-beings will have figured out how to escape to such a neighboring universe.)

Furthermore, other theories, such as the steady-state theory, which postulates that the universe was not created from the big-bang, and is not expanding, is plagued with problems that do not agree with scientific observations. Indeed, most cosmologists have already abandoned this theory (Greene 2011: 122). It can therefore be concluded that nature is not as stagnant as steady-state theory is.[40]

It has also been suggested that the universe could be set to a never-ending "bouncing" loop (i.e., oscillatory universe or cyclic multiverse). When dealing with such a theoretical query, we must start by looking for clues that are pointing toward a specific direction. In this case, we must again consider the second law of thermodynamics, which constrains what can be achieved by any

---

38 The Planck length is the scale in which the known laws of gravity, space, and time break down. Likewise, Plank time is the point at which no further subdivisions can occur.

39 This negates the "quilted universe" theory.

40 Recommended reading for the case that we do not live in a static universe: Alfred North Whitehead's *Process and Reality* (1929).

cyclical process—that is, at least in a singular universe (Davies and Gribbin 1992: 123-124). Furthermore, according to the calculations of astrophysicist Herman Zanstra, the cyclical model would have had a beginning as well because entropy would have decreased in the past until there was no cycle, only a big-bang (Greene 2011: 122-123). These findings indicate that the universe most likely does not operate on an infinite motion loop of expansion and compression. Indeed, for these reasons and others, many cosmologists believe that the universe is indeed finite and expanding (Davies and Gribbin 1992: 124).

However, I propose that the type of limits that we face in our own universe do not exist in the Infinite Energy Circuit. Therefore, before the big-bang time did not exist—that is, at least in our particular universe.

We must therefore differentiate between the finite space-time continuum that was generated by the big-bang, and the energies that illuminate and animate all activity within. This energy can now be examined by scientists on a quantum level. It is telling, for instance, that quantum particle waves are essentially able to transcend time and space (1992: 215). Indeed, the nature of photon energy is now known to have infinite properties (1992: 243). It is also telling that an electric current that is streaming in a superconductive state can flow in a ring forever without losing energy (1992: 206). It can be contended that a similar type of activity is occurring in the IEC. The conservation of energy law (i.e., the first law of thermodynamics), which states that an energy system can neither be destroyed, nor created, but rather only transformed, is also pointing back toward the original transcendent super-force. Therefore, it is not the universe that is infinite in space and time, but rather the energy behind it that is most likely endowed with such transcendent faculty.

However, it is also possible that the energy that is broken down in a black-hole may also perhaps be reinvested to other points in our own universe—that is, besides the energy that is released in the form of Hawking radiation. Newly emerging findings are indicating that there may be different kinds of black-holes. One type may contain singularities (such as the Schwarzchild black-hole), which may be ingress interfaces into the IEC. The other type is most likely a worm-hole (such as rotating Kerr black-holes), which may be nexus points to other bubble universes that exist alongside our own—or perhaps to other locations in space within our own universe (Kaku 2014: 292). It therefore seems that the multiverse[41] and what I am referring to as the Infinite Energy Circuit, might be creating and sustaining its projections via worm-hole arteries in the space-time continuum. Indeed, the big-bang black-hole connection is predicted by general relativity. This is also the same type of theory presented by physicist Nikodem Poplawski. Poplawski's findings predict that at the other end of the worm-hole is not another black-hole, but rather a white-hole—which also agrees with the laws of general relativity. According to Poplawski, worm-holes would also help to explain gamma-ray bursts, which are most likely associated with supernova star explosions at the edge of the universe. Therefore, it may be possible that these mini big-bangs could be the discharges of energy either from our own or from other universes (Than 2010).[42]

However, many astrophysicists are skeptical of this worm-hole theory. They argue that the mathematical models that predict it fail to consider disruptive factors related to radiation, material, and gravity. However, it should be taken into account that this contention is based only on current understandings. Indeed, it might be possible that quantum activity, that is not fully understood at this time, could remove the need for a singularity in worm-holes—that is, if these points were of a particular condition (Davies and Gribbin 1992: 272).

* * *

---

[41] The multiverse paradigm that I am suggesting is not related to the quantum so-called "many worlds" parallel universe theory, which holds that split realities can emerge based on the actions of beings in this world, and that these split realities contain doppelgangers. I contend that this theory would most likely only apply to hypothetical time-travel scenarios. Therefore, a more accurate designation may be pluriverse.

[42] The idea that black-holes could be worm-holes was also presented in a study conducted by Thibault Damour of the Institut des Hautes Etudes Scientifiques in Bures-sur-Yvette, France, and Sergey Solodukhin of International University Bremen in Germany (Shiga 2007).

The ultimate question that we are now left with is: could the act of the original energy source to extend itself into what has become our universe have been caused by a spontaneous fluctuation, or by the intention of an energy system that is endowed with sentient faculty?

I propose that the IEC is not static, but rather is alive and pulsing with a never-ending and never-beginning amount of super-force energy. Such a dynamic state might allow for just enough entropy (i.e., unknown factors related to disorder) for fluctuations to occur. In this case, the IEC can not only be envisaged as an unending fractal set, but as super-force energy that is in a state of perpetual movement through itself. Indeed, when physicists look into the quantum world they do not always see the type of predetermined clock-work order that is asserted by the theists; but rather they see evidence of a balance between order and disorder. This may be the type of randomness that sparked the big-bang. If this is true, then the presence of a "Creator" that is referred to in the religious tradition is unnecessary.

Judeo-Christians will sometimes cite scientific findings in order to argue that the conditions for this universe to exist had to be of such precise conditions that only a Supreme Being could have planned it. However, fluctuations that occur in the IEC could be happening all of the time, but only something like one in an astronomical amount are precise enough to lead to the creation of an entire universe.

It is true that physicists acknowledge that conditions related to the primary forces of nature are of such fine-tuned proportions that if anything was even slightly different life would not be possible. However, it may also be possible that other types of life-forms would be able to develop in different types of conditions. Furthermore, it can also be contended that all of the forces that make life possible can be traced back to a single source; namely, the Infinite Energy Circuit, and therefore balance each other out through a natural holistic process. Indeed, it seems that it is the inherent nature of the primordial super-force energy of the IEC to naturally gravitate toward order and life. It can be contended that this phenomenon is more influenced by the energeia[43] of its own intrinsic entelechy,[44] rather than by magical intervention by a humanoid deity. Indeed, when scientists examine the universe, they observe a natural stochastic[45] process that has occurred over the course of eons—as opposed to "days" (Genesis 1.1-31, 2.1-3). An increasing amount of scientific findings are pointing toward a universe that is not governed by rigid predestination, but rather by natural attractor dynamics. In this case, there is no need for premeditated "intelligent design"—at least on a macrocosmic scale. (The microcosmic level will be examined later.)

However, it must also be acknowledged that the primordial energy that is emitted by the IEC also seems to be endowed with metaphysical characteristics that incline towards beauty and sublimity. Indeed, not everything in the universe can be reduced to a cold mathematical mechanism. These supramundane characteristics enable it to be associated with the concept of the Godhead.

However, if consciousness exists within the IEC/Godhead it is obvious that it does not exist like that of our own. Therefore, for human-beings to fully comprehend the nature of the IEC/Godhead might be like a chimpanzee attempting to comprehend quantum mechanics. In this case, it is likely that the IEC/Godhead exists on a level that is beyond the cognitive range of the current stage *Homosapien* primate. Moreover, there is no plausible reason to believe that this energy system manifests itself as any one particular humanoid individual—although there are certainly mystical and religious

---

43 *Energeia* is a word that was used by Aristotle to denote action or the act of functioning. In a Judeo-Christian sense, it is used to denote the power of God.

44 *Entelechy* is a term that was used by Aristotle to denote that which actualizes what is otherwise only potential (e.g. form from matter). This is a reference to the underlying force that directs growth from a fundamental state into a complete and functional form (which is related to Aristotle's theory of *hylomorphism*). The philosopher Leibniz referred to entelechy as a primitive active force that is latent within the monad. I use this term here to refer to intrinsic life-endowing attractor dynamics.

45 The stochastic process evolves over time according to natural probabilities, as opposed to stringent deterministic laws.

figures who can claim to have some sort of special connection with it. Such individuals could be mistaken as the omnipotent creator force itself by the layman. Indeed, it can be contended that it is more likely that the infinite energy source itself is much more transcendent, universal, omnipresent, and phenomenal, and is not limited to any singular, familiar, and Earthly form.

* * *

What can be extrapolated from the available data, is that we are living in the aftermath of a tremendous explosion. When this explosion cooled down what was left in its wake was a zone that can be described as a "bubble." However, the term bubble may not be an accurate depiction. This is because scientific discoveries are pointing toward the conclusion that the universe is not spherical like a bubble, but rather flat.[46] Although this finding is not yet considered to be settled beyond all doubt, until future evidence indicates otherwise, the universe should be envisaged as flat, rather than spherical.

Within this horizontal explosion, the dimensions of space and time were generated. Therefore, it can be hypothesized that the universe was not created for our specific anthropic requirements—as the theists assert, but rather the other way around. Indeed, this is why there is nothing extraordinary about our planet's position in the universe (Davies and Gribbin 1992: 113, 115; Greene 2011: 146). It can therefore be stated that to assume that human-beings and the planet Earth are the ultimate *raison d'être* of the entire universe is an unfortunate combination of ignorance and arrogance.

What is therefore being proposed is not the traditional picture of a bearded and white-robed man, sitting on a throne in the clouds, personally judging every single human-being, but rather a singular but all-pervading universal life-force. A transcendent energy system

---

[46] map.gsfc.nasa.gov/universe/uni_shape.html; Krauss, Lawrence M. "Lawrence M. Krauss; A Universe from Nothing; Radcliffe Institute," Uploaded by Harvard University, *Youtube.com* (July 17, 2013) Accessed January 9, 2017: youtu.be/vwzbU0bGOdc

that does not manifest itself as any particular gender, or race, or individual, and is not confined to any one particular creed, but rather exists as more of a unified but dualistic energy system that is both feminine (*Yin*) and masculine (*Yang*). It is this balance of forces that gives birth to galaxies, solar systems, planets, and all living things down to a subatomic level. The ancient Chinese Taoist sage, Lao Tzu (i.e., Laozi), described this energy as having a type of profound and paradoxical characteristic. Therefore, when one is in harmony with the spirit of the IEC, one is in harmony with "The Way":

> Without form or image, without existence, the form of the formless, is beyond defining, cannot be described, and is beyond our understanding. It cannot be called by any name. Standing before it, it has no beginning; even when followed, it has no end in the now, it exists; to the present apply it, follow it well, and reach its beginning.
> —Lao Tzu, Tao Te Ching

The mystics believe that our connection to the transcendent source is located within the soul (i.e., psyche). The soul can be defined as the individuated "spark"[47] of the monad entity; while the spirit can be defined as an energy field that is subjected to the vibrational resonance of the entity. (More on this subject later.)

The goal of the devotee is to establish a connection with this original and higher state and employ it as a means in which to evolve as a co-animator. The Hindu personal god Krishna, for instance, is said to be (in the Bhagavad Gita) the "personality of the Godhead." Therefore, it is not that Krishna *was* God, but rather he was a manifestation of an individual who was in a perfect state of harmonic attunement with the "Super-soul" of the universal Godhead source.

> The Lord, as Super-soul, pervades all things.
> —Bhagavata Purana (Srimad Bhagavatam)

---

[47] This is a term that has been used by mystics.

In this case, it can be deduced that the Elohim were not actually themselves the original universal creative source, but rather were individuated monad manifestations *of* this phenomena.

Nevertheless, we are still left with the persisting question: who were the Elohim?

CHAPTER III

THE INTER-WORLDLY REALITY

According to the ancient texts, not only did the gods exist, and not only did they appear in humanoid form, but they were also said to have had the ability to fly. The vehicles of the gods are sometimes poetically referred to as rising boats, or barges, that sail through the sky:

> [. . .] the holy barge which traverses the sky, Nanna, the lord [. . .]
> —A Praise Poem of Sulgi (Sulgi D)[48]

> Thou shalt come forth into heaven, thou shalt sail over the sky, and thou shalt hold loving intercourse with the Star-gods. Praises shall be made to thee in the boat.
> —Book of the Coming Forth by Day (Book of the Dead), The Osiris Ani, Whose Word is Truth, In Peace, the Truth-Speaker, Saith

The designation "Star-gods" refers to the cosmic characteristics of those otherworldly beings. Indeed, the original Sumerian cuneiform

---

[48] Also in A Balbale to Suen (Nanna A) text.

sign that was used to indicate the word for god was shaped like a star (van der Toorn 1999: 356).

References to the stars and the gods also appear in the Bible:

> You have lifted up the shrine of your king, the pedestal of your idols, the star of your gods—which you made for yourselves.
> —Amos 5.26

Gods who are represented by the symbol of the star are most likely the same beings who are referred to elsewhere in the Bible as the "starry hosts" (i.e., "hosts of heaven").

> In the two courts of the temple of the Lord, he built altars to all the starry hosts.
> —2 Kings 21.5

It is commonly assumed that the hosts of heaven refer to stars, planets, and the forces of nature (Mettinger 1999: 920; Niehr 1999: 429); however, the word that is translated as "host" derives from the Hebrew word *sabat/sabaoth/tsebaoth*, which literally means troops or army (Betz, 1999: 268; Cross 1997: 70). This is a reference to celestial beings, not celestial bodies. Indeed, in the following passage, the starry hosts are not associated with inanimate objects, but rather with the "gods":

> [. . .] they burned incense on the roofs to all the starry hosts and poured out drink offerings to the gods.
> —Jeremiah 19.13

This definition is contextually congruent with the Ugaritic texts (KTU 1.10 i:3-4), in which celestial beings who are referred to as the "sons of god" in the "assembly of stars" are reported (Parker 1999: 798). Indeed, Ugaritic religio-culture played a part in influencing the West Semitic milieu.

The association between gods and planets most likely derived from Babylonian astrology. However, in this system the gods were not the planets themselves, but rather the planets were named after the gods—just like these same planets have been re-named after the Roman gods in the present era. For example, the god Marduk was associated with the planet Jupiter. However, we know that Marduk was not literally a planet because texts and iconography portray him with humanoid features. The same can be said for the other gods and their respective planets (e.g., Venus with Ishtar; Saturn with Ninurta; Mercury with Nabu; Mars with Nergal) (Jastrow 1911: 111, 217-218, 224-226).

There may also be a conflation occurring between ordinary stars and the type of Unidentified Flying Objects that appear as stars from far away. Indeed, reports of UFOs that look like stars in the present era are numerous (Davis 2014: 104, 143; Kean 2010: 86, 310; Sturrock 1999: 11). In this case, when the ancient people witnessed those sidereal objects moving in the night-time sky in an intelligently guided manner, they must have believed that they were being monitored by celestial beings. Offerings and extolments were extended to the star gods to curry favor. Indeed, this is what the previously cited verse from the book of Jeremiah is referring to.

The belief in the "star people" extends throughout the world (Mack 1999: 6, 156, 162). The Mesoamerican Maya of the pre-Columbian Americas, for instance, based their religio-cultural creed on the existence of those celestial beings (Milbrath 1999: 1, 198). According to one African account, the star people descended to the Earth in "sky boats" (Mack 1999: 192). These reports coincide with a Delaware native American legend of a "Great Man above," who descended to Earth in ancient times (Mayor and Sarjeant 2001: 156).

Another object that is commonly associated with the gods of the ancient world is the winged-disk. Throughout the ancient buildings and monuments of the Middle East, the image of the winged-disk is found (Figure 2; Figure 3). These disks are commonly referred to by scholars as "solar disks"; however, this is not an accurate interpretation. Because these objects are depicted with wings indicates that they had the ability to fly. It is commonly believed that the wings represent the natural orbital movement of the sun; however, not only are the gods sometimes shown inside these flying disks (Figure 4) but they are also shown being taken up into the air by these same objects as well (Figure 5). Furthermore, the solar disks that actually do represent the sun are depicted with light rays emanating from them, as opposed to wings (Figure 6).

(Fig. 2)

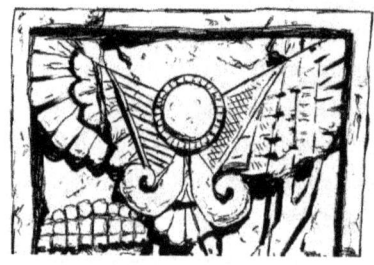

(Fig. 3)

(Fig. 4)

(Fig. 5)

(Fig. 6)

If the winged-disk was intended to represent the orbital movement of the sun, we should then expect to find the very same wings attached to depictions of the moon; however, such images remain elusive. Furthermore, neither is the moon god, Nanna-Sin, ever seen inside or on the crescent symbol of the moon.

In the following passages, the god who dwells in the disk is Horus.

> Homage to you, O ye gods, who live in your Hall of Maati, who have no taint of sin in you, who live upon truth, who feed upon truth before Horus, the dweller in his disk.
> —Litany from the Book of the Dead

> Then Thoth told Isis not to fear, but to put away all anxiety from her heart, for he had come to heal her child, and he told her that Horus was fully protected because he was the Dweller in his disk, [. . .]
> —The Legend of the Wanderings of Isis

What is exceptional about these reports is that Horus was not a sun god. Why then would he be associated with the disk? This next passage is even more revealing:

> Horus took the form of a great winged disk, which flew up into the air and pursued the enemy, and it attacked them with such terrific force that they could neither see nor hear, and they fell upon each other, and slew each other, and in a moment not a single foe was left alive. Then Horus returned to the Boat of Ra-Harmakhis, in the form of the winged disk which shone with many colours [sic] [. . .]
> —The Legend of Horus of Behutet and the Winged Disk

Why would the sun be described as flying around and maneuvering around in battle? And why would the sun be described as having many colors? In these instances, interpreting the winged disk as the sun cannot be justified.

One of the problems that serves to obscure the original meaning pertains to translators of the hieroglyphics themselves who may be influenced by conventional interpretations. In these cases, the assumption is made that the disk is referring to the sun, and the word "sun-disk" is erroneously inserted into the translation.

Conversely, it is acknowledged that there are other instances in the artworks and the texts where the disk is described as the sun. Indeed, this seems to be what happened during the reign of Pharaoh Akhenaten. Therefore, the challenge is to differentiate between disks that actually do refer to the sun, versus disks that refer to the flying vehicles of the gods. Due to the poetic language and abstract cultural imagery of the time, discerning between the two requires careful examination. In these cases, contextual indications must be considered.

However, there may be another reason why the sun and the flying disk of the gods were conflated. It may also be possible that the sun was believed to be a giant fiery flying disk that was piloted by the gods, in much the same way that the smaller fiery disks that were also seen in the sky were controlled by the gods. Indeed, it was believed by many in the ancient world that the sun god caused the Sun to rise and set each day.

In the following passages, the disk of the gods is described as an object that could be entered by the gods:

> I enter in by the Disk, I come forth by the god Ahui. I shall hold converse with the Followers of the Gods.
> —Book of the Coming Forth by Day (Book of the Dead), The Chapter of Advancing to the Tchatchau Chiefs of Osiris

What is also significant about this report is that the disk does not refer to the sun god, Ra. Another reference appears in the following passage:

> Hail, Power of Heaven, Opener of the Disk, thou Beautiful Rudder of the Northern Heaven. Hail, Ra, Guide of the Two Lands, thou Beautiful Rudder of the Western Heaven. Hail, Khu, Dweller in the House of the Akhemu gods, thou Beautiful Rudder of the Eastern Heaven. Hail, Governor, Dweller in the House of the Tesheru Gods, thou Beautiful Rudder of the Southern Heaven.
> —Book of the Coming Forth by Day (Book of the Dead), The Addresses of the Northern Heaven (The Addresses of the Four Rudders)

The "Power of Heaven" was not the Sun god Ra, because this deity is mentioned in the very next sentence as a being who was separate from the "Opener of the Disk." It is also telling that the opener of the disk is associated with the "Northern Heaven," since the Sun is not usually associated with the north, but rather with the east and the west. This most likely indicates that the author was not referring to the Sun, but rather to some other "Power" who was associated with a disk that had been witnessed in the northern sky.

There are also references to gods who dwell in arc-shaped enclosures—i.e., domes—that "belong to the sky":

> [. . .] that you might say that this Pepi [the pharaoh] will be among them, the gods in the sky, for you have assembled those in the (sky's)[49] arcs and banded together those who are the Imperishable Stars.
> —The Pyramid Texts of Pepi I, Spells for the Entering and Leaving the Tomb

The word "sky's" appears to have been inserted by the translator, perhaps due to damaged text. Some might argue that this is a reference to the arch of the firmament of the sky; however, this is

---

[49] The insertion in parenthesis appears in the following translation: Allen, James P. *The Ancient Egyptian Pyramid Texts* (Society of Biblical Literature, 2005).

unlikely because the word "arcs" that is next to it is plural. Therefore, it can be maintained that the information most likely refers to the arch-shaped objects that the gods, and perhaps even some selected humans, ascended to the sky in. In the following passage, the arch-shaped object is referred to as an "enclosure":

> [. . .] he [the pharaoh] is the one who will go up to Horus's enclosure that belongs to the sky.
> —The Pyramid Texts of Pepi I, Commending the Spirit to the Gods

In some instances, the image of the winged disk is found next to symbols that appear to be depictions of the Sun and the Moon (e.g., the stele of King Nabonidus). This seems to confirm that the winged disk cannot be associated with the Sun. However, mainstream scholars claim that these other sun-like images are not the Sun, but rather the planet Venus (Ascalone 2005: 93, 110, 221). However, just like the symbol of the Moon, the Venus goddess is never depicted inside such disks. The Venus interpretation also does not explain why the "sun disk" has wings and Venus does not. Nor does it explain why the winged disk is depicted taking people up into the air. Nor does it explain accounts of flying disks that are not associated with either a Sun or a Venus deity.

One possible reason why the winged disk is sometimes shown alongside images of the Sun and the Moon is because the artists were conveying the message that the observant gods oversaw human activity from a vantage point that was next to the Sun and the Moon (see figure 7).

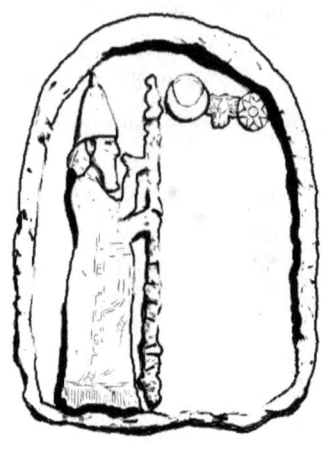

(Fig. 7)

What is also significant about this particular image is that the symbol that is next to the flying disk and the moon bears seven points; however, the star symbol that was associated with Venus was not a seven-pointed star, but rather an eight-pointed star (Benard and Moon 2000: 26, 69). This is a small but significant difference.

Nevertheless, it must also be acknowledged there may have indeed been instances when the sun-like image of the star (i.e., the eight-pointed star) did represent Venus (e.g., Kudurru Melishipak stele). Therefore, it is possible that the winged disk and the Sun were originally separate images that, in some instances, were conflated. This conflation began to occur in ancient times and has been further promulgated by modern-day scholars. Indeed, it must be understood that these images were rendered in different regions and in different times. They were also created in an era in which absolute definitions were not firmly established and maintained by academically-based institutions. Consequently, complete consistency should not be expected. This is due to the natural fact that observers tend to interpret such images according to their own understandings. Therefore, the best that an objective examiner can do is to consider each depiction on a case-by-case basis.

* * *

Similar types of phenomena that are associated with the gods were documented in the Hindu Mahabharata text. In this poetic epic, that was composed between 400 BCE and 400 CE, it is reported that the flying vehicles (*vimanas*) were piloted by the gods:

> And the gods in cloud-borne chariots came into view the scene so fair [. . .] Bright celestial cars in concourse sailed upon the cloudless sky.
> —The Mahabharata, Book II. Swayamvara. Part IV, the Suitors

A section of the Mahabharata, called the Bhagavad Gita, reports that an early group of mystical progenitors, i.e., the "Manus," were interplanetary beings:

> The seven great sages and before them the four other great sages and the Manus come from Me, born from My mind, and all the living beings populated the various planets descend from them.
> —Krishna, The Bhagavad Gita, chapter 10, The Opulence of the Absolute

In the ancient Hindu epic, the Ramayana, the following descriptions of flying illuminated vehicles of the gods are also found:

> The Gods themselves from every sphere,
> Incomparably bright,
> Borne in their golden cars drew near
> To see the wondrous sight.
> The cloudless sky was all aflame
> With the light of a hundred suns
> Where'er the shining chariots came
> That bore those holy ones.

—Ramayana, Canto XLIV. The Descent Of Ganga.

With Indra near him, to the sky
On a bright car, with flame that glowed,
—Ramayana, Canto LXIV. Dasaratha's Death.

Vibhíshan set upon the throne;
The flying chariot Pushpak shown.
How Brahmá and the Gods appeared,
And Sítá's doubted honour [sic] cleared.
How in the flying car they rode
To Bharadvája's cabin abode.
The Wind-God's son sent on afar;
How Bharat met the flying car.
—Ramayana, Canto III. The Argument.

Then, at his hest, the car rose high
And sailing through the northern sky,
—Ramayana, Canto CXXIX. The Meeting With Bharat.

There are also repeated references to supernatural beings who dwell in the sky:

Lord Vishnu slew his demon foes,
Amid the dwellers in the skies
—Ramayana, Canto XII. The Heavenly Bow.

In the Bhagavata Purana (i.e., the Srimad Bhagavatam) there is an account of a great war in which flying vehicles were used by both the gods and a warrior king who was associated with the gods. In the following passage from this text, it is reported that one of the flying vehicles was also referred to as a "Saubha," which translates as aerial city. This large mother-ship was able to move so fast that it seemed to defy the laws of physics:

> Moving hither and thither like a whirling firebrand, from one moment to the next seen on the earth, then in the sky, on a mountain top and then in the water, remained that Saubha airship never in one place.
> —The Bhagavata Purana, (Srimad Bhagavatam) Canto X, chapter 76

Of course, these accounts became increasingly embellished over the centuries, until cultural aesthetics and devotional aggrandizement overtook literal descriptions. This explains why the flying vehicles of the gods were portrayed as ornate palaces. This is also why some accounts depict the flying vehicles of the gods as being drawn by magical horses.

It must also be understood that literacy was not common in the ancient world. The reason why we do not have direct first-hand reports is because those who witnessed such phenomena were not able to directly document and therefore preserve a literal account of their experience. By the time these oral reports reached those who were qualified to document it, they had clearly been affected. These accounts were then altered even further by scribes who put these reports into a relevant religio-cultural context. Indeed, this same type of process occurs in many present-day historically-based Hollywood movies. Such depictions often tend to stray from a realistic portrayal of events in order to heighten dramatic effect.

Furthermore, it is typically assumed that when the gods were described by the scribes as coming down from the "heavens" that this was an indication that they had descended from the invisible world of spirit; however, in the ancient world, the term heavens originally referred to the sky. Likewise, the term "worlds" may not have been a reference to heaven and Earth, but rather to planets. Indeed, a reference to planets appears in the following translation:

> O mighty-armed one [Krishna], all the planets with their demigods are disturbed at seeing Your great form.
> —Arjuna, The Bhagavad Gita, chapter 11, The Universal Form

* * *

Inside the Great Pyramid of Egypt, there are narrow shafts that lead out from both the king's and queen's chambers. Egyptologists suggest that these shafts could have been used for ventilation purposes; although it is also believed that they could have also served some sort of spiritual/religious function. It is therefore compelling that calculations reveal that in the year 2500 BCE—which is the time in which the Great Pyramid was constructed—the northern shaft of the king's chamber inside the Great Pyramid was aimed toward the star Alpha Draconis, and the southern shaft was aimed at Zeta Orionis. In the queen's chamber, the northern shaft was aimed toward the star Beta Ursae Minoris, and the southern shaft was aimed toward Sirius (Verner 2007: 200, 202). What is also especially significant about this situation is how these locations concur with the information that is provided by the Pyramid Texts, which report that the ancient Egyptians were interested in these specific points in outer-space:

> With Orion in the eastern arm of sky shall you go up, with Orion in the western arm of the sky shall you go down. Sothis [i.e., Sirius], whose places are clean, is the third of you two: she is the one who will lead you two in the Marsh of Reeds to the perfect paths in the sky.
> —The Pyramid Text of Pepi I, Address to the Spirit as Osiris in the Duat

The Pyramid Texts report that these shafts were what the Egyptians of antiquity referred to as "sky windows," or "perfect paths in the sky." These were channel tracks that were plotted out for the course of the transmigrating soul (*Ka* or *Ba*) of the pharaoh and his wife, after the expiration of their physical encasement. It appears that the objective of the procedure was for the king and queen, who was thought to be a direct descendant of the "star gods," to return to the planetary realm of their divine ancestors and live forever in their eternal abode.

> A gate to the *Akhet* [horizon] will be opened for you in the sky, the god's hearts will be welcoming at meeting you, and they will take you to the sky in your *ba* [soul], you having become *ba* as one of them.
> —The Pyramid Text of Pepi I, Sending the Spirit to the West

> (You) shall (live like those in the sky live, you shall evolve) more than those in the world evolve. Raise yourself by your (own) force. When you go forth to the sky, the sky shall give you birth like Orion.
> —The Pyramid Text of Pepi II, Sending the Spirit to the Sky

> Orion is his brother, Sothis is his sister, and he will sit between them in this world forever.
> —The Pyramid Text of Queen Neith, Boarding the Sun-Boat

It is therefore most likely that the Great Pyramid was intended to be not only a tomb but a gateway to the world of the spiritual and interstellar gods.

> The *mastaba* [royal tomb] has been opened for you, the sarcophagus's (lid has been pulled back for you), the sky's door has been pulled open for you [. . .]
> —The Pyramid Texts of Pepi II, Preparing to Leave the Duat

In his book *Seeds of Knowledge, Stone of Plenty*, biological engineer John Burke and Kaj Halberg present findings from experiments that were conducted at various sacred sites around the world, including the Meso-American and Egyptian pyramids. Using instruments such as magnetometers, Burke and Halberg detected unusual electromagnetic fluctuations at these locations. The findings indicate that the man-made stone structures may have been built to harness the natural telluric energies of the area, which somehow aided in the convergence between the higher realm of the gods and the lower physical plane. A similar concept is found in the records of the pre-Columbian civilizations of Meso-America:

> [. . .] after his death, King Pacal was transformed into a god linked with Jupiter, an apotheosis that carried the ruler to heaven. Other rulers were transformed into Venus after death. They traveled on the soul's road, the Milky Way, to reach their celestial abode. The Precolumbian Maya, like other great civilizations, believed their stars were gods, and their rulers derived power from their connection with the cosmos in life and in the afterlife.
> —Susan Milbrath,[50] *Star Gods of the Maya*

Some scholars interpret the records to mean that the gods had transformed themselves into stars and planets. However, it is more likely that it was believed that the celestial gods were from or returned to these locations.

The collected information from around the world is pointing toward the conclusion that our planet has been visited and is continuing to be visited by beings who are not indigenous to our world. It is also evident that our ancestors referred to these advanced non-human intelligences (NHIs)[51] as the "gods"—i.e., the "Elohim."

* * *

---

[50] Susan Milbrath, Ph.D., is an affiliate professor of anthropology at the University of Florida, and a curator of Latin American Art and Archaeology at the Florida Museum of Natural History.

[51] I will mostly be using the term "non-human intelligence" (NHI) as opposed to "extraterrestrial" (ET), due to the connotations that the later term has with popular preconceptions and misconceptions about the topic.

It must be understood that according to scientific probability, it is actually *less* likely that Earth could be the only living planet in an active universe. Astronomers estimate that in our galaxy alone there are approximately a 150 to 200 billion stars. The astronomer Frank Drake, at the University of California at Santa Cruz, calculated the probability of solar-systems that could contain Earth-like planets capable of supporting intelligent life to be in the range of 200,000 (Kaku 1994: 283). Although this calculation cannot be considered to be absolute, at no times do the probability equations ever reach zero (Denzler 2001: 75-76). Moreover, it is now calculated that there are even more planets orbiting stars than even Drake determined (Kaku 2014: 301). Using a special microlensing technique, a team of researchers (i.e., Probing Lensing Anomalies NETwork, i.e., PLANET) concluded that planets that orbit stars are the rule, rather than the exception (Cassan, et al. 2012: 167-169). An approximate estimate from this survey indicates the probable existence of more than 10 billion terrestrial planets across our galaxy alone!

Astronomers are beginning to detect the presence of these other planets. In 1995, two astronomers at the University of Geneva discovered a planet that was orbiting a Sun not unlike our own (Borenstein 2007). These discoveries set off a planet hunting race. By the end of 1996 six more planets were discovered. In 2007, a team of European astronomers discovered a planet ("581c") orbiting a red dwarf sun, which exists in a circumstellar habitable area (i.e., "goldilocks zone") that is considered possible for life to exist (Than 2007). In 2015, astronomers discovered the closest earth-like planet yet that is circling its own sun-like star, in the constellation Cygnus. This planet (Kepler-452b) orbits at approximately the same distance, and is also approximately the same size as Earth. It is also interesting that it is well over 1 billion years older than our own planet. The discovery of Kepler-452b was announced together with the report of 521 other planetary candidates, which brings the total number at the time of writing to over 4,696. What is more, is that a dozen of those planets are within habitable zones (Khan 2015). In 2016, astronomers discovered three planets orbiting an ultracool dwarf star (Trappist-1) that are similar to both the size and temperature of Earth (Chu 2016).

Likewise, in 2016 a potentially life-supporting earth-like planet (Proxima b) was discovered next door to our own solar-system (Escude, et al. 2016: 437-440). It can therefore be surmised that not only is life on other planets possible but even likely.

> Furthermore, the Hubble Space Telescope has given us an estimate of the total number of galaxies in the visible universe: one hundred billion. Therefore, we can calculate the number of earthlike planets in the visible universe: one hundred billion times one hundred billion, or one hundred quintillion earthlike planets.
>
> This is a truly astronomical number, so the odds of life existing in the universe are astronomically large, especially when you consider that the universe is 13.8 billion years old, and there has been plenty of time for intelligent empires to arise—and perhaps fall. In fact, it would be more miraculous if another advanced civilization did not exist.
>
> —Michio Kaku,[52] *The Future of the Mind*

A planet in our own solar-system has even come close to qualifying. Mars may seem like a dead planet, but an increasing amount of evidence is beginning to suggest otherwise. In 1976, the Viking orbiter took pictures of what appear to be river-bed channels running along the surface of Mars, as well as a polar ice-cap. In 2008 NASA's Phoenix Mars Lander also uncovered water in the form of ice that was uncovered in the Martian Soil (Brown, et al. 2008). In 2013, NASA's Mars rover Opportunity also discovered signs of water, which indicates that at least microbial life on Mars may have once been possible (Wall 2013). These findings were confirmed by Mars rover Curiosity (Wall 2013). In 2015, Mars Reconnaissance Orbiter

---

[52] Michio Kaku, Ph.D., is a professor in theoretical physics at the City College of New York and a visiting professor at the Institute for Advanced Study at Princeton and New York University.

provided evidence that water still continues to flow on Mars (Dunbar 2015).

The signs are pointing toward the existence of a living universe. It can therefore be posited that the refusal by some to comprehend its capacity is in fact only a reflection of that individual's own limited mentality, and not that of reality itself.

* * *

Where then did these so-called "star gods" originate? It may be telling that the ancient Egyptians were especially interested in Sirius, as well as Orion, Alpha Draconis, and Ursae Minoris. Another location that is being reported by present-day contactees from around the world is the seven-star cluster of the Pleiades (Bader 2007: 296; Marciniak 1992). It is therefore interesting that according to some Lakota and Dakota Cherokee native American accounts, human beings are said to be descended from star people from the Pleiades (Mack 1999: 170). It is also compelling that representations of the Pleiades have been found in some early Mesopotamian cylinder seals. In the following image, for instance, the seven stars of the Pleiades are depicted in the upper left corner. Directly below the Pleiades, a flying craft descends from the sky. On the right side, the god (believed to be Shamash) meets with a servant (Figure 8).

(Fig. 8)

However, astronomers calculate that the Pleiades is only about a 100 million years-old. This is problematic since our own Sun/star is approximately 4.6 billion years old. It is therefore unlikely that advanced beings could have come from a system that is younger than our own. The same can be said of Sirius (200 to 300 million years old) and Orion (20 million to only 3 million years old).

Another location that is being reported by contactees[53] is Venus (Tumminia 2007: xxviii). However, not only is Venus approximately the same age as the Earth but its dense carbon-dioxide atmosphere and sulfuric acid cloud cover is incompatible with the requirements for life to exist.

However, it may be the case that the NHIs are not originally from either Venus, Sirius, Orion, or the Pleiades, but rather have established what contactee George Van Tassel referred to as "substations" (Van Tassel 1952: 45-46). Contactees[54] explain that the NHIs are not effected by harsh conditions on Venus because (according to Luis Fernando Mostajo Maertens)[55] the NHIs dwell in subterranean constructions. If this is true, it indicates that this remote location may be a sort of base of operations that the NHIs use for the purpose of monitoring events on Earth. Indeed, the planet's inhospitable atmosphere would mean that it is unlikely that they would ever be discovered and disturbed by their human subjects.

It is therefore intriguing that the planet Venus was so especially regarded by the pre-Columbian Maya.[56] For example, their primary calendar (the *Tzolk'in*) was based on the position of Venus. Likewise, the El Caracol observatory at the Mayan site of Chichen Itza was

---

[53] e.g., Howard Menger, George Adamski, and Luis Fernando Mostajo Maertens. However, I have been unable to validate the authenticity of their reports. Indeed, Menger's films and Adamski's photographs in particular are suspicious, and some of the claims made by others, such as Luis Fernando Mostajo Maertens, are difficult to confirm.

[54] *I Rode a Flying Saucer: The Mystery of the Flying Saucers Revealed.* (New Age Publishing Co., 1952).

[55] archive.org/details/GalacticDiplomacy.LuisFernandoMostajo

[56] See the Dresden Codex.

specifically aligned with the transit of Venus. Although these orientations may pertain to time-keeping (e.g., the *Sheaf* interval cycle), it might also be possible that this specific location was chosen as a base reference point because the Maya and the Aztecs were seeking to align themselves with what they believed was the home world of the gods—specifically, the divine being Quetzalcoatl/Kukulcan, who was the primary deity who was associated with this planet (Aveni 2001: 26, 84, 275; Milbrath 1999: 35-36). Indeed, it is known that Mayan/Aztec astronomy was practiced not solely for mundane reasons, but rather primarily for divination purposes, and that celestial bodies and astrological mythology were linked with the existence of the gods, who were thought to be able to influence inter-planetary events (Aveni 2001: 186; Milbrath 1999: 1). It is also compelling that, although some of the other planets in the solar-system were known to the early Meso-Americans, no single planet was as especially regarded as Venus (Aveni 2001: 26, 38; Milbrath 1999: 34-36). Mainstream scholars surmise that Quetzalcoatl transformed himself into this planet after he departed from this world (Aveni 2001: 145). However, another possibility is that Quetzalcoatl did not become Venus itself, but rather he transported himself and dwelled on (or inside) this planet. This accords with the information that is recorded in the Florentine Codex,[57] which reports that the Aztecs believed that the gods came from the heavens.[58]

Nevertheless, it must be acknowledged that the Meso-American record is markedly abstract, ephemeral, complex, culturally affected, and incomplete—especially due to the fact that many records were destroyed by the Pauline Christian Conquistadors. Indeed, there is much that is not known, which renders arriving at a single deduction, at least in regard to the Meso-American extraterrestrial deity

---

[57] The Florentine Codex (or: The General History of the Things of New Spain) is a sixteenth-century manuscript written by the Franciscan friar Bernardino de Sahagun and his partners.

[58] Book XII, chapter 8. This also accords with other Mesoamerican legends that report that the god Ome tecutli also came from the heavens (Book X, chapter 29).

hypothesis, unlikely. Nevertheless, it can be maintained that preternatural beings who were associated with outer-space and other planets in pre-Columbian Meso-America remains pertinent.

In his book *ULOs: Unidentified Lunar Objects Revealed in NASA Photography*, (which was the basis of the 2014 documentary *Aliens on the Moon*) Alan Sturm presents NASA photographs of peculiar structures on the moon. These structures seem to indicate the presence of substations on the Moon. My own interpretation of these images is that most can be explained as naturally occurring formations that only appear to have symmetrical features. Nevertheless, there were at least a couple of images that I was not able to dismiss. Unfortunately, due to the low resolution of the photographs, as well as the disjointed and faint coarseness of the structures themselves, it is difficult to be certain. One possible reason for the roughness of the structures might be because these are not recent constructions, but rather were built and abandoned long ago—perhaps in ancient times. In this case, at least a few of these photographs should not be dismissed.

According to some reports, the original planet that the humanoid NHIs originated from is located in the constellation Lyra (Huntley 2002: 59-63; Stone 1995: 4-6, 20). It is therefore compelling that in 2013 it was announced by NASA's Kepler space observatory mission that at least two of the discovered planets in Lyra (Kepler 62e and Kepler-62f) exist in habitable zones (Kaku 2014: 297; Overbye 2013). Moreover, the age of the planets were calculated to be in the 7.4 billion-year-old range, which would make these objects nearly 3 billion years older than Earth (Borucki, et al. 2013). It is also compelling that this discovery occurred *after* claims were made by contactees of a Lyran origin. However, at the present time, the Lyra contactee reports seem especially unreliable. Moreover, I have not been able to verify any ancient records that make any substantial reference to the gods and Lyra—that is, besides the Grecian lyre of Orpheus folktale, which does not apply.

In regard to the general subject of origin, it must be acknowledged that we are simply missing enough credible information at this time.

Hopefully, reputable contactees will be able to provide details in the future.

One of the problems that we must deal with pertains to the problem that one faces when directly in the midst of the *mysterium tremendum*. During such boundary-breaking encounters, it is not always possible for the experiencer to be thinking in terms of practicalities and the scientific method.

* * *

In his popular and controversial book *The Twelfth Planet* (and the subsequent editions of the Earth Chronicles series), Zecheria Sitchin, presented the theory that the god-like beings who are referred to in the ancient Mesopotamian texts as the "Nefilim" (i.e., Nephilim), as well as the "Anunnaki" (i.e., Anunna), were a race of extraterrestrials from an undiscovered planet in our own solar-system called Nibiru. Note: Sitchin will be referred to frequently throughout this work due to how influential his contribution has been to the extraterrestrial deity hypothesis. However, no evidence exists that indicates that Nibiru was any planet in our solar-system that has not yet been discovered. Moreover, in this modern age of advanced technology (e.g., the Hubble telescope), it is extremely unlikely, if not utterly impossible, that another life-supporting planet in our own solar system has not yet been discovered. Furthermore, in the Babylonian Enuma Elish text, Nibiru seems to be referred to not as a planet but rather as a star. Sitchin also asserted that Nibiru was referred to as "the planet of crossing," which he believed was a reference to its unique orbit (Sitchin 1976: 237, 240). However, according to the Babylonian Mul-Apin text, other planets, such as Venus, Mars, and Saturn, were also referred to as planets that "crossed the sky" (Watson and Horowitz. 2011: 65, 67, 189). Furthermore, these texts are younger than the older Sumerian reports, and therefore most likely do not reflect the original definition. The original Sumerian texts all describe Nibiru (i.e., Nibru, or Neburu) in local terms; namely, as a city. It was also described as a location that could be reached by boat,[59] and where the "black-headed people" (i.e., Earthlings) also dwelled.[60] If these descriptions are accurate, they nullify an extraterrestrial location. However, all of the different descriptions could be telling us that the Mesopotamian's understanding of Nibiru was not much better than our own. Moreover, if the city really was as grand as the Sumerian texts indicate, then it is certainly peculiar that it has never been found on Earth—especially when other major cities of the same era and location have been located. Therefore, even though Sitchin's Nibiru theory is most likely incorrect, until the city is located on our own planet, it cannot be ruled out that it existed on some other. If Nibiru does exist in our own solar-system, the most likely location for such a city would not be on an undiscovered planet, but rather Venus. In this case, Nibiru might have been an underground substation. This would explain why it was described in the Sumerian texts as a sanctuary of the gods.

In 2016, another planet in our solar system was allegedly discovered. I say "allegedly" because at the time of writing, only its indirect effects have been detected. Of course, an undiscovered planet in our own solar-system would seem to fit with Sitchin's Nibiru theory. However, this hypothetical planet, which at this time is being referred to as "Planet Nine," is out of the immediate range of the Sun, which means that life on Planet Nine would not be possible. Sitchin asserted that the Nephilim were able to evolve in such a sunless environment because the planet was warmed from an interior source (Sitchin 1978: 254). However, I have not been able to locate any text, both inside or outside of Mesopotamia, that indicates this. Indeed, Sitchin does not cite a source for this. Moreover, even if lifeforms could exist in such an inhospitable environment they most certainly would not appear like life that has evolved on a sun-effected planet. Scientists at the California Institute of Technology who are studying Planet Nine, believe that *if* this planet does exist (or any other undiscovered planet in our own solar system) it would most likely be another cold dead "ice giant" (e.g., Uranus and Neptune). If the planet

---

[59] Enki and the World Order
[60] Lament for Nibru

had some sort of interior heat source, and because heat is usually associated with light, it should be easier to detect. However, neither Planet Nine nor Sitchin's twelfth planet have been directly observed. Indeed, this purported planet is so far out in the dark abyss of space that its surface is not even reflecting any Sun light at all. It can be concluded that Planet Nine cannot be Nibiru, and therefore Sitchin's theory pertaining to this particular subject cannot be correct.

* * *

It must be understood that sightings of unexplained aerial phenomena have been occurring for centuries.[61] [62] In 776 CE, for example, two fiery "shields" appeared over the Saxon army (Vallee and Aubeck 2009: 68).[63] In 1580 CE, a round and fiery object was seen rising over the Straights of Magellan, before it transformed into the shape of a "lance" (Vallee and Aubeck 2009: 171-172).[64] Likewise, the following is reported to have occurred in Attica Greece, in 404 BCE. In this account that was recorded by the second- and third-century CE

---

[61] Another document that is commonly adduced by "ancient astronaut" theorists is the Tulli Papyrus, in which it is reported that fiery disks were seen flying over ancient Egypt. However, the history of this document is uncertain, indeed even dubious, and therefore will not be submitted as evidence.

[62] Many more examples can be found in Harold T. Wilkin's book, *Flying Sauces on the Attack* (Citadel Press, 1954), as well as *Wonders in the Sky: Unexplained Aerial Objects from Antiquity to Modern Times: and Their Impact on Human Culture, History, and Beliefs*, by Jacques Vallee and Chris Aubeck (Tarcher Penguin 2009). However, it must be noted that in some cases I was unable to validate the source information that was provided. Therefore, further research and confirmation is needed.

[63] Original source: *Carolingian Chronicles: Royal Frankish Annals and Nithard's Histories*. Bernhard Walter Scholz (trans.). The University of Michigan Press, 1972: 53-55.

[64] Original source: Pedro Sarmiento Gamboa. *Viajes Al Estrecho de Magallanes Por el Capitan Pedro Sarmiento de Gamboa, En los Annos de 1579. y 1580. Y Noticia de la Expedicion Que despues hizo para poblarle*, 1768: 205.

theologian Clement of Alexandria, there is a reference to an aerial "pillar":

> When Thrasybulus was bringing back the exiles from Phyla, and wished to elude observation, a pillar became his guide as he marched over a trackless region ……..
> The sky being moonless and stormy, a fire appeared leading the way, which, having conducted them safely, left them near Munychia, where is now the altar of the light-bringer.
> —Clement of Alexandria, Stromata (Book I, chapter 24)

This same imagery appears in the Old Testament/Tanakh, where we are told that Yahweh guided the Israelites in a "pillar of fire and cloud":

> By day Yahweh went ahead of them in a pillar of cloud to guide them on their way, and by night in a pillar of fire to give them light, so that they could travel by day or night.
> —Exodus 13.21

> During the last watch of the night Yahweh looked down from the pillar of fire and cloud at the Egyptians and threw them into confusion.
> —Exodus 14.24

> When Moses entered the tent, the pillar of cloud would descend and stay at the entrance of the tent, while Yahweh would speak with Moses.
> —Exodus 33.9

In Numbers 12.5, Yahweh is also described descending in a "pillar of cloud." However, it may have not been the pillar that "stood" at the entrance of the tabernacle, but rather Yahweh himself, after having emerged from the object. Indeed, the passage does not indicate that Yahweh himself was the pillar, but rather that he was "in" the pillar:

> Yahweh came down in a pillar of cloud, and stood at the door of the Tent [. . .]
> —Numbers 12.5

The accounts indicate that it was an object could both descend (Exodus 34.5; Numbers 12.5), and ascend (Exodus 40.36; Numbers 9.17). Furthermore, Moses was also able to "enter into" this object (Exodus 24.18).

We are also told that this craft was associated with something that is referred to in the original Hebrew texts as the *kabod* of Yahweh. The meaning of the word kabod is uncertain (Cross 1997: 166). This is because the word can be used in different ways. The kabod is most commonly translated as "glory"; however, this definition is not always applicable in regard to the context in which it appears. According to Exodus 24.17, for instance, the physical form kabod appears as a "consuming fire." The kabod is often described as something that appeared inside a "cloud" (e.g., Exodus 16.10). However, the cloud reference may actually be a primitive description of the luminous nimbus that surrounded the kabod. It is therefore likely that there may be a conflation occurring between the glorious physical appearance of this illuminated object and the glorious presence of Yahweh himself—who is the being who controlled that cylinder-shaped pillar of flame and cloud.

Some scholars have suggested that the fire and the smoke that was seen at the top of Mount Sinai/Horeb during the contact event with Moses (Exodus 19.18, 24.16-17) was nothing more than a volcano (Cross 1997: 167, 169). There are numerous reasons why this interpretation is untenable, indeed even preposterous; which is why I do not even feel the need to expend time and effort into dispelling this conjecture. Indeed, eminent scholar F. M. Cross also dismisses this interpretation in his book: *Canaanite Myth and Hebrew Epic*.

Although we might tend to interpret the image of a pillar as a vertical column, the reports, including the biblical accounts, do not specifically describe it this way. If the pillar of fire, or pillar of cloud, was horizontal, it would match with what witnesses in our own time more commonly describe as cylindrical or "cigar-shaped" UFOs (Kean 2010: 54, 75, 195; Sturrock 1999: 11, 90).

Likewise, in the year 597 CE, a fiery pillar was seen by numerous witnesses rising from the ground in Ireland, before it disappeared into the sky (Vallee and Aubeck 2009: 61-62).[65] In 1110 CE, a fiery pillar was seen moving over a Russian monastery in an intelligently-guided manner. The residents believed that it was an angel of God (Vallee and Aubeck 2009: 97-98).[66] In 1428 CE, a flaming "column" was seen rising into the sky over Forli Italy (Vallee and Aubeck 2009: 122).[67] In 1662 CE, a group of thirty to forty people in England "beheld a great pillar of fire" flying up and down in the sky. The object proceeded to turn into a "blazing star," before changing again into three separate "moons." These smaller luminescent objects then morphed together into a bent "bow"-like structure.[68] In 1721 CE, a slow moving "pillar of fire" advanced toward the city of Bern Switzerland, before it silently burst into three "globes of fire" that flew off in separate directions (Vallee and Aubeck 2009: 249).[69] In 1950, villagers on Nunivak island off the coast of Alaska reported seeing a "pillar of flame," "hundreds of feet in length," rising out of

---

[65] Original source: Adamnana, William Reeves, (ed.) *Life of Saint Columba, founder of Hy.* Book III (Edmonston & Douglas, 1874).

[66] Original source: Samuel H. Cross (trans., ed.), and Olgerd P. Sherbowitz-Wetzor (trans., ed.). *Russian Primary Chronicle: Laurentian Text.* Mediaeval Academy of America. (no date given): 10, 14-15, 204-205. The work (also known as the "Tale of Bygone Years") is believed to have been compiled and written by the monk Nestor. The original book was published in Keiv in about the year 1113 CE (Encyclopedia Britannica Online, s.v. "The Russian Primary Chronicle").

[67] Original source: Filippo Guarini. *I Terremoti a Forli: In Varie Epoche.* Stabilimento, 1880: 12, 13.

[68] *Mirabilis Annus Secundus* (The Second Year of Prodigies), 1662 (LIII): 21.

[69] Original source: Thomas Crawford. *An Account of Terrible Apparitions and Prodigies Which Hath Been Seen Both Upon Earth and Sea, In the End of Last, and Beginning of this Present Year, 1721.* Thomas Crawford, (Publishing date not given): 6.

the Bearing sea. The fiery object disappeared and reappeared three times on its way up before it morphed into an oval shape. The object then changed course and flew northward.[70] A skeptic might argue that this was nothing more than a missile launched from a submarine; however, this incident was reported five years before the first ballistic missile was launched from a submerged submarine (Encyclopedia Britannica Online, s.v. "rocket and missile system"). Furthermore, not only did the flame disappear and reappear three times, and not only did the object transform into an oval shape, and not only did it change course, and not only was the actual physical missile portion itself not described in the account, but no smoke trail was reported either. Therefore, a submerged missile explanation is not plausible—nor does such an explanation account for similar types of sightings before the twentieth century.

In 1952, a fiery cigar-shaped UFO was seen by numerous witnesses in Copenhagen. This same type of luminescent and cylindrical object was also reported in France that same year.[71] Indeed, 1952 was a momentous year in terms of sightings of this particular type of UFO. Formerly classified CIA documents released to the public under the Freedom of Information Act reveal that most of the sightings that were reported during that visitation wave occurred in Africa and the surrounding region, including some parts of Europe. In some reports, the cigar-shaped objects were reported emitting a luminous cloud in their wake. Although, in other instances, the "shinning" crafts were described as not leaving behind any smoke trail at all.[72] It is telling that these cigar-shaped objects were also sometimes seen together with disk-shaped UFOs,[73] and were not always seen flying in a consistent speed or direction,[74] which negates the possibility that these were nothing more than meteors, human-

---

70 bluebookarchive.org/page.aspx?pagecode=NARA-PBB92 -459&tab=1

71 bluebookarchive.org/page.aspx?PageCode=MISC-AFOSR4-133, bluebookarchive.org/page.aspx?PageCode=MISC-AFOSR4-136

72 www.cia.gov/library/readingroom/docs/DOC_0000015467.pdf

73 www.cia.gov/library/readingroom/docs/DOC_0000015470.pdf

74 www.cia.gov/library/readingroom/docs/DOC_0000015470.pdf

made rockets, or missiles. Furthermore, one of these cigar-shaped objects that was seen by numerous people flying along the Algerian coastline during the 1952 wave was described as "enveloped in orange flames."[75] This is indeed very much similar to the UFO that is reported in the Bible.

Indeed, throughout the Old Testament/Tanakh, accounts of an other-worldly individual who flew in the sky are reported:

> Sing to the one who rides through the skies, which are from long ago. He speaks with a thundering voice. Announce that God [Elohim] is powerful. He rules over Israel, and his power is in the skies.
> —Psalm 68.32-34

Among the words that are used to describe the object that Yahweh flew through the skies in was "cloud."

> Yahweh is riding on a fast-moving cloud and is coming to Egypt [. . .]
> —Isaiah 19.1

Likewise, in 1707 CE a "long dark cloud of a cylindrical figure which lay horizontally," was seen on the southern coast of England. The stationary object swelled with brightness and became fiery, before it returned to its original state (Vallee and Aubeck 2009: 246).[76]

In these cases, it is likely that the object was most likely not a literal so-called "cloud," but rather something that appeared like a hazy light-hued object from the ground—which should not be confused with the clouds that are used by UFOs to conceal themselves. Indeed, some of the UFOs emit vaporous smoke that can also be confused for a cloud-like appearance. One of these objects

---

[75] www.cia.gov/library/readingroom/docs/DOC_0000015470.pdf

[76] Original source: John Morton. *Natural History of Northamptonshire: With Some Account of the Antiquities*. 1712: 350.

was described as being "enveloped in smoke trails."[77] For example, in 1952 the citizens of Oloron and Gaillac France witnessed a cylinder-shaped craft that emitted exhaust that was described as a "cottony cloud" (Trench 1966: 31). That same year, a luminous cigar-shaped craft that emitted a silvery smoke trail was also reported by numerous witnesses in Morocco and Tangier.[78] It was during that same wave of sightings that a flying saucer was seen that emitted white smoke over Morocco. Witnesses reported that the object came to a stop before it darted out of sight,[79] which indicates that it was neither a meteor nor any other known human-made object. Similarly, in 1966 an army officer stationed off the coast of Wake Island reported seeing a "cloud" expand with a light inside of it. The object sped away into the sky after hovering for several minutes.[80] In 1984, pilots of two Russian Tu-134A airplanes witnessed an object that initially appeared as a yellow star before it morphed into the shape-shifting green luminescent "cloud." Thousands of people also witnessed the object from the ground. One of them (retired Colonel A. Kovalchuk) described it as a cigar-shaped craft that was enveloped in a green halo (Stonehill 1998: 88). It is therefore compelling that in the ancient Egyptian Pyramid Texts, it is reported that one of the objects that transported the pharaoh up into the air appeared as a "cloud":

> Pepi Neferkare has gone up on a cloud and descended
> [...]
> —The Pyramid Text of Pepi II, Ascending to Nut

Descriptions of so-called "clouds" were not the only objects that the gods were seen moving through the skies in. There are also references to flying objects that made a whirring commotion and stirred up dust as they launched and landed. For example, in the Vedic Hymn to Vata

---

[77] www.cia.gov/library/readingroom/docs/DOC_0000015475.pdf
[78] www.cia.gov/library/readingroom/docs/DOC_0000015469.pdf
[79] www.cia.gov/library/readingroom/docs/DOC_0000015465.pdf
[80] www.nicap.org/bluebook/unknowns.htm

text, we read that the flying "chariot" of the gods whirled dust upon the earth when it thundered up into the heavens:

> Now for the greatness of the chariot of Vata. Its roar goes crashing and thundering. It moves touching the sky, and creating red sheens, or it goes scattering the dust of the earth.
>
> —To Vata, Vedic Hymn, Mandala X, Hymn 168

These descriptions are similar to the "whirlwind," "cloud," and "chariot of fire" imagery that is found in the Bible. It can be contended that in both of these cases, the so-called "chariot" may simply refer to a transportation craft, not to a literal two-wheeled carriage that is pulled by horses. Therefore, it is possible that the so-called "horses" that pull the "chariot of fire" that appear in 2 Kings 2.11 was a cultural interpretation that was inserted by a pious scribe who was not present to personally witness the events. Indeed, it should also be understood that people tend to describe UFOs in the shape of objects that are familiar with. This explains why UFOs in the past were more often described as shields, torches, spears, cannon balls, pillars, etc. (Vallee and Aubeck 2009), while UFOs in the present era are more often described as cigars, spheres, and saucers.

In 2 Kings 2.1, Elijah is taken up into heaven in something that is referred to as a "whirlwind."

> When Yahweh was about to take Elijah up by a whirlwind into heaven, Elijah went with Elisha from Gilgal.
> —2 Kings 2.1

These accounts imply that the whirlwind had the ability to transport an individual up into the sky. This may be because the flying objects may have not only kicked up a whirlwind of dust and smoke as it ascended and descended but also may have even made the sound of a whirling wind type of noise as it did so. Indeed, some modern-day UFO witnesses also report the sound of whirring, rushing, or humming sounds that emanate from UFOs (Sturrock 1999: 272; Trench 1966: 153, 159). In his book *Intruders*, author Budd Hopkins presents interviews that he conducted with people who have come in to close proximity to extraterrestrial crafts; some of whom describe ships that made a "humming or whirring" sound, like that of a "spinning top." Other witnesses report UFOs that rapidly rotate (Davis 2014: 91; Kean 2010: 66; Sturrock 1999: 226) and UFOs that stir dust up into the air as they depart from the ground (Sturrock 1999: 264). Indeed, accounts of UFOs that rotate or spin and that emit sound are numerous. In 1955, for example, a woman in Switzerland witnessed a large oval metallic object in the sky that was maneuvering in an intelligently-guided manner. The object emitted sparks and smoke that began to spin around the object as the craft ascended.[81] In 1952, "fiery disks" were seen that emitted buzzing and hissing sounds over the Belgian Congo. A fighter pilot who was dispatched to intercept the flying objects reported seeing a craft that was enveloped in spinning fire, while its inner metallic core remained completely still.[82] A similar type of UFO was reported by separate witnesses on the Ivory Coast that same year.[83] Likewise, in East Germany, also in 1952, two witnesses, one of which reported in a sworn testimony before a judge that he and his daughter had seen a rotating conical craft that was surrounded by a "ring of flames" launch from a remote forested area where it had landed. The craft at first made a humming sound before it lifted from the ground, at which point it changed to a higher-pitched whistling sound. This object was also reported by other witnesses in the area, who erroneously assumed that they must have seen a "comet"—despite the fact that it could not have been a comet due to the fact that comets traverse outside of Earth's atmosphere. The initial witness was at first reluctant to come forward and testify because he thought that he must have seen a top-secret Soviet military project and believed that he might be in danger for "knowing too much." What is also interesting about this

---

[81] bluebookarchive.org/page.aspx?pagecode=MISC-AFOSR4-325&tab =2
[82] www.cia.gov/library/readingroom/docs/DOC_0000015463.pdf
[83] www.cia.gov/library/readingroom/docs/DOC_0000015469.pdf

case is that before the object departed, two humanoid men in "shiny metallic clothing," who were standing near the craft, were also observed.[84] In more primitive times, such advanced beings were perceived as gods.

A similar type of rotating and fiery UFO was recorded in the sixteenth century by the Franciscan friar Bernadino De Sahagun. In this account, a fiery "whirlwind" is reported to have descended from the sky in an intelligently-guided manner:

> And when night had fallen, it thereupon rained at intervals; it sprinkled at intervals. It was already deep night when a fire appeared. As it was seen, as it appeared, it was as if it came from the heavens like a whirlwind. It went continually spinning about; it went revolving. It was as if the blazing coal broke into many pieces, some very large, some only very small, some just like sparks. It arose like a coppery wind. Much did it rustle, crackle, snap. It only circled the ramparts at the water's edge; it went toward Coyonacazco. Then it went into the middle of the water. There it went to disappear.
> —Bernardino De Sahagun, *Florentine Codex (General History of the Things of New Spain)* (Book XII, chapter 39)

* * *

Witnesses report that these objects sometimes not only display multicolored lights and a fiery appearance, but when they are seen in daylight they usually appear to be constructed out of a metallic substance. These types of descriptions concur with the biblical account that is reported in the book of Ezekiel, in which a deity is reported to have descended from the sky in a luminescent object:

---

[84] www.cia.gov/library/readingroom/docs/DOC_0000015464.pdf

> I looked, and I saw a windstorm coming out of the north—an immense cloud with flashing lightning and surrounded by brilliant light. The center of the fire looked like glowing metal.
> —Ezekiel 1.4

Here we have the first description of a metallic object that was used by Yahweh to travel through the sky in, which is also related to a whirlwind. In some translations, this metal is reported having a golden "amber" color; and in others it is called "electrum," which is a pale golden- and silver-colored alloy. These descriptions comport well with modern-day UFO reports.

Much has already been written and said by both ancient astronaut theorists and skeptics concerning this incident that is reported in the book of Ezekiel. Skeptics point out that the Ezekiel account is consistent with other motifs from the same time and region and that there is nothing "alien" about it; while others deny the legitimacy of the book of Ezekiel altogether.

In order to understand what is happening in this biblical book, it is necessary to understand the history of the book of Ezekiel itself. The fact is that the history of the book of Ezekiel is especially dubious, and is therefore most likely not a literal account of what actually happened. This is because it does appear that the author, or more likely, authors, employed the use of Babylonian iconography, which the priest-scribes had been exposed to during the captivity period. Indeed, it is understood by scholars that the book of Ezekiel was written during that time, and that it was most likely not written by Ezekiel himself, but rather by members of a priestly school in Babylon who were influenced by him (Blenkinsopp 1996: 167). It was most likely these priests who were responsible for the stylistic embellishments that appear. This may have been done not only to impress upon the reader the supernatural nature of Yahweh but was also used to put him into a relevant context—which was greatly needed during that time.

It must be understood that witnesses tend to interpret these types of extraordinary events according to their belief system.

> Both my own research and the preceding historical review suggest that AAN (Alien Abduction Narratives) Tellers bring their own personal and social histories into the events, "experiences," and testimonies surrounding their "alien abduction." From a psychosocial perspective, they are influenced by a number of factors, including their religious backgrounds, personal spiritual styles, quasi-religious movements around abduction, leadership and integration into specific groups, narratives of conversion, and rituals of conversion.
>
> When religious contexts are already operating within an individual's social network, they can influence the process by which a personal narrative develops. Just as seemingly ordinary events can acquire new meanings during a process of religious conversion, it is not difficult to see how a series of attentional changes and anomalous perceptions can be integrated into an AAN. In addition, AANs can influence the personal religious context by adding a new subjective component to an existing religious framework.
>
> —Scott R. Scribner,[85] Alien Abduction Narratives and Religious Contexts[86]

It cannot be emphasized enough that most people in the ancient world were illiterate. Indeed, it is calculated that only five to ten percent of

---

[85] Scott R. Scribner studied astronomy and cultural anthropology at Harvard Universities Mount Hermon Liberal Studies Program; physics and sociology at Rensselaer Polytechnic Institute, where he was a member of a NASA development group; philosophy, hypnosis, and cognition at the University of New Hampshire; psychology and theology at Fuller Theological Seminary, and theoretical psychology at Greenwich University.

[86] From the book: Tumminia, Diana G. (ed.) *Alien Worlds: Social and Religious Dimensions of Extraterrestrial Contact* (Syracuse University Press, 2007).

people were able to read and write (van der Toorn 2007: 10). Those who experienced sighting or contact events did not have the ability to document their account themselves. Consequently, these stories were passed on orally for years and centuries before they were documented by a qualified scribe. Therefore, not only did the accounts clearly change over time but it is evident that the scribes inserted their own interpretations on top of the older accounts.

The book of Ezekiel begins with an account of a man by a river. The account is suspicious because it begins in first-person, before switching to second-person in the second paragraph. This second author then proceeds to repeat redundant information, which clearly indicates that this document has been affected by a second person. The account then switches back to first person before the vision of Yahweh takes place.

The book of Ezekiel is not a literal description of what actually happened; however, like most other ancient accounts, it was most likely based on a genuine experience that was embellished at a later time. Therefore, in order to understand this arcane report, it is necessary to deconstruct it down to its primary components. In this case, what we are left with is a man who witnessed phenomenal lights in the sky; a luminescent and metallic flying craft that made a turbulent sound as it descended, and a preternatural being who piloted the craft. It is also telling that the flying craft was described as capped with a "shining" "dome":[87]

> A voice came from above the dome over their [cherubim's] heads as they stood still with their wings lowered. Above the dome over their heads was

---

[87] The word "dome" is translated from the Hebrew word *raki'a*. In some translations it is interpreted as "firmament," which is also a reference to an expansive arch shape. Indeed, the dome translation appears in many versions, such as the Common English Bible, the Complete Jewish Bible, the Contemporary English Version, the Expanded Bible, God's Word Translation, the Good News Translation, the Names of God version, the New Century Version, the New Revised Standard, and others.

> something that looked like a throne made of sapphire.
> On the throne was the figure that looked like a human.
> —Ezekiel 1.25-26

The dome is said to have been made out of "crystal"; however, the reference to crystal may have been used to describe a smooth, hard, and glass-like material that was "shinning." Here is how the word appears:

> Something like a dome was spread over the heads of the living creatures. It looked like dazzling crystal.
> —Ezekiel 1.22

This account is also very similar to sightings that occur in the present era, in which fiery and metallic flying saucers are seen that appear to have glass-like domes on top of them. Since domes mostly exist in a circular shape, it is possible that the object that Yahweh was seen upon may have been what we would refer to in our own time as a saucer. In this case, it can be maintained that the account that is reported in the book of Ezekiel is indeed consistent with modern-age descriptions of UFOs.

Not only did Yahweh travel through the skies in what we would today refer to as an Unidentified Flying Object but he referred to his servants as "earthling man" (Isaiah 5.15, 13.12). Although the term for a man from earth is sometimes translated as "man," "people," or "humanity," etc., such translations are not the most accurate interpretation of the original word. The word for man is translated from the original Hebrew word *Adam*, which derives from the word for earth—i.e., *Adamah* (Singer, et al. 1901: 178; Gruenwald 2012: 61-62). Therefore, "Adam" was not originally a name, but rather a generic term for a man from the earth—i.e., Earthling. Indeed, the New World Translation Bible translates the word Adam as "earthling man." This is the most accurate translation. It is also possible that this could be a reference not only to soil but to the planet Earth itself.

Unlike his human followers, Yahweh himself was able to dwell up in "heaven":

> Elohim rides through the ancient heaven, the highest heaven. Listen! He makes his voice heard, his powerful voice.
> —Psalm 68.33

However, in these cases, the word translated as "heaven" or "the heavens" (*shamayim*) originally referred to the firmament of the sky, not to the blissful after-life "Kingdom of Heaven" (*Basileia tou Ouranou*) that was referred to by Jesus centuries later. Indeed, the original Mesopotamians did not distinguish between heaven and the sky (Hutter 1999: 388). The following biblical passage confirms this:

> There's no one like your Elohim, Jeshurun! He rides through the heavens to help you. In majesty he rides through the clouds.
> —Deuteronomy 33.26

Indeed, the Hebrew Bible indicates that Yahweh dwelled in the sky:

> Elohim looks down from heaven on Adam's descendants to see if there is anyone who acts wisely, if there is anyone who seeks help from Elohim.
> —Psalm 53.2

> Then from heaven, your dwelling place, hear their prayer and their plea, and uphold their cause.
> —1 Kings 8.49

In the following passage, Yahweh indicates his concern for the presence of the other Elohim, who, like himself, also had the ability to fly through the sky and perform works that appeared to be supernatural:

> Do not learn the ways of the nations. Do not be frightened by the signs in the sky [. . .]

—Yahweh, Jeremiah 10.2

Here we find another example of Yahweh—or rather, the agent of Yahweh, who was speaking on his behalf—showing concern for the ability of his rivals in the council of the Elohim. And here again, it is reported that the foreign gods were not mere wooden carvings and golden idols, but rather were actual living beings who also had the ability to show "signs in the sky," just as Yahweh did. A reference to these other gods appears in the following Mesopotamian text:

> When Anu had gone up to the sky, and the gods of the Apsu had gone below, the Anunnaki of the sky made the Igigi bear the workload.
> —Atra Hasis

Likewise, in the ancient Greek texts (e.g., Hymn to Demeter), there are also references to "gods who dwell in the sky."

Both the biblical and the coinciding ancient regional texts report that these "signs" were the fiery cloud-borne whirlwind vehicles (i.e., so-called "chariots" or "boats") of the Elohim/gods. The airborne NHIs are also indirectly referred to in the following passage:

> Who in the skies can compare with Yahweh? Who among the heavenly beings [sons of Elohim] is like Yahweh?
> —Psalm 89.6

Strange objects in the sky are not only related to the Old Testament world but in the New Testament book of Acts we are told of an encounter that the apostle Peter had with a flat-shaped object that came down from out of the sky. In this account, the four-cornered object was compared to a "sheet"; however, we know that it was not a literal sheet, because as it got closer he was able to look "into it" and see animals inside. He was told telepathically (i.e., in a "trance" state) that the animals were there for him to eat:

> I was in the city of Joppa praying, and in a trance I saw a vision: a certain container descending, like it was a great sheet let down from heaven by four corners. It came as far as me. When I looked intently at it, I considered, and saw the four-footed animals of the earth, wild animals, creeping things, and birds of the sky. I also heard a voice saying to me, 'Rise Peter, kill and eat!"
>
> —Peter, Acts 11.5-7

This description of a trance-like telepathic state of communication also correlates with present-day alien contactee reports (Jacobs 1992: 25, 87; Mack 1999: 10, 15, 56).

A similar encounter is reported to have occurred in the year 675 CE. According to this account, a luminous object that was described as "a light from heaven, like a great sheet," descended from the sky while a group of nuns were praying. The object maneuvered around the area before returning to the sky (Vallee and Aubeck 2009: 64).[88]

Of course, the other instance of an Unidentified Flying Object appearing in the New Testament occurred when a shining white orb appeared in the sky to the three wise-men and directed them to the birth place of Jesus. This object is compared to a "star" by the author; however, we know that it was not a literal star because it had the ability to navigate through the sky in an intelligently-guided manner and direct the wise-men to the manger. Indeed, numerous reports of UFOs that are described as "stars" or spheres of light, are reported in the present era as well.

* * *

According to Zecheria Sitchin, the extraterrestrial presence on this planet can be traced back almost half a million years (Sitchin 1978:

---

[88] Original source: Bede, *Ecclesiastical History of England*, (Book IV, chapter 7) Henry G Bohm, [reprint 1867]. Originally completed in the eighth century CE.

253). It is an age that some might associate with the mythological civilizations of Atlantis, Lemuria, and Mu (which is a subject that will be examined further ahead). These are civilizations that are said to have lasted in some form or another right up until a cataclysm that some would say corresponds with the biblical flood catastrophe—which Sitchin dates to somewhere between the eleventh- and ninth-centuries BCE (Sitchin 1976: 401-402, 409). In this case, does any evidence exist that could support such incredible claims?

At this time, no confirmed physical evidence exists that can be dated that far back in time—at least in the form of cities. However, we must also consider the reports of humanoid footprints that have been found embedded into stone surfaces around the world. The prints could only have been made when the material was still soft, thousands and even millions of years ago. Some of these tracks seem to even appear alongside dinosaur tracks. However, it is now known that most of these cases are either misidentifications or even outright hoaxes (e.g., the Paluxy River bed and the Carson City Nevada tracks). Nevertheless, it is worth questioning if we have received a full, unbiased, and irrefutably proven conclusion for every case.

In ancient times, sightings of such anomalous footprints were apparently more common than they are today (Mayor and Sarjeant 2001: 144-145, 148). The ancient Greek historian Herodotus,[89] for instance, reported seeing one of these giant prints that had been impressed into a rock in Scythia. Perhaps some of these prints were either collected or unintentionally destroyed, or covered up, or eroded by natural processes. One such set of humanoid prints that were embedded into the surface of a rock was discovered near the American town of Braystown (i.e., Braytown) in 1883 (Mayor and Sarjeant 2001: 155-156).[90] The barefoot footprints, sixteen inches long, were purportedly of an abnormally large humanoid with six toes. A large humanoid with six toes coincides with the biblical account of a giant with "six fingers on each hand and six toes on each

---

[89] The Histories, Book IV

[90] Original source: *American Antiquarian and Oriental Journal*, Vol. 7. Stephen D. Peet (ed.), (F.H. Revell, 1885): 365-366.

foot" (2 Samuel 21.20). In this case, could the tracks have been carved into the rock by a Christian? Although this is a possibility, conclusive evidence to support this explanation remains elusive. In the late nineteenth century, "good sized" humanoid footprints were discovered embedded into a rock on a mountain trail in the state of Delaware (reported by Professor J. Brown of Berea College) (Mayor and Sarjeant 2001: 156-157).[91] These types of findings should not be automatically disregarded just because of the mere possibility of hoaxes, or because of the existence of a few higher profile erroneous straw-man cases, or because they do not conform to the standard paradigm. If even just one such track is found to be credible the significance must be acknowledged. Moreover, it may also be possible that these are not the tracks of human-beings, but rather of the humanoid NHIs and their terrestrial progeny (more on this subject later). Indeed, what is also interesting about the larger prints is how they correlate to modern-age reports of "tall" humanoid NHIs (Davis 2014: 44, 49; Mack 1999: 67, 75, 194; Trench 1966: 150, 158; Tumminia 2007: 314).

Another anomaly that is in need of unbiased examination are the abnormally large humanoid skeletons that have also been uncovered (Dewhurst 2014; Fowke 1902: 73, 358). Many of these seven- to eleven-foot-tall skeletons were discovered in the nineteenth century inside North-American earthen burial mounds that were hundreds and in some cases thousands of years old (Fowke 1902: 145, 394, 458).[92] In 1897, for example, in the state of Wisconsin the skeleton of a man that was over nine-feet-tall was discovered in one the mounds.[93] In 1886, other skeletons of "gigantic" size were unearthed in

---

[91] Original source: *The American Antiquarian and Oriental Journal*, Vol. 7. Stephen D. Peet (ed.), (F.H. Revell, 1885): 39.

[92] Many more of these cases are reported in Richard J. Dewhurst's book, *The Ancient Giants Who Ruled America: The Missing Skeletons and the Smithsonian Cover-Up*. (Bear & Company, 2014).

[93] query.nytimes.com/gst/abstract.html?res= 9B02EED61330E333A25753C2A9649D94669ED7CF

Cartersville Georgia.[94] What is more, is that these finds correlate with the native American Iroquois, Kansa, Omaha, Osage, Pawnee, and Shawnee tribe legends, that all tell of an early time in which a race of giants inhabited their land (Mayor 2005: 40-41, 54-55, 192-94; Peet 1885: 107-108). In most cases, artifacts that were found next to those skeletons included jewelry, weapons, and other items; many of which were made out of copper[95] (Fowke 1902: 345, 725; 456; Powell 1894: 24, 302-303, 324-325). Some of the artwork found in the mounds display pre-Columbian Mexican motifs (Fowke 1902: 705; Powell 1894: 306). This has caused some to propose that the mound builders may have derived from the Aztec or Toltec cultures (Powell 1894: 599). These are compelling locations because of how well it accords with the South and Meso-American legends that also tell of an early age of giants.[96] According to these reports, both the Mexican Tlaxcaltecas and the South American Inca went to war against those beings (Mayor 2005: 74-78, 80-81). Similar accounts also appear in the legends of the Nahua, the Olmecs, and the Xicalancs (Helmolt 1901: 250, 264). It was believed that the city of Teotihuacan was originally inhabited and therefore presumably built by those tall-statured people (Mayor 2005: 88). It is therefore significant that the early ancient history of the city of Teotihuacan is unknown (Encyclopedia Britannica Online s.v. Teotihuacan; Hearn 2015). According to the Aztec Legend of the Suns, before the present age there were four previous ages that are referred to as "Suns." The first Sun was the age of giants (Aguilar-Moreno 2006: 139). The Aztecs believed that it was these beings who constructed the first pyramids (Mayor 2005: 88-89). Likewise, according to the Ixtlilxochitl people, Cholula (site of the Great Pyramid) was also originally inhabited by

---

[94] query.nytimes.com/gst/abstract.html?res= 9C07E1D71330E533A25756C0A9629C94679FD7CF

[95] In some cases, iron, silver, and gold were also found inside the mounds; however, it is likely that these metals had been acquired by the Europeans in a later era.

[96] Some of the reports can be found in the *Florentine Codex*, the *Ixtlilxochitl Codex*, The *Chronicle of Peru* by Pedro de Cieza de Leon, and the records of the Catholic priest and historian Juan de Torquemada.

giants—that is, before they were defeated by the Olmecs (Mayor 2005: 89).

The sixteenth-century Franciscan friar Bernadino De Sahagun reported in the *Florentine Codex* (i.e., *General History of the Things of New Spain*), that the original "Tolteca" (i.e, Toltecs) were a tall, wise, and peaceful people. These early inhabitants of Mexico not only built pyramids but were said to have been proficient in the use of plants and medicine. According to the indigenous reports collected by De Sahagun and his associates, the Toltecs also sought out "mines of silver, gold, copper, tin, mica," and "lead."[97]

What is also notable about the Toltecs is their relationship with the legendary being Quetzalcoatl, who they regarded not as a mythological flying serpent, but rather as a divine humanoid patriarch.[98] Skeptical scholars surmise that this anthropomorphic depiction was nothing more than a human priest or king who had assumed the role of the god. The claim is based on the apparent fact that some prominent humans in a later period were bestowed with the name of this deity. However, these different identities do not negate the possibility of an original anthropomorphic deity. Moreover, the serpent was most likely the animal symbol of the humanoid god. Indeed, a similar tradition appears in the Mesopotamian pantheon (more on this subject later).

It is likely that it was not the Toltecs themselves who were the giants, but rather they seem to have been influenced by those earlier foreigners. It may also be possible that contact between the natives and the foreigners occurred in an earlier age. In this case, the Toltecs were not the NHIs themselves, but rather were the offspring of those beings. The individual who came to be known as Quetzalcoatl may have been responsible for this intermingling. Unfortunately, in regard to these matters, the record is so deficient that it is difficult to delineate a clear picture of events. Hopefully, future archaeological discoveries will provide further elucidation.

---

[97] Book X, chapter 29

[98] Book X, chapter 29

Ancient accounts of giant humanoid beings who were the builders of massive stone structures was also reported by the Samaritans of the Levant (which is an ethnic group that is related to the Israelites). According to their account, the giants who survived the biblical flood were the original builders of the tower of Babylon (Mussies 1999: 345). A similar account is also reported in the ancient Grecian tale of Otus and Ephialtes (i.e., Aloadae). According to this classic narrative, the two titans attempted to invade the heaven of the gods by piling up mountains on top of each other.

It appears that race of tall-statured people were eventually driven out of South and Meso-America. They seem to have migrated into North America, all the way up to the copper-rich regions of the Great Lakes, and then back down into mid-western America (Fowke 1902: 51, 88). A South and Meso-American origin may also explain why some of the North American mounds were constructed in the shape of truncated and terraced pyramids (Fowke 1902: 51, 60, 71; Powell 1894: 23, 235, 297). One of the largest mounds (i.e., Monk's Mound at Cahokia in Illinois) has a larger base circumference than both the Great Pyramid in Egypt and the Pyramid of the Sun in Mexico (Speirs 2014: 1).

Of course, the current scholarly consensus maintains that the mound builders were nothing more than an ancient native American race from the archaic Woodland Period—most likely the Adena and Hopewell cultures. However, it may actually be the case that those indigenous people were only imitating the works that had been constructed by their neighbors in the region. Indeed, this not only explains the existence of South and Meso-American artwork but the existence of abnormally large skeletons.

Another alleged example of the existence of an ancient race of tall-statured people in America that is usually cited in books and websites that are dedicated to these subjects, are the red-haired giants of Lovelock cave in Nevada. According to Paiute native American legends, a red-haired people called the *Si-Ti-Cah* (i.e., *Sai'i*) once inhabited their land. In 1911, two bat guano miners discovered material remains in Lovelock cave. Among the numerous objects recovered were red-haired mummies. According to the testimony of one of the miners (James H. Hart), one of the mummies was a "giant" (Loud and Harrington 1929: 169). However, this corpse was mentioned along with the discovery of a mummified man who was "six feet six inches tall." It does appear in this account[99] that the mummified corpse and the giant were intended to be descriptions of the same individual. However, six and a half feet is not an extraordinarily abnormal height. Perhaps the tall corpse appeared like a giant compared to the other smaller human remains that were found in the same cave. Indeed, smaller mummies were mentioned. Therefore, the giant interpretation, in this particular case, is most likely not accurate.

In her book *Fossil Legends of the First Americans*, Adrienne Mayor[100] also dismisses the legends of the Lovelock cave giants. However, she claims that the skeletons that were thought to be giants were probably nothing more than mammoth and/or cave bear bones; however, mammoth and bear bones were *not* found in the Lovelock cave. Mayor is aware of this but assumes that because mammoth and bear bones were found a hundred miles away that they could have become conflated. However, this is extremely unlikely. Furthermore, her explanation is plausible only in instances where single bones or teeth are found. Although in Lovelock cave, a fully intact humanoid corpse was discovered; not indiscriminate bones (Loud and Harrington 1929: 168). What is also concerning about Mayor's interpretation is that she uses this explanation for *all* giant remains; while simultaneously omitting instances where entire humanoid skeletons and mummified remains were uncovered. Although she does cite one instance in which giant humanoid skulls were found, she assumes that they only "seemed human" (Mayor 2000: 81). However, human skulls are distinct from animal or dinosaur skulls— including even our nearest primate ancestors. Mayor's interpretations seem to be based on a provisional Occam's razor type of technique, in which the examiner, who may be seeking to maintain a pre-

---

[99] Llewellyn Lemont Loud and Mark Raymond Harrington. *Lovelock Cave*, Vol. 25, (University of California Press, 1929).

[100] Historian and scholar of natural history folklore at Stanford University.

established conclusion, assumes that whatever is the easiest, or standard, or preferred explanation, must be the truth.

Based on the accumulated reports, it is apparent that isolated mammoth and dinosaur bones were indeed *sometimes* confused for the bones of giant human-beings by the native Americans, and these bones were most likely erroneously conflated with the earlier legends of giant men. However, the two should not be confused.

Although the Lovelock cave discovery cannot be used to evince the existence of giants in North America, there are plenty of other cases that do. One such discovery was unearthed in the same Nevada area just seven years earlier. In 1904, workers on a construction project in Winnemucca Nevada purportedly uncovered part of a skeleton of an abnormally tall man.[101] According to the report that was published in the *Saint Paul Globe*, the remains were examined by a "Dr. Samuels" who "pronounced them to be the bones of a man who must have been nearly eleven feet in height." In this case, it is extremely difficult to believe that a modern doctor was unable to tell the difference between human and animal/dinosaur remains.

Furthermore, in 1921, the full skeleton of a "man" between eight- to nine-feet-tall was discovered in a mound in Pennsylvania that had been excavated by Doctor W. J. Holland, curator of the Carnegie Museum, and his assistant, Doctor Peterson.[102] This is another case where the remains were examined by qualified professionals, and were not determined to be the remains of either an animal or a dinosaur. Indeed, according to the article that was published in the Stanstead Journal: "the curators believe the man whose skeleton they secured belonged to the mound builder class." (Note: I contacted the Carnegie Museum to inquire about this discovery, but my message was never responded to. There are several ways that this can be interpreted. Without hearing an explanation from the museum for

---

[101] Reported in the *St. Paul Globe*, St. Paul, Minnesota, January 24, 1904: 28, under the title: "May be Related to the Cardiff Giant: Bones of a Human Skeleton Eleven Feet High are Dug Up in Nevada."

[102] *The Stanstead Journal*. February 3, 1921: 4.

their silence on this matter, I will not attempt to assume which is the most likely.)

Moreover, I have also not been able to find any indications that any of these discoveries were used by anyone in order to make money for themselves—which, of course, is the primary indication of a hoax. These cases are not limited to North America. In 1890, the bones (humerus, tibia, and femoral mid-shaft) of a humanoid who would have stood approximately ten- to eleven-feet-tall was discovered by the anthropologist Georges Vacher de Lapouge in a Bronze Age burial grave in France that dated back to the Neolithic period.[103] The remains were unearthed in a cemetery in which regular size human skeletons were also found. The bones of the "giant of Castelnau" were studied by specialists at the University of Montpellier and their authenticity were confirmed. In this case, we must inquire if the abnormal size was the result of some biological condition, such as a pituitary gland disorder. However, ten to eleven feet is still an extremely anomalous height, even for an individual suffering from a gland disorder. Indeed, most individuals inflicted with this disease do not grow to such a height. The tallest living human-being ever recorded (according to the Guinness Book of World Records) was eight feet eleven inches. However, the Castelnau giant was approximately almost two feet taller. Moreover, the pituitary disorder explanation still does not account for the existence of numerous giant skeletons that were discovered in North America. The possibility that all of these remains were nothing more than individuals who were suffering from gland disorders is, of course, preposterous.

What then happened to all of the remains? It is documented that some of the skeletons were so far decayed that they could not be preserved (Fowke 1902: 116; Powell 1894: 84, 405, 588), while other skeletons simply disappeared (Fowke 1902: 73). It might also be possible that at least some of these remains were repatriated back to

---

[103] Published in the French journal *La Nature*, Vol. 18, Iss. 888, June 7, 1890: 11-12, under the title: "Le Géant Fossile De Castelnau" by G. de Lapouge. Also reported in *The Popular Science News and Boston Journal of Chemistry and Pharmacy*, Vol. XXIV, No. 8, August 1890.

the Native American community, who may have buried the evidence. Another possibility is that material that is considered to be too far out on the "fringe" is unlikely to be displayed in a mainstream establishment. Indeed, as will be made evident further ahead—when the subject of parapsychology is examined, people tend to reject, either consciously or subconsciously, information that conflicts with their beliefs. A clear example of this problem (e.g., confirmation bias; experimenter expectancy effect; inattentional blindness) occurs in political scenarios, when a report that favors one particular political party is seen as legitimate by members of that party, while the same report is seen as illegitimate by members of the opposing party—and vice versa when the facts are reversed. Indeed, as will be made apparent further ahead in this investigation, the belief that mainstream science is free from bias is a consequential misconception.

Richard J. Dewhurst presents a strong case for the existence of these anomalous skeletons in his book: *The Ancient Giants Who Ruled America: The Missing Skeletons and the Smithsonian Cover-Up*. Dewhurst contends that archaeological evidence of a giant race of mound builders in ancient America has been marginalized, ignored, and even deliberately repressed by some in the scientific community—specifically by those at the Smithsonian Museum. Dewhurst presents evidence in the form of a report to the secretary of the Smithsonian titled, "On Limitations to the Use of Some Anthropologic Data." The document was composed by the original director of the museum, John W. Powell. Powell was a committed denier of what he referred to as the "romantic fallacy" of a lost race of ancient Americans (Powell 1894: xli). Powell made it clear in the letter that evidence that would seem to prove the existence of "races of antiquity," and what he referred to as "the lost tribes"—which is a reference to Mormon creed, were not to be exhibited. Consequently, the abnormally large skeletal remains that were sent to the museum were never seen again. Indeed, this policy explains why references to skeletons of abnormally large size are inexplicably absent from

Powell's book on the subject on the mounds[104]—and yet, such references do appear in the works of his colleagues (e.g., Gerard Fowke).[105]

It was around that same time that other persons who were affiliated with the Smithsonian also repudiated any theory that might work to subvert the standard interpretation that they were committed to retain. Foremost among the critics was the militant polemicists Ales Hrdlicka. Although, it is now known that some of his biased beliefs, such as a "short history" of America, have been proven to be wrong (Silverberg 1968: 229).

It is likely that the cover-up that was initiated by members of the Smithsonian Museum was a part of a secular effort to undermine anything that might provide credibility to the biblical report of the existence of "giants" (Deuteronomy 2.11, 2.20; 3.11-13; Numbers 13.33; 1 Samuel 17.4; etc.). By suppressing the evidence, Powell and his associates must have believed that they were doing humanity an evolutionary favor.

The list of all of the anomalous skeletons, footprints, and objects that have been discovered around the world and documented in scientific publications, such as The *American Journal of Science* and the *Scientific American*, are numerous and would require the writing of another book altogether in order to fully examine. Therefore, I must refer the reader to the works of Richard J. Dewhurst (The *Ancient Giants Who Ruled America*), Brad Steiger (*Worlds Before Our Own*), Karl Shuker (*Unexplained*), and Michael A. Cremo, and Richard L. Thompson (*Forbidden Archeology* and *The Hidden History of the Human Race*). I submit these examples not as proof, but only as material that deserves further examination by *unbiased* researchers. Furthermore, *all* of the cases must be honestly studied before any verdict is reached; as opposed to just a few easy straw-

---

[104] *Report on the Mound Explorations of the Bureau of Ethnology. The Twelfth Annual Report of the Bureau of Ethnology to the Secretary of the Smithsonian Institution*, 1890-'91. 1894.

[105] Gerard Fowke. *Archaeological History of Ohio: The Mound Builders and Later Indians*. (Ohio State Archaeological and Historical Society, 1902).

men hoax cases (e.g., "The Cardiff Giant")—which is the technique that is often resorted to by the so-called "debunkers."

What can be stated for certain is that the belief in the existence of giant humanoid beings is world-wide. The coherency of this matter is significant and undeniable. The ancient Greek writer Hesiod, for example, wrote about these beings in the Theogony (eighth century BCE), in which he associated the giants with the demi-god "Titans" of ancient Greece. Similarly, in the Indian Ramayana epic[106] are more reports of so-called "giant" humanoid beings. Likewise, in the biblical book of Genesis, we are told of a class of beings known as the "Nephilim":

> The Nephilim were on the earth in those days—and also afterward—when the sons of God [Elohim] went to the daughters of humans and had children by them. They were the heroes of old, men of renown.
> —Genesis 6.4

Sitchin claimed to have translated the word Nephilim down to the original Semitic root *n-ph-l*, which he concluded could be interpreted to mean "to be cast down." (Sitchin 1978: 171). According to his interpretation, this could mean that these were beings who descended to the Earth from the skies. However, this is unlikely because, according to Genesis 6.4, the Nephilim were the hybrid offspring of human females and the "sons of Elohim." It is therefore difficult to believe that the Nephilim would have descended to Earth if they had been born on Earth. Indeed, the Bible reports that these beings dwelled on Earth.

Another more common interpretation stems from the root *naphal*, which could also be interpreted as "fallen" (Coxon 1999: 619-620). This definition posits that the Nephilim were the "fallen ones." Judeo-Christians attribute this definition to the fallen angels who were cast out of heaven, along with Lucifer. However, this too is not an accurate definition. This is because the Nephilim themselves were not

---

[106] Canto XXII, Dasaratha's Speech, etc.

the so-called fallen angels, but rather they were the alleged progeny of those illicit copulations. The fallen designation would seem to be a derogatory reference to such offspring—perhaps the equivalent of the present-day term "bastard." However, this too is unlikely because the Nephilim were described as "mighty men," and the "men of renown." In other words, they were described in honorable terms.

It is therefore likely that the translation stems from the Aramaic noun *npyl/naphil*, which means "giants" (Brown, et al. 1906: 658; Jastrow 1903: 923; Coxon 1999: 619-620). Indeed, this is how the word appears in a majority of the original translations. This Aramaic term may have been appealing to the Hebrews because the word could have a double meaning, in that it could denote a race of *giants* who were causing the *downfall* of the world (Jastrow 1903: 923). However, this interpretation does not rule out a non-human genus, since—as will be explicated further ahead—the reason for their so-called "giant" stature may be attributed to their extraterrestrial genetics.

The link between the Nephilim and the giants is confirmed in the following biblical passage:

> [. . .] They said, "The land we explored devours those living in it. All the people we saw there are of great size. We saw the Nephilim there (the descendants of Anak come from the Nephilim). We seemed like grasshoppers in our own eyes, and we looked the same to them.
> —Numbers 13.32-33

The "sons of Anak" is a reference to the name of a giant who is called Anak[107] (Joshua 21.11). An etymological conflation between the name Anak, and the sons of Anak—who are also referred to as the

---

[107] According to *Peake's Commentary on the Bible* (edited by M. Black, Routledge, 2001), and *The Hebrew and English Lexicon of the Old Testament* (edited by Francis Brown, Oxford University Press, 1906), the word Anak means "long neck"; which is most likely a reference to his giant stature.

Anakim—and the Mesopotamian Anunnaki is unlikely; although it is possible that there was indeed a relationship between these beings nevertheless.

The Anunnaki/Anunna were a race of divine beings in Mesopotamia. They are sometimes associated with the underworld (e.g., Nergal, Ningishzida, etc.); however, most of them cannot be limited to this classification. Indeed, the word Anunnaki itself means "those of princely blood," or "royal offspring" of the patriarchal god, Anu (Gwendolyn 1998: 7; Leeming 2005: 21).

Although it is also likely that not all of the so-called "giants" were of royal lineage, it is possible that the mysterious beings known as the Igigi were of the same or perhaps similar genus, and yet not of direct aristocratic descent—which is why they are referred to as laborers in the Atra Hasis text. It is also possible that the beings who are referred to in the Judean/Israelite records as the sons of Anak, as well as the Rephaim and the Emites, were also of direct Anunnaki descent.

> The Emites used to live there. These people were strong, as numerous, and as tall as the people of Anak. They were thought to be Rephaim, like the people of Anak, but the Moabites called them Emites.
> —Deuteronomy 2.10-11[108]

Some of these archaic beings were portrayed as a race of giants. The giant called Goliath, who fought with David in the popular biblical story, was most likely one of these beings. Indeed, Mesopotamian pictorial representations show what appears to be a race of giants towering over their terrestrial human servants (Figure 9, 10, 11).

---

[108] Also repeated in *Deuteronomy* 2.20.

(Fig. 9)

(Fig. 10)

(Fig. 11)

Mainstream scholars have devised their own explanation for these depictions. They surmise that this was the artist's way of aggrandizing the god or king. While this may be true in the case of human kings who were striving to link themselves with the gods (e.g., the Victory of Naram Sin stele), such an interpretation does not explain the textual descriptions of literal giant beings. For example, the biblical character Goliath was never described as a god or even a king; nor are the non-biblical characters depicted in some of the visual images of the same approximate era and region presented in a god-like context (e.g., Figure 11). Therefore, when the human kings were depicted as giants this was actually a way in which to promote the perception that the king was a direct descendant—i.e., Nephilim—of the original titan lords—i.e., the Anunnaki.

All of these different textual reports and archaeological discoveries from around the world are pointing toward the literal existence of an ancient and possibly prehistoric race of extraordinarily tall-statured people. One possible reason for this anomalous genetic trait may have been because those giant human-beings bore a closer genetic connection to the "tall" NHI progenitors.

However, it should also be noted that not all of these sightings of humanoid NHIs are of a tall stature. According to some contactees, their appearance is indistinguishable from our own. Perhaps this is who the Igigi were. This would seem to indicate the existence of not only heterogeneous races among the NHIs but also a common genetic link with *Homo-sapiens*.

According to some of the previously cited South and Meso-American legends, the first pyramids were constructed by the giants. What is also compelling about those immense stone structures is how similar they are to the ziggurats of Mesopotamia. In this case, we must ask if there could there be a connection between the Nephilim, the giants, and the ziggurat/pyramid temples in both the Old and New Worlds? If this is true, then before the arrival of those tall-statured people from the other side of the world, the indigenous tribes of the New World lived not unlike some remote Amazonian indigenous tribes do to this day. When the native people came across those anomalous foreigners who were constructing massive stone complexes on their land, they may have felt threatened, and attacked the intruders. This explains the legends of a war between the natives and the giants. The foreign beings seem to have chosen a pacifistic stance, and abandoned the area, rather than engage in violent warfare with the indigenous people. When the natives constructed their own stone complexes, or in the case of Teotihuacan and perhaps Cholula as well, took over the settlements that had been abandoned by the foreigners, they may have been imitating what they had seen those people do before them. This is because although the natives may have felt threatened by those mysterious intruders, they seem to have been impressed by them as well. In this case, perhaps the legend of Viracocha/Kon Tiki derived from not only sightings but perhaps even some peaceful contact experiences. This theory explains why the Inca and the Aztecs mistakenly believed that the Spanish Conquistadors were those returning advanced beings. Indeed, the Florentine Codex[109] confirms that not only did the Spaniards have "white" skin and "long beards" but some had "yellow" hair. It is also true that the European Spanish are mostly taller than the indigenous races of South and Central America. Indeed, the height difference between these two racial stocks can still be seen to this day.

This theory not only explains the similarities between the pyramids of the New World and the Mesopotamian ziggurats, and it not only explains the presence of "giants," but it also explains the legend of Viracocha—and perhaps even the legendary deity, Quetzalcoatl as well.

* * *

The information that derives from numerous sources is pointing towards the conclusion that the Elohim who appeared in physical form—as opposed to spiritual form—were a race of preternatural beings that included persons who were referred to in Mesopotamia as the Anunnaki/Anunna. It was these beings who may have engendered a hybrid sub-species who were referred to as the sons of Anak, the Rephaim, the Emites, and the Nephilim.

The Elohim who has come to be known to the world as Yahweh/Jehovah was allegedly not only a primary figure in the council of the Elohim but he too was of similar genus. Indeed, the only extant data that indicates that Yahweh was unique comes to us from claims that were made by Yahweh's devoted followers

---

[109] Book XII, chapter 7.

themselves. However, when concurrent ancient texts are examined and compared it becomes obvious that there were others in the council assembly of the gods—which the Old Testament/Tanakh admit existed—who not only bore similar abilities but were also attributed with the very same elite status:

> [. . .] father Enki, you are king of the assembled people.
> —Enki and the World Order
>
> Elohim rules the nations. Elohim sits upon his holy throne.
> —Psalm 47.8
>
> O lord of wide understanding, who is as wise as you? Enki, the great lord, who can equal your actions?
> —Enki and Ninmah
>
> Who in the skies can compare with Yahweh? Who among the heavenly beings is like Yahweh?
> —Psalm 89.6
>
> O Marduk, thou are chiefest among the gods, they fate is unequaled.
> —Enuma Elish (The Epic of Creation)
>
> God [Elohim] presides in the great assembly; he renders judgment among the "gods" [Elohim].
> —Psalm 82.1

In great majority of cases, the Mesopotamian texts predate the biblical record. Therefore, this is not a case of so-called "false gods" attempting to imitate the original real God. Moreover, it is evident that Yahweh and his retinue of priest/prophet/scribes were carrying on in the same way as others had done before them.

I alone am the creator. When I came into being, all life began to develop. When the almighty speaks, all else comes to life.
—Atum, Hymn to Atum

See now that I myself am he! There is no god besides me. I put to death and I bring to life, [. . .]
—Yahweh, Deuteronomy 32.39

It can therefore be concluded—as will be explicated in upcoming chapters—that the only attribute that was unique about Yahweh and his movement was not only its zealous adherence to tradition but its realization of the power of the written word.

* * *

I feel compelled once again to reference the George Van Tassel case, which I consider to be a possible legitimate contactee. I say "possible" because of the proliferation of hoaxes in recent times that has corrupted the field.

George W. Van Tassel was an aircraft mechanic and owner of the Giant Rock Airport in the southern California Mojave Desert area in around the mid-twentieth century. According to Van Tassel, the NHIs channeled information about future events on this planet to him telepathically (the scientific veracity of this claim will be examined further ahead), which was sometimes accompanied by light phenomena in the sky; although, according to Van Tassel, the NHIs did also apparently meet with him in person on at least two occasions. In biblical times, Van Tassel would have been referred to as a "prophet," and the extraterrestrial beings who he was in contact with would have been referred to as "angels," and/or the "gods," or "God." What is also compelling about the Van Tassel case, is that when the names of some of the NHIs who he was in contact with are put side by side with the names of the gods of the ancient world, a significant similarity occurs (Van Tassel 1952): e.g., Ashtar Anshar; Molca

Milcom; Singba Shiva; Tolta Tonatiuh; Klacta Kartikeya; Noot Nut; Rea Rhea.

Another possible legitimate modern-day contactee is James Gilliland. Gilliland is the owner of the ECETI (Enlightened Contact with Extraterrestrial Intelligence) Ranch in Washington state. Gilliland emphasizes the spiritual relationship with beings who he refers to as "off world visitors." Thousands of visitors to his ranch have witnessed UFOs flying around the nearby Mount Adams area. Military aircraft have also been observed attempting to intercept these ships. Some of these sightings were photographed and videotaped by witnesses and are available to be viewed at his website, eceti.org. Not only are anomalous lights in the sky frequently observed around his property but some, such as Gilliland himself, have experienced direct personal contact events with humanoid NHIs. It is Gilliland's aim to facilitate a "Galactic Exchange Center" and thereby create a positive environment for peaceful interaction with the visitors. One message that has been consistently coming through via the contactees is that the NHIs do not—or perhaps, no longer—wish to be regarded as "gods."

* * *

Another reoccurring aspect that continues to appear in many contact cases are reports of a spiritual or inter-dimensional element (Mack 1999: 17, 64, 68). It has therefore been speculated that these otherworldly visitors are not extraterrestrial at all, but rather are inter-dimensional "ultraterrestrials."[110] The collective information is indicating that they have evolved to a point where they are both. Indeed, this spiritual element accords well with the ancient reports of the interdimensional star gods:

---

[110] According to Chinese Taoist philosophy, these higher beings are called *xian* (transcendents).

An important aspect of the [Egyptian] gods is their *ba*. The *ba*, often translated 'soul', is an hypostasis[111] of the gods (or the dead) in their capacity to move from one realm (one reality, one plane of being) to another. Thus the dead are present among the living as *ba'u* (the plural of *ba*), [...]
—Karel van der Toorn,[112] "God," *Dictionary of Deities and Demons in the Bible*

Indeed, the similarity between the NHI contact event and the religious theophany are significant. Such experiences have even given rise to so-called "UFO cults" in the modern times—e.g., Aetherius Society, The Raelians, etc. (Tumminia 2007). It can be contended that many of the ancient religions began as what we would refer to today as UFO cults.

Contactees often have imminent metaphysical messages for humanity, which include moral injunctions, apocalyptic pronouncements, and themes of being chosen by a higher power. Such people often classify their own experiences as spiritual and transformational, and their messages become new mythological material for anyone who will listen.
—Anna E. Kubiak,[113] *Aliens from the Cosmos*[114]

---

[111] Hypostasis refers to an underlying reality, substance, or state of being. It usually denotes the soul/spirit.

[112] Karel van der Toorn, Ph.D., is a professor of religion and society at the University of Amsterdam.

[113] Anna E. Kubiak, Ph.D., is a cultural anthropologist and an associate professor at the Institute of Philosophy and Sociology of the Polish Academy of Sciences.

[114] From the book, Tumminia, Diana G. (ed.) *Alien Worlds: Social and Religious Dimensions of Extraterrestrial Contact.* (Syracuse University Press, 2007).

It should be understood that despite conditions that we modern-age human-beings consider to be primitive, ancient human-beings had the same cognitive capacity that we do. Therefore, there is no reason to believe that ancient people fabricated god-like humanoid beings entirely from their imaginations. Rather, the signs are pointing toward the conclusion that the ancient people based their descriptions on actual sightings and encounters. This explains why they put so much time and effort into not only sacred religious functions that were dedicated to these other-worldly beings but to the construction of enormous temple complexes. The purpose of the ziggurat towers (as opposed to the Egyptian pyramids) may have been to provide an inviting location for the star gods to land and bless the people with their guidance and protection. The massive effort that it took to build those structures was based on the absolute certainty that those advanced celestial beings literally existed. Therefore, such monumental endeavors cannot be attributed to misconceptions concocted by unevolved minds. Indeed, the pyramid/ziggurats that were built with such extraordinary precision attests to the exceptional genius of those people.

* * *

The images rendered by the artists of Mesopotamia and the Levant indicate that the gods were not only humanoid in appearance but were also clad in the cultural attire of the time and region. However, if the extraterrestrial deity hypothesis is true, then we should expect to see the gods clothed in apparel that is more befitting of advanced interstellar beings—as opposed to wool robes and bull-horned tiaras. Similarly, why would such advanced beings be shown wielding primitive weapons, such as a bow and arrow, and swords?

There are two possible reasons for this: It is likely that the artists were depicting events that they themselves were not present to witness. These events may have occurred many years earlier and developed over the course of time as these legendary accounts expanded and became infused with subjective interpretation, religious devotion, and cultural motifs. The purpose of these images may have been to put the deity into a relevant context for the community, and thereby inspire faith in a system that the priests and the kings were attempting to establish and to maintain. Indeed, we find a similar practice employed in most major world religions in the present era. For example, the common visual conception that the public has of Jesus as a light-complexioned Caucasian man with long brown or blond hair is an altered depiction that was formulated during the European Medieval/Renaissance age. The truth is that most men in Jesus's time and region bore darker Semitic features and wore their hair short, except for the sides of their head (Leviticus 19.27)—which is why the apostle Paul condemned long hair (1 Corinthians 11.14). Indeed, these very same features can be seen today in Hasidic Jewish males.

Another possible explanation is that at least a segment of beings who were regarded as gods were not the NHIs themselves, but rather were the progeny of sexual relations with human-beings. Such "Nephilim" could have become deified. However, this explanation is unlikely due to the fact that if this was the case then such gods, such as Enlil, Inanna, etc., should appear in the practical records, such as the Sumerian King List. However, they do not. Indeed, the gods are depicted, for the most part, as being separate from humankind. It is therefore likely that the artistic renderings of the gods are not literal depictions. Therefore, the indications are pointing toward a more minimalist interpretation.

* * *

We are now left with the following question: To what extent did the Elohim/gods literally exist in our world, and to what extent did they exist only in the aggrandized legends of man?

Over the course of years that I have spent researching and contemplating this question, my view on this matter has slightly changed. I originally suspected that although the ancient reports were obviously embellished, they were still, for the most part, based on actual events and literal individuals. However, based on the data that I have seen in the ensuing years, I now feel that it is much more likely that direct personal interaction between terrestrials and extraterrestrials was, for the most part, minimal. Indeed, this conclusion coincides with the reports of brief encounters with NHIs in current times as well. What seems to have happened is that over the years these events became affected to such an extent that they essentially took on a life of their own.

Another possible albeit less likely explanation, is that at least a small group of Anunnaki, or perhaps even a single individual, settled the Earth and directly acclimated into the terrestrial environment for a longer period of time in an early age. Indeed, this may be what happened during the Enki/Ea period—which will be examined further ahead.

* * *

Of course, present-day skeptics of this type of phenomenon do not accept the vague and extraordinary nature of these reports. Such critics adhere to the maxim: "extraordinary claims require extraordinary proof." The problem with this stance is that it creates a double standard. In this situation, investigators are not concerned about objectively searching for indicators that point toward a possible truth, but rather are more concerned about devising ways of disregarding such indicators in order to maintain a pre-established conclusion. This type of circular reasoning (i.e., informal fallacy) is not only unjustifiable but unscientific.

In regard to the question of evidence, the truth is that UFOs have not only been seen, photographed, videotaped, and filmed by multiple witnesses, but have also been simultaneously tracked on radar (Kean 2010: 1, 37, 58; Sturrock 1999: 28-32, 74, 173-215). Some of the evidence, such as the photographs examined by the 1998 Sturrock panel (more on this subject later), have also passed rigorous scientific examination and show no signs of having been mistaken for natural or man-made objects, or having been tampered with by a hoaxer (Sturrock 1999: 197-233). It is also significant that some of the reports also come from credible sources, such as high-ranking military personnel and police officers (Sturrock 1999: 25, 298-330).

There are also many cases of sightings that were simultaneously witnessed by both witnesses on the ground as well as pilots in the air. Investigations revealed that the multiple witnesses in these cases did not know each other, and there were no indications of collusion (Sturrock 1999: 91, 371). Moreover, biochemical analysis of landing sites conducted by organizations such as GEIPAN (i.e., GEPAN) and SEPRA[115]—who were charged with the task of not simply concocting disingenuous ways of discrediting the extraterrestrial hypothesis but with uncovering the truth—have revealed ground traces in the form of soil compression and landing gear tracks, as well as thermal, magnetic, radioactive, and other biochemical molecular physiological anomalies that had significantly effected vegetation and soil (Kean 2010: 126, 132-134; Sturrock 1999: 93-100, 225, 371). It is also compelling that objects, including human-beings, that have come near UFOs also bear high radiation readings (Kean 2010: 51). How a civilian hoaxer could have implemented this life-endangering feat is unknown. Indeed, some witness who have had close encounters were afflicted with physiological effects; some of which pertained to radiation exposure (Kean 2010: 188; Sturrock 1999: 101-104). Some of the findings of the French GEPAN/SEPRA investigations were consulted in the COMETA report (1999). The COMETA report;[116] (published under the title "UFOs and Defense: What Must We Be Prepared For?") is a document researched, written, and compiled by a committee of scientists, military generals, and other high-level experts. The study concluded that not all cases could be dismissed as mere hoaxes, nor as natural known phenomena, and that the extraterrestrial hypothesis, in some cases, was the only explanation that they were left with (Kean 2010: 1, 120). Likewise, the American "Project Sign" investigation (1948) arrived at this same conclusion (Kean 2010: 104, 193, 210). Furthermore, James E. McDonald, a professor of meteorology at the university of Arizona and a senior physicist at the Institute of Atmospheric Physics, personally

---

[115] GEIPAN (formerly known as GEPAN) (1977-1988) and SEPRA (1988-2004) are French UFO research organizations.

[116] archive.org/details/TheCometaReport

investigated the UFO question and testified in a Congressional hearing in 1968 that UFOs were of "a matter of extraordinary scientific importance." Moreover, he leaned toward an extraterrestrial explanation for the phenomena (Kean 2010: 111-112).

One of the first major scientific research projects regarding the UFO question was conducted by the "Condon Committee" (i.e., The University of Colorado UFO Project) (1966-1968), which was a scientific group that was funded by the United States Air Force. After examining hundreds of cases, the Condon Report dismissed the extraterrestrial hypothesis and concluded that further investigations would be unlikely to produce any findings of scientific value. However, subsequent investigations of the study itself uncovered major flaws. Scientists (e.g., J. Allen Hynek; James E. McDonald; Thorton Page; David Saunders; Peter A. Sturrock, and the investigators at the American Institute of Aeronautics and Astronautics) criticized the report for its unscientific methodology and misleading assumptions. Closer examination revealed that the investigators (primarily Edward Condon himself) had simply disregarded cases that could not be explained. Furthermore, some of the cases that were interpreted were downright laughable. Case 6, for example, examines an incident in which a metallic disk-shaped UFO that emitted flashing red, green, and white lights descended from the sky and stopped at a distance of twenty to thirty feet above one of the several witnesses on the ground. Two police officers who were called to the scene also witnessed the same "vehicle"-sized object. According to the Condon investigation, what the witnesses had seen was nothing more than the planet Jupiter! (Condon Report: 266-70; Sturrock 1999: 25)

It is therefore evident that the true aim of the Condon investigation was not to discover the truth, but rather to produce conclusions that would be the most comfortable to the general public and to the skeptics themselves. Indeed, Condon project coordinator Robert J. Low expressed his concern about the subject in a 1966 memo to two university deans. In the letter, Low revealed his personal opinion that undertaking an investigation into such a controversial topic would jeopardize the prestige of the scientific community. Instead, he advised that the investigation should only "appear a totally objective study"[117] (Dean 1998: 38; Kean 2010: 110-111). Therefore, the actual unscientific aim of the Condon investigation was to dismiss the subject.

In 1998, a panel of scientists examined UFO cases in which physical evidence was collected. The event was organized by Peter A. Sturrock, emeritus professor of applied physics at Stanford University. Although the panel was not convinced of unquestionable proof of the existence of extraterrestrial life, it did acknowledge that some of the phenomena could not be explained by conventional means (Sturrock 1999: 132) and that the subject is worthy of further study[118] (Sturrock 1999: 123, 153, 372). Moreover, the panel concluded that the Condon Report should be rejected (Sturrock 1999: 153).

Another problem that is serving to prevent much-needed serious study of the topic pertains to the lack of funding for high-level scientifically-based investigations due to an erroneous belief that UFOs are not a respectable subject, and that there is no data worth examining (Sturrock 1999: 153).

Of course, the skeptics would have the public believe that only slack-jawed kooks believe in such things as "aliens"; however, not only are there scientists who are open to idea that we are not alone in the universe, but some have even experienced their own sightings:

> Surveys of engineers and scientists in 1971 and again in 1979 showed that anywhere from 18 percent to 22 percent had sighted something that could have been a UFO. Most of those individuals (88 percent) had discussed it only with family and friends. A 1973 poll of members of the American Institute of Aeronautics and Astronautics found that most of the group would not report a UFO sighting unless they were guaranteed anonymity. In 1975 a poll of the American

---

[117] www.nicap.org/docs/660809lowmemo.htm
[118] news.stanford.edu/pr/98/980629ufostudy.html

Astronomical Society (AAS) offered astronomers an opportunity to do just that: It revealed that 5 percent of the respondents had witnessed some kind of puzzling phenomenon in the sky. The conventional wisdom among UFO skeptics, however, was that scientists (and especially astronomers) never had UFO sightings and never made UFO reports. That wisdom was wrong. Not only did a fraction of the AAS report sightings anonymously, some well-known astronomers came forward to report that they, too, had seen a UFO.

—Brenda Denzler,[119] *The Lure of the Edge*

* * *

Another problem pertains to the problem of witnesses who have been instructed to keep quiet. Some witnesses who were interviewed by CIA agents were told that the events that they experienced had "never happened" (Kean 2010: 227)—which were more like instructions than factual deductions. Indeed, other witnesses in the military were also given similar orders (Davis 2014: 61, 142, 148; Kean 2010: 230). Not only have organizations, such as the Air Force, often been suspiciously uncooperative with investigators (Sturrock 1999: 47) but some witnesses were instructed by Air Force OSI (Office of Special Investigations) agents not to talk publicly about their sightings (Kean 2010: 188). Air Force Regulation 200-2 (1954)[120] instructs Air Force personnel on the proper procedure for reporting "Unidentified Flying Objects." According to these guidelines, information pertaining to UFOs can be released to the public only when it is "positively identified as a familiar object." However, "objects which are not explainable" are not to be "revealed."

This policy of denial and suppression compelled a former CIA director, Navy vice admiral and member of the board of governors of

---

[119] Brenda Denzler received her Ph.D. in religious studies from Duke University.

[120] www.cufon.org/cufon/afr200-2.htm

the National Investigations Committee on Aerial Phenomena (NICAP), to issue the following statement:

> It is time for the truth to be brought out in open Congressional hearings. [. . .] Behind the scenes, high-ranking Air Force officers are soberly concerned about the UFOs. But through official secrecy and ridicule, many citizens are led to believe the unknown flying objects are nonsense. To hide the facts, the Air Force has silenced its personnel.
> —Roscoe H. Hillenkoetter[121]

Critics of the extraterrestrial hypothesis cite a lack of evidence, however, complaining about a lack of evidence when evidence is being suppressed is not a legitimate argument. Even Condon himself admitted that "where secrecy is known to exist, one can never be absolutely sure that he knows the complete truth" (Sturrock 1999: 49).

An example of an admitted cover-up by a government official is the 1997 "Phoenix Lights" case. This event that took place in the American state of Arizona and the surrounding region, is regarded by UFOlogists to be one of the most significant mass sighting cases ever. In a press conference after the sightings had become world-wide news, Governor Fife Symington downplayed the sightings, and even mocked the idea of the presence of space aliens; however, years later Symington admitted in an interview that he too had seen the UFO (Kean 2010: 253-261). He stated that the reason he withheld this information was because he felt that, as a public official, that it was his duty to calm apprehensions.

A NASA report (in a contract with the Brookings Institute) titled "Proposed Studies on the Implications of Peaceful Space Activities for Human Affairs" (1959-1960) addressed the "implications of a

---

[121] "Air Forge [sic] Order on 'Saucers' Cited; Pamphlet by the Inspector General Called Objects a 'Serious Business.'" (*Nytimes.com*, Feb 28, 1960). Accessed Oct 14, 2015.

discovery of extraterrestrial life." The conclusion that full disclosure of the reality of "intellectually superior creatures" who are not indigenous to this world could provoke "unpredictable" and "anxiety provoking" reactions (Davis 2014: 155; Strange 2007: 244). Indeed, those who have had a close encounter with this phenomenon will often experience profound ontological shock (Mack 1999: 210). It is this type of disturbance that is preventing the truth from being confirmed to the public. Therefore, the reason for the cover-up is related to the concerns pertaining to mass panic (Davis 2014:148; Kean 2010: 227). Public officials, including even the president himself, are informed about the reality of the existence of extraterrestrial life on a need-to-know basis only (Kean 2010: 236).

This cover-up was reported by a former NASA contractor, Donna Hare, who disclosed that she personally witnessed a NASA employee using an airbrush to cover-up a picture of a flying saucer in a NASA photograph.[122] According to Hare, there were some NASA employees who knew about this policy, and others who did not. This information may have been confirmed by the hacker Gary McKinnon (alias "Solo"), who, in 2001-2002, infiltrated NASA computers. During one of his sessions, McKinnon allegedly discovered un-retouched NASA photographs of extraterrestrial spacecraft.[123] His activities were discovered and he was identified and indicted. (He has so far been able to fight extradition to the United States, where he faces up to sixty years in prison.)

The threat of mass hysteria may be especially manifested in the Judeo-Christian community, who are likely to interpret the presence of otherworldly beings as a sign of the apocalyptic End Times that is described in the biblical book of Revelation. Indeed, there are many Judeo-Christians who not only believe in extraterrestrials but perceive them as being related to demons who have a role to play in the foretold Armageddon (Denzler 2001: 124, 149-152; Scribner 2007:

---

[122] *Aliens on the Moon: The Truth Exposed*. Written, directed, produced by Robert C. Kiviat. (Robert Kiviat Productions, Inc., 2014).

[123] "Hacker Fears 'UFO cover-up'." (*bbc.co.uk*. May 5, 2006). Accessed Sept. 15, 2016. news.bbc.co.uk/2/hi/programmes/click_online/4977134.stm

145; Strange 2007: 244). Disclosure of such an alien reality may even result in unfortunate consequences. Indeed, the Brookings Institute study[124] that was commissioned by NASA determined that the "consequences for earth attitudes and values" for the discovery of extraterrestrial life "may be profound." The report also issued the following warning:

> Anthropological files contain many examples of societies, sure of their place in the universe, which have disintegrated when they have had to associate with previously unfamiliar societies espousing different ideas and different life ways; others that survived such an experience usually did so by paying the price of changes in values and attitudes and behavior.
> —Donald N. Michael,[125] et al., Proposed Studies on the Implications of Peaceful Space Activities for Human Affairs

It is now possible to see how the government groups that were formed to investigate the UFO situation felt obligated to take steps necessary for the preservation of the stability of society by downplaying and even covering up information related to the fact that we are not alone in the universe. For example, in 1952 the chairman of a CIA commissioned UFO investigation committee and Assistant Director for the Office of Scientific Intelligence, Marshall Chadwell, wrote about his concerns regarding mass panic in a memorandum that was addressed to the Director of Central Intelligence (General Walter Bedell Smith) (Fawcett and Greenwood 1992: 125; Haines 2007; Kean 2010: 105):

---

[124] Proposed Studies on the Implications of Peaceful Space Activities for Human Affairs. www.nicap.org/papers/brookings.pdf

[125] Donald N. Michael, Ph.D., was a twentieth-century American professor of psychology and planning and public policy at the University of Michigan. He was a former member of the Joint Chiefs of Staff, the National Science Foundation, the Peace Research Institute, and the Brookings Institution.

> The public concern with the phenomena [of UFO activity] [. . .] indicates that a fair proportion of our population is mentally conditioned to the acceptance of the incredible. In this fact lies the potential for the touching off of mass hysteria [. . .] In order to minimize risk of panic, a national policy should be established as to what should be told to the public regarding the phenomena.
>
> —H. Marshall Chadwell

In another memo, Chadwell admitted that out of 1,500 reported flying saucer cases studied, 20% percent were unable to be debunked. He conceded that the UFO activity was "not attributable to natural phenomenon or known types of aerial vehicles" (Haines 2007; Kean 2010: 105). Chadwell believed that the "flying saucer" problem was real enough that he recommended that the issue be brought to the attention of the National Security Council. However, he was reluctant to conclude that the phenomenon was extraterrestrial in nature; and instead suggested possible Soviet activity[126]—although, it is now known that the Soviets were dealing with their own UFO problem.

Fortunately, thousands of UFO related documents have been released to investigators in America through the Freedom of Information Act. Although, some of the more sensitive data was blacked out, we now know just how seriously the American government, as well as others from around the world, have considered this activity. After investigative teams, such as Project Sign concluded that UFOs were not Soviet, nor any other known terrestrial craft, these types of government-affiliated groups were given new directions; that is, not to investigate but to put the public at

---

[126] At the time of writing, transcripts of these declassified CIA documents are available at the CUFON website: www.cufon.org/cufon/cia-52-1.htm. It is also noteworthy that on its own website, the CIA does not deny the validity of the Chadwell memos: www.cia.gov/library/center-for-the-study-of-intelligence/csipublications/csi-studies/studies/97unclass/ufo.html

ease by down-playing UFO reports (Dean 1998: 36). Although these front groups explained away a majority of the sightings as such improbable things as weather balloons, sunspots, planets, and mass hallucinations, a significant amount of cases remained unexplained (Davis 2014: 47, 58, 60; Kean 2010: 118). Despite this anomalous gap, the public front groups were publicly shut down. Meanwhile, behind the scenes, other groups continued to investigate (Kean 2010: 146-147).

* * *

If the skeptic finds this explanation for the cover-up difficult to believe, then such a person should consider the fact that a panic related to the presence of alien beings has already happened. In 1938, the writer, director, actor, producer, Orson Welles, broadcast a radio play version of the H.G. Welles' classic *War of the Worlds*. The story depicted an invasion of Earth by Martians. The performance was presented as an actual news report in order to heighten dramatic effect. Many of the listeners who tuned in late did not know that they were listening to a fictional enactment, and as a result an actual hysteria ensued. Even though the panic was primarily caused by the report that the so-called "Martians" were using weapons of mass-destruction in order to annihilate the human-race, a following investigation revealed that a substantial portion of the panic was related to the mere existence of aliens.

In his book *The Invasion from Mars*, Hadley Cantril[127] posited that the *War of the Worlds* incident provides insight into the social sciences and mass psychology. Mr. Cantril's research concluded that the root of the hysteria was to be found in the "complete inability of the individual to alleviate or control the consequences of the invasion." Cantril deduced that the aliens not only represented a threat to the security of the world but to people's most valued belief

---

[127] Hadley Cantril, Ph.D., was a psychologist and chairman of the Department of Psychology at Princeton University, and founder of the Institute for International Social Research and the Office of Public Opinion Research.

systems. From a psychological point of view, the aliens were not only destroying our world but our reality.

> When an individual believes that a situation threatens him he means that it threatens not only his physical self but all of those things and people which he somehow regards as a part of him. This Ego of an individual is essentially composed of the many social and personal values *he* had accepted. *He* feels threatened if his investments are threatened [. . .]
> —Hadley Cantril, *The Invasion from Mars*

What Cantril is referring to is related to what Sigmund Freud classified as the s*uper-ego*. This condition is related to the intrinsic human need for a sense of security through the comfort of traditional morality, normality, order, and cultural regulations.

A significant factor related to both the super-ego and the presence of extraterrestrials is their relation to religion. This is an issue that was acknowledged by Cantril's research as well. It was also discovered that some of the victims of the *War of the Worlds* debacle did actually believe that they were in the midst of the apocalyptic Armageddon that is predicted in the Bible. This may be an issue that the NHIs themselves are aware of, which may be why they have chosen not to officially reveal themselves. Perhaps they too are concerned about the possibility of creating a disruptive and dangerous situation. According to some reports by witnesses who have experienced personal contact events, some of the beings do not want to be discovered—that is, by the general public (Jacobs 1992: 309). Indeed, this was the feeling that I myself received from my own encounter. It is therefore apparent that they do not feel that we are ready in our present state of development for contact to take place on a public level.

Furthermore, the presence of the NHIs in ancient times and their relation to the religious doctrines and history of our world presents a formidable predicament, since it is a revelation that is incompatible with traditional and present-day norms. It is evident that there were others beside Cantril who were observing and considering this situation when they were formulating a course of policy in regard to these matters.

It was approximately one decade after the *War of the Worlds* broadcast that the American public faced its first wide-scale wave of public UFO sightings. Some have attributed this to the fact that the NHIs interest was rekindled when we discovered the ability to split the atom when the first nuclear bomb was detonated in 1945; and/or, it may have been due to the fact that this was also the time in which human-beings were discovering the ability that would propel them into space via the development of rocket technology. Indeed, this is what was conveyed to contactee George Van Tassel.

Besides the mass hysteria scenario, other information has also come to light that indicates that a full disclosure might also jeopardize the stability and profitability of certain established economic systems as well. This incompatibility between our two worlds is allegedly due to the existence of extraterrestrial technology that operates on some type of free-energy system that we have not yet discovered. This may have been the same type of technology that was discovered by the nineteenth- and twentieth-century scientist Nikola Tesla. Tesla stated that a free energy system could be derived from the energy that is present throughout the universe. He believed that the ability to attach machinery to the very "wheel-work of nature" could be possible. Conspiracy theorists believe that government agents confiscated his notes after his death in order to prevent the secret of free-energy from being released; and thus, preserve the more lucrative fossil-fuel driven economic system. It is a theory that should not be dismissed.

* * *

What, then, are the intentions of these beings? Are they benevolent or malevolent?

One thing is apparent: that is, if they had wanted to eliminate us it would seem from the reports of their superior technology that they could easily have done so; and yet, they have not. In fact, not only have they not done so but they appear to have taken intentional steps not to directly interfere in the affairs of our world—that is, at least not in any significant way that we know of in the present age. According to George Van Tassel, the NHIs confirmed to him that they "have no reason to destroy" (Van Tassel 1952: 21). Furthermore, contactees also report that the NHIs are concerned about the welfare of our planet, and warn of careless human influence on the environment (Greer 2013: 83; Mack 1999: 88).

It is evident that world governments have concluded that even if the unexplained cases are indeed evidence of the existence of extraterrestrials they do not indicate malicious intent, and therefore do not present a threat to national security. The American director of Air Force Intelligence, General John A. Samford, publicly addressed the topic of UFOs after the 1952 sightings. Although his team were supposedly able to explain a majority of the reports as ordinary objects, he admitted that they were not able to explain all of the cases. Nevertheless, he assured the public that they had no reason to believe that these unexplained objects presented "any conceivable threat to the United States" (Kean 2010: 105). Indeed, the American investigations have concluded that these unidentified flying objects do not represent a serious threat to national security (Fawcett and Greenwood 1992: 2). This conclusion is not based on the premise that the NHIs do not exist, but rather that the NHIs have not indicated hostile intent.

Some might argue that their appearance over military and nuclear missile facilities indicates hostile intent. This is a reference to the 1967 Malmstrom Air Force Base incident, in which ten intercontinental ballistic missiles suddenly became inoperative at the same time that a UFO was observed in the sky above the base (CBS News 2010).[128] Power was restored after the UFO departed. No practical cause of the failure was ever found. This indicates that the NHIs are aware of our military abilities and have the power to shut our defenses down if they so desire. The intention of that display

---

[128] For a full account of this incident see: www.cufon.org/cufon/malmstrom/malm1.htm; www.nicap.org/babylon/missile_incidents.htm

seems to have been to prove to us that if they had wanted to destroy us they could easily have done so.

Skeptics argue that if such beings exist why then would they not announce themselves by landing in a public venue. The reality is that such an incident would be perceived—or rather misperceived—as a threat to national security. It must also be understood that the NHIs themselves would also be putting themselves in harm's way by doing this. To assume otherwise is simply naive.

It is evident that the NHIs themselves do not wish to frighten the public by officially revealing themselves to those who are not ready to receive them. Instead, it appears that the benevolent NHIs—as opposed to the Grays, who seem to be the ones who are abducting people against their will—prefer to make contact peacefully, in controlled private settings, with individuals who they believe are mentally mature. Indeed, it is evident that this situation was examined by both terrestrial and extraterrestrial leaders, and the conclusion that was reached by both parties was the same; namely, that the public is not ready to know the truth.

Although the NHIs have chosen not to officially land in a public venue and ask to be taken to our leader, like a scene from an old Hollywood movie, they have revealed their presence in the skies over public places. Furthermore, as will be explicated further ahead, according to contactees, it is apparent that the NHIs exist in a higher vibrational energy state, and are also endowed with a higher intellectual ability. In this case, it may be difficult for them to lower themselves to our lower earthly plane. Indeed, this is what is being reported by contactees (Mack 1999: 56, 63, 66).

On April 1997, a closed congressional briefing in Washington DC, sponsored by the Center for the study of Extraterrestrial Intelligence (CSETI), as part of its "Project Starlight" program, moved to petition congress to bring about open hearings on the subject of UFOs and extraterrestrials. (See www.disclosureproject.org). According to CSETI director Steven Greer, there are individuals in government and military positions who feel that the public has a right to know the truth, but are obligated not to disclose this information. If the truth were to ever be publicly revealed its impact would surely be felt deeper than only the shock at the confirmation that we are not alone in the universe, and that our extraterrestrial neighbors are far more advanced than us. The other aspect of this revelation is a historic one, in that it could lead to the realization of the historic role that these inter-planetary and inter-dimensional beings have had in our world.

The critics who confidently argue against the extraterrestrial hypothesis are often simply unaware of the facts. Those who become more aware are more likely to accept the reality that we are not alone in the universe. Indeed, this is precisely what happened to astronomer and professor J. Allen Hynek, who began his investigation into UFOs as a skeptical scientific consultant to American governments investigations, such as Project Sign (followed by Project Grudge and Project Blue Book). However, after decades of investigation, he was compelled by the evidence to change from a staunch so-called "debunker" to an advocate for disclosure of the truth of the reality of UFOs. Indeed, Hynek also came to believe that there was even enough evidence to defend the extraterrestrial hypothesis.

Unfortunately, the public has been inundated with Hollywood movie depictions of so-called "aliens." This phantasia has effected mass perception and has caused many to regard the entire topic as nothing more than outlandish concoctions. However, such critics are more influenced by fiction than by truth; just as other reactionary skeptics are more influenced by hoaxers.

Unfortunately, in this twenty-first-century reality, hoaxes are more easily able to be perpetrated by unscrupulous individuals who have become proficient in digital technology that is able to mimic real-life phenomena. The proliferation of not only hoaxes but of the ability of skeptical watchdogs to expose these hoaxes have served to create an overall environment of distrust. Just like the fable *The Boy Who Cried Wolf*, even when genuine phenomena is captured on video, etc. it is usually no longer taken seriously. Even if the skeptics cannot successfully disprove the evidence, the assumption is made that the phenomena must still be the result of mistaken identification or fraud nevertheless. I therefore suspect that even if the NHIs were to land in some public venue and officially announce themselves to the world, it would still be treated as nothing more than a prank or a conspiracy concocted by a so-called "sinister secret government" (which is a term that is frequently used by conspiracy theorists).

The fact is that a vast majority of this modern-day society has become mentally accustomed to this man-made construction that we refer to as civilization. In so-called "developed" countries, it is a work-based and consumerist world of shopping malls, television, internet, alcohol, and drugs—both legal and illegal; as well as academic and religious institutions that do not always provide a complete and accurate account to its pupils. In the case of secular institutions, extraterrestrial beings and the supernatural are subjects that are simply incompatible with the environment they have constructed. However, it must be understood that this was not true in the ancient era. Consequently, assumptions are often made and information disregarded for the purpose of maintaining a familiar and preferred condition. The truth is that the truth cannot always be found inside the rooms of man's construction, and that genuine reports of fantastic events and preternatural beings are, in some cases, indicative of a greater universal reality that exists outside the bounds of our limited everyday world. It is a reality that our ancient ancestors were aware of, but that we have forgotten.

Skeptics, who are usually in the physicalist camp, adamantly relegate themselves to only one level, and expect—if not outright demand—that others do the same. However, it is a position that cannot always be justified. Indeed, it can be contended that the sardonic scoffers who confidently argue against the existence of transcendental phenomena and the extraterrestrial hypothesis will one day be proven to be on the wrong side of history.

* * *

The fact that we are not alone in the universe is not something that should cause us to panic. This is because it is evident that these advanced beings understand the precious value of life and do not wish to disrupt it. Even the Grays, who are apparently the ones who are abducting people against their will, are at least making the effort not only to return their human subjects unharmed but, in most cases, cause their subjects to be unaware of the event; thereby reducing the chance of inducing mental trauma—at least, that seems to be the intention.

Unfortunately, due to the less than ideal condition of present-day society, it is understandable why those in positions of power have concluded that the common public is not ready to know the truth. Indeed, it is a conclusion that I am unable to completely refute.

CHAPTER IV

THE CREATION OF EVOLUTION

Although the medieval Islamic historians were the first to investigate the civilizations that preceded them in the Middle East, the western world did not begin to discover this buried civilization until the current age. The objects that were uncovered in what was once the lands of ancient Mesopotamia—a region that stretches over what is now present-day Iraq, Syria, and some parts of Iran and south-eastern Turkey—have provided valuable insight into early human history. What makes this region of the world and these findings so especially significant is that it is a location that is not only the anthropological "Cradle of Civilization" but the birthplace of three of the major religions of the world.

It was an archaeological team from Britain that uncovered the library of King Ashurbanipal, in what was once the Assyrian city of Nineveh. The find proved to be the largest collection of cuneiform texts ever discovered. Epigraphers were able to translate thousands of pictographic and cuneiform inscriptions by comparing them with writings that were found at the Rock of Behistun in the mountains of Iran. This carved rock provided a side-by-side comparative rendering of three cuneiform languages (i.e., Elamite, Babylonian, and Old Persian). By cracking the cuneiform code, hundreds of thousands of letters, hymns, incantations, law codes, legal records and stories, were able to be translated from inscriptions that were found on various plaques, steles, cylinder seals, and bas-relief carvings. Besides the pottery, statuary, and tools that were uncovered, entire buildings, including over thirty pyramid-like ziggurats were also revealed. These discoveries provided not only valuable insight into the world of the Babylonians and the Assyrians but especially the earlier Akkadian and Sumerian civilizations. Radio-carbon dating of the artifacts indicate that civilization began earlier than previously thought. Most historians now believe that the early Ubaid period began sometime around the sixth or seventh century BCE. Moreover, the traditional time-line of the history of man that was originally formulated by the Roman Catholic Church (i.e., Bishop Ussher), which was thought to not go back further than 4004 BCE, was proven to be wrong.[129]

What was also discovered were earlier Mesopotamian versions of the biblical accounts of the creation of man and the flood story. In regard to the flood account, evidence of a deluge in the form of strata-layers of disturbed soil were uncovered by excavators in the Middle East (Montgomery 2012: 150). Archaeologists discovered the deposits at the sites of the Sumerian cities of Ur and Kish. However, the Ur flood was dated to around 3500 BCE, while the Kish flood was dated to over 500 years later (Kramer 1967: 18; MacDonald 1988: 15-16). Evidence of a second flood that occurred in around 2600 BCE was also discovered. Flood stratum was also uncovered at the city of Shuruppak. This is an intriguing location since this was where the original Sumerian Noah (i.e., Ziusudra) was reported to have lived.[130] The Shuruppak flood was dated to somewhere between 2950 to 2800 BCE, which is within the approximate date of the Kish deluge (Kramer 1967: 18; MacDonald 1988: 16; Montgomery 2012: 154). However, no evidence of significant flooding was found elsewhere in Mesopotamia; not even in the nearby city of Eridu (MacDonald 1988: 19). Based on calculations of the Sumerian King List, scholars have concluded that a cataclysmic flood occurred sometime between 2900 to 2800 BCE (MacDonald 1988: 18).

---

[129] For more on this subject, I recommend the book *The Rocks Don't Lie*, by David R. Montgomery (W.W. Norton & Co. 2012).

[130] Sumerian Flood Epic.

Although this matches close enough with the Kish and Shurruppak flood, it does not concur with the Ur event.[131] In the case of the Ur deluge, the waters never even reached a high enough level to cover the entire city (Montgomery 2012: 152-153). Moreover, in the Kish and Shurrupak floods, the waters did not exceed more than fifteen inches! (MacDonald 1988: 18, 20) It can therefore be concluded that the tale of the deluge became grossly exaggerated over the years, until the waters were said to have covered the entire world—which is another example of an embellished biblical account that was based on a real event.

The discovery of the archaeological evidence must cause us to re-evaluate not only the traditional interpretations that pertain to biblical history but its related doctrines as well. Armed with this knowledge, we must return to the beginning: to the book of Genesis.

* * *

While most of the world is familiar with the biblical creation story, the actual literal meaning of the objects and characters that are found in the book of Genesis have remained, for the most part, unsolved—that is, despite what some zealous traditionalists may assume.

The first symbol in the garden of the Elohim is the "serpent." The account tells us that the serpent persuaded the first woman to eat from the forbidden fruit of the "tree." She does so before persuading "Adam" to do the same. But the Lord God discovered their transgression and sentenced them to banishment. The second symbol in the garden is the tree of knowledge. The third symbol is the mystery "fruit" of the Tree of Life, which allegedly had the ability to make one become as immortal as God himself.

---

[131] This data contradicts Sitchin's theory that the Great Flood was caused by the melting of ice during the end of the last Ice Age. We know that this is not true because of not only the findings already presented but because the end of the last Ice Age occurred over 10,000 years ago, when civilization did not yet exist.

> Then the Lord God [Yahweh Elohim] said, "The man has now become like one of us, knowing good and evil. He must not be allowed to reach out his hand and take also from the tree of life and eat, and live forever.
> —Genesis 3.22

In order to understand what all of this means, it is necessary to understand that the author of Genesis was, of course, not actually present to witness those alleged events, which is why the use of allegorical descriptions were used. Because the events that were described in the book of Genesis were symbolic, this would seem to indicate that the creation of the first human-beings was also not a literal event.

Of course, the current debate between the religious creationists and scientific evolutionists has been waging since Charles Darwin published his monumental work *On the Origin of Species*. The theory of evolution seemed to be proven after a series of archaeological discoveries that validated the link between humans and animals (e.g., *Australopithecus*, etc.). What was discovered was physical evidence of a transitional primate/human anthropoid. In this case, could the creationists forever be discredited?

One observation that is apparent, is that present-age *Homo-sapiens-sapiens* are so far removed from the animal world that it is understandable why many believe that we human-beings could not share a direct genetic link with animals. Traits that are unique to human-beings not only include upright posture/bipedalism and advanced cognitive functioning, but also emotional crying with tears, blushing, prolonged post-natal growth in infants, absence of a penile bone (i.e., *baculum*) in males. Indeed, it seems that we human-beings are an anachronism. The sizable and inconsistent jump in the genetic time-line would seem to leave open the possibility of some type of extraordinary occurrence that influenced the genetic progression of our species.

Concerning the question of divine intervention and the evolution of life on Earth, one possible explanation was originally proposed by the ancient philosopher Anaxagoras. Anaxagoras theorized that life on this planet may be the result of seeding from a non-earthly source. This can be related to the modern theory that is referred to as "directed panspermia." However, it should be noted that the developers of this theory (Crick and Orgel) did not attribute this act of seeding by the traditional Judeo-Christian God, but rather by extraterrestrials (Crick and Orgel 1973: 341-346). Furthermore, in their book *Intelligent Life in the Universe*, scientists I.S. Shklovskii and Carl Sagan agree that either intentional or unintentional seeding of our planet, in the form of micro-organism contamination from visitation events, by extraterrestrial beings is not something that can be ruled out (Shklovskii and Sagan 1966: 211-212). Directed panspermia also relates to the theories of British astronomer Frederick Hoyle.[132] Hoyle deduced that the spontaneous formation of life (i.e., *abiogenesis*) from a Darwinian "primordial soup" alone was not possible.[133] Although it has been suggested that such a spark could have been ignited by lightning, or perhaps even by a meteorite containing life-bearing seed elements—perhaps from another planet (i.e., *exogenesis*).

In 2015, Milton Wainwright[134] and a team of researchers from the University of Sheffield and the University of Buckingham Centre for Astrobiology, examined dust and minute matter that was gathered by a balloon in the stratosphere. What they discovered was a microscopic metallic sphere that was composed mostly of titanium, with traces of vanadium, that had "filamentous life on the outside and a gooey biological material oozing from its center" (Rao 2015; Wickramasinghe and Tokoro 2014: 5). What is more, is that there is no explanation for how this anomalous object could have been found

---

[132] Fred Hoyle was a twentieth-century professor of astronomy and experimental philosophy and a fellow of the Royal Society.

[133] Fred Hoyle. *The Intelligent Universe: A New View of Creation and Evolution*. (Holt, Reinhart and Winston, 1984).

[134] Milton Wainwright, Ph.D., is a British microbiologist, honorary professor at Cardiff University, King Saud University, Centre for Astrobiology at the University of Buckingham, Megunaroden Slavjanski Institute, and senior lecturer in the Department of Molecular Biology and Biotechnology at the University of Sheffield.

at such a height. Wainwright hypothesized: "One theory is it was sent to Earth by some unknown civilization in order to continue seeding the planet with life." Indeed, one of the world's leading experts on interstellar material, Chandra Wickramasinghe,[135] stated that the discovery is evidence for the existence of extraterrestrial life (Rao 2015).

For years, Professor Wickramasinghe has been avowing the validity of the panspermia hypothesis. According to Wickramasinghe, physical life did not originate on Earth, but rather was born in an interstellar cloud of organic dust in outer-space. Evidence of this hypothesis exists in the form of fossilized microbial life, such as bacteria, that have been found embedded into meteors (Wickramasinghe and Tokoro 2014: 4-5). Indeed, bacteria is the foundation of all organic life. Furthermore, scientific experiments have proven that microbes can exist in the harsh conditions of outer-space, and are even able to travel from one to planet to another (i.e., *lithopanspermia*) (Druyan, et al. 2014: S1, E11; Gaskill 2014). Wickramasinghe and his colleagues have even declared that this cosmic genesis theory has been "overwhelmingly verified" (Wickramasinghe and Tokoro 2014: 2, 4). It can therefore be concluded that not only is there life outside of Earth's atmosphere but life actually originated in outer-space.

However, despite Wickramasinghe's and his team's credentials, and despite the scientific validity of the findings, they have been mostly ignored and dismissed by the dominating elite in the scientific establishment. This, of course, is due to the problem of skeptics, who, tend to dismiss and in some cases even censor information that does not conform to the doctrine that they are committed to retain—even if

---

[135] Chandra Wickramasinghe, Ph.D., Sc.D., is an award-winning astrobiologist, astronomer, and mathematician. He was a professor of applied mathematics and astronomy at the University College, Cardiff, and the University of Cardiff, at Wales; director of the Cardiff Centre for Astrobiology and the Buckingham Centre for Astrobiology; honorary professor at Glamorgan and Buckingham Universities, and former staff member at the Institute of Astronomy at the University of Cambridge.

such information meets the standards required for scientific justification. This is especially true when such information might endow credibility to anything pertaining to paranormal or extraterrestrial life.

> If a jury comprised of 12 impartial men and women were presented with the full range of evidence on the existence of extraterrestrial life, and the cosmic origins of life, there is scarcely any doubt that the verdict will be positive. So overwhelming is the totality of the evidence we have discussed. Ingress of extraterrestrial life to the Earth would appear to have been established beyond a shadow of doubt. The fact that this conclusion is not widely known or publicised is in the authors' view entirely a function of state control of scientific paradigms, of a kind reminiscent of the behaviour of totalitarian political regimes. Refusal to conform with the strictures of authority is met with serious consequences that are particularly damaging for young scientists at the start of their careers in science. For them the award of grants to support their work, approbation by peers, or even their very livelihood is threatened. Under such repressive constraints progress toward any form of objective truth is virtually impossible.
> —Chandra Wickramasinghe and Gensuke Tokoro, Life as a Cosmic Phenomenon

Wickramasinghe and his associates contend that the existence of extraterrestrial life not only represents an existential threat to the established scientific orthodoxy but also a socio-economic hazard to other entrenched institutions as well. The sociological repercussions pertain to fears that relate to the existence of extraterrestrial life that may be superior to our own, and therefore presents a security threat. The economic ramifications pertain to scientific funding programs in place that support the established paradigm (Wickramasinghe and Tokoro 2014: 6).

* * *

When confronting the types of questions that pertain to the differences between the creationists and evolutionists, one factor emerges that serves to provide a possible explanation, as well as a bridge between these two diametrically opposed schools of thought. That factor is the reoccurring presence and influence of intelligent beings who are not indigenous to our world.

People who claim to have been contacted by NHIs in the present era, have been told by humanoid beings that they have been a presence on this planet for millennia. There also seems to have been occasions when they exerted a direct influence upon events in this world. Indeed, this is the same type of conclusion that was reached by Timothy Good, in his ground-breaking book *Above Top Secret*:

> I believe that man's progress on this planet has been monitored by beings whose technological and mental resources make ours look primitive and theirs "supernatural" in comparison. The fact that many of the visitors are similar to us physiologically indicates that we share a genetic link. Could it be that some of the have had a hand in our evolution?
> —Timothy Good, *Above Top Secret*

According to the ancient texts, the answer to Good's question is a definitive *yes*.

Both the Old Testament/Tanakh as well as the preceding Mesopotamian versions, report that human-beings were created by the Elohim/gods. The following eighteenth-century BCE Akkadian Atra Hasis text explains that this was done because the gods required workers:

> When gods instead of man did the work, bore the loads, the god's work was too great, the work too hard, the trouble too much.
> —Atra Hasis

> "Create a mortal, that he may bear the yoke! Let him bear the yoke, the work of Ellil [Enlil], Let man bear the load of the gods."
> —Enki/Ea, Atra Hasis

Of course, these archaic accounts are highly affected. Therefore, in order to understand what these texts (e.g., the Enki and Ninmah text; the book of Genesis; etc.) are actually attempting to report, it is necessary to consider the underlying essence of the account, as opposed to its idiosyncratic details.

According to these reports, the first being that could be classified as human was conceived by combining Anunnaki and terrestrial "flesh and blood"—which may have been an ancient laymen's term for DNA.

> Nintu shall mix the clay with his flesh and blood. Then a god and man will be mixed together in clay.
> —Atra Hasis

In the following passage, there is a reference to a "creature" who already "exists." This is a reference to a hominid primate that was apparently used in this operation:

> Oh my mother, the creature whose name you uttered, it exists. Bind upon it the [image?][136] of the gods; [...]
> —Enki, The Creation of Man[137]

---

[136] The word "image" is a rendering by Samuel Noah Kramer that is derived from the damaged word in the original text (Kramer 1963: 150).

[137] Also known as the Nippur Tablet.

Likewise, in the book of Genesis, we are told that human-beings were created in the image of the gods:

> Then Elohim said, "Let us make man in our image, in our likeness [. . .]
> —Genesis 1.26

Sitchin surmised that the terrestrial that was used in this *in vitro* experiment was the hominid *Homo-erectus* (Sitchin 1978: 341-344). However, I contend that *Heidelbergensis* (i.e., *Homo-rhodesiensis*) is a better candidate. This new conjecture posits that the genetic line of Heidelbergensis could have separated into two distinct branches. The first could have naturally evolved into Neanderthals, while the second could have become the first artificially engendered *Homo-sapiens* human-beings.

> Nintu shall mix with his [an Anunnaki] flesh and blood. Then a god and a man will be mixed together in clay.
> —Atra Hasis

> Then the Lord God [Yahweh Elohim] formed a man from the dust of the ground and breathed into his nostrils the breath of life, and the man became a living being.
> —Genesis 2.7

In the biblical version, the substance that the first man was created from is usually described as "dust." The word can also be translated as "soil" (Contemporary English Version), "dirt" (The Message), and "slime" (Wycliffe Bible). The original Sumerian term that was used is translated as "clay."

> Ninmah threw pinched-off clay from her hand on the ground and a great silence fell. The great lord Enki said to Ninmah: "I have decreed the fates of your creatures

and given them their daily bread. Come, now I will fashion somebody for you, and you must decree the fate of the newborn one!" Enki [. . .] said to Ninmah: "Pour ejaculated semen into a woman's womb, and the woman will give birth to the semen in her womb. Ninmah stood by for the newborn [. . .] and the woman brought forth [. . .]
—Enki and Ninmah

In the following passage, the designation Earthlings is translated as "aborigines":

Ea [Enki] formed the aborigines from Kingu's blood, Marduk put the aborigines to work [. . .]
—The Enuma Elish

According to these texts, the mother goddess carried the hybrid to term in her own womb until giving birth by natural means.

This brings up questions related to inter-species breeding compatibility. However, it may be possible that this operation was executed through a scientific process that is not yet known to us in our present scientific stage.

By mixing together these two genetic pools, the Elohim were allegedly able to engender a being that was strong enough to carry out work-related functions, as well as able to understand basic language—i.e., "commandments," but not intelligent enough to rival them and challenge their authority—at least, that was the intention. The result may have been early *Homo-sapiens*—i.e., "thinking man." This theory not only explains the anomalous gap between the human and earlier primate world but it also explains why some of the NHIs bear humanoid characteristics. In other words: we were made in their image (Genesis 1.27).

According to the Mesopotamian records, the Anunnaki individual who assisted in the birth process was a female who is referred to as Ninti—i.e., Nintu, Ninmah, or Ninhursag. What is common to all three names is the Sumerian root *Nin*, which could either mean god/goddess/queen/lady or lord. The Sumerian word *ti* (which is affixed to the designation *Nin-ti*) can either mean "life" or "rib" (Kramer 1963: 149). Indeed, these two definitions can be used interchangeably in Sumerian. Therefore, *Nin-Ti* was the goddess of the rib of life. This coincides with her alternate appellation *Nin-tu*, which most likely means "birth lady"; as well as her epithets: the "womb goddess" (i.e., *sassuru*) and "midwife of the gods: (i.e., *tabsut ili*) (Dalley 1989: 326). It is therefore interesting that the term "rib" is also found in the Old Testament/Tanakh, where it is written that the first woman was created from the "rib" of Adam. However, it was not Eve who was the first mother, but rather an Anunnaki female who is "the lady of the rib" (i.e., the lady of life) (Kramer 1963: 149). The scribe who inserted this into the book of Genesis was most likely influenced by this earlier Sumerian description. Therefore, the "life" of Adam may be a reference to his DNA.

It is also compelling that among Ninhursag's names and epithets were Ama/Amma and Mami (Dalley 1989: 324, 326; Dukstra 1999: 603). What is common to these words is the root *ma*, which is nearly universal among the Indo-European languages. This Sumerian cognate most likely evolved into the words *mama/matar* (Persian), *mama/mater* (Latin), *mome/modor* (Middle and Old English), and the modern-day English words mother and mom.

> There is one race of man, and another of gods; but
> from the same Mother do we draw our breath.
> —Pindar, Nemean Odes

In the following gnostic Christian tractate, a dialogue is recounted between Jesus and the disciple James, in which Jesus instructs the disciple on the nature of the "alien things." Note that the word that is translated as *"Acamoth"* is a reference to the Greek term for the creator goddess (i.e., Sophia)—who may have been based on the Sumerian creator Goddess, Ninti:

> They [the humans] are not entirely alien, but they are
> from Acamoth, who is female. And these she produced

as she brought down the race from the Pre-existent One. So then they are not alien, but they are ours [. . .] At the same time, they are alien because the Pre-existent One did not have intercourse with her, when she produced them.

—Jesus, Nag Hammadi Codices, The First Apocalypse of James

According to this account, the female Elohim did not produce the first human-beings sexually, but rather through another means; which concurs with the Sumerian account.

It is therefore compelling that scientists who have examined the genetic history of the human-race have discovered evidence of a single mother by examining mitochondrial DNA—which is a type of cell material that is passed from mother to child. The analysis indicates that all humans share at least one common female ancestor who lived in Africa thousands of years ago. The geneticists have named this woman "Eve" (Avise 1998: 38-40; Ghose 2013). Different studies have produced different results; therefore, a precise date cannot be confirmed at this time. Nevertheless, the approximate range of the mitochondrial evidence, at the present time, is somewhere between 99,000 to 148,000 years ago (Poznik, et al. 2013: 562-565).

Other studies that are based on the genome sequencing of the male Y chromosome have produced similar results. The findings indicate that all male human-beings share a common ancestor who also lived in Africa, sometime between 120,000 to 254,000 years ago (Ghose 2013; Karmin, et al. 2016: 459-464; Poznik, et al. 2013: 562). This places the timeline within the approximate range of the mitochondrial evidence and the emergence of Homo-sapiens. However, it must be noted that these findings do not prove that these two individuals were the very first human-beings, or that they knew each other, but rather they were the first that bore verifiable lineages that remain unbroken. Nevertheless, it is apparent that human chronology begins to narrow considerably when it approaches the 200,000-year range. It could therefore be possible that it was in around that era that some transition occurred that altered the course of primate evolution; and it was this transition that permanently separated human-beings from the animal world.

If a genetic intervention event took place as far back as these dates suggest, and civilization does not go back further than approximately 7000 years, this would seem to prove that the original Sumerian accounts, from which the biblical account is based, regarding the creation of human-beings cannot be true. However, before we can arrive at this conclusion we must first inquire if this discrepancy be explained. A possible explanation will be offered further ahead.

* * *

The Sumerian texts report that the first human-beings were created because the gods required workers.

> [. . .] they [the human-beings] took hold of [picks and spades][138] .... Made new picks and spades, made big canals to feed the people and sustain the gods.
> —Atra Hasis

This was done because the minor gods (i.e., the Igigi) were tired of the labor.

> [. . .] the senior gods oversaw the work, while the minor gods were bearing the toil. The gods were digging the canals and piling up the silt in Harali. The gods, dredging the clay, began complaining about this life.
> —Enki and Ninmah

It was in a later age that a demand also developed for warriors and worshipers.

---

[138] Interpretation of missing text is based on extant information in the same text.

According to these accounts, the African birth-mother goddess generated other hybrid offspring as well. Most of the hybrids were destined to take up either the "pickax"[139] or the "hoe,"[140] and dig the irrigation canals that directed water from the Tigris and the Euphrates rivers into the orchard plantation farms of the colony settlements of the Anunnaki/Elohim.[141] Of course, the most well-known of these plantation farms is referred to in the Bible as "Eden" (Genesis 2.10-14).

In the original Sumero-Akkadian language, *Edin* (i.e., *Edinnu/Edinna*) was a fertile steppe plain where the animals grazed and where vegetables and fruit were cultivated (Hamblin 1987: 130). The cognate also appears in an ancient bi-lingual Akkadian-Aramaic text, where it also translates as a fruitful, plentiful, and well-watered location (Berlin 2011: 228-229). Although we may tend to think of the Garden of Eden as a beautiful heavenly paradise, according to the biblical report itself, the garden was more of a workplace environment:

> Then Yahweh Elohim took the man and put him in the Garden of Eden to farm the land and to take care of it.
> —Genesis 2.15

This passage concurs with the Mesopotamian account, which states that the first human-beings were created to be workers.

In this case, the biblical so-called "Adam" and "Eve," which should not be conflated with the original *Homo-sapiens*—as the author of Genesis did, may have been among the first to be placed in the Mesopotamian settlements. Indeed, the biblical account tells us that the man was subsequently relocated to Eden (Genesis 2.15). This coincides with the anthropological evidence, which indicates that Homo-sapiens migrated—or perhaps, in some instances, were transported—out of Africa.

---

[139] Creation of the Pickax
[140] Song of the Hoe
[141] Atra Hasis

However, it is likely that the activity that occurred in Africa, in the earlier prehistoric era, was not based on farming, but rather on mining. The ancient land of Nubia (modern-day northern Sudan) was a region known as the "land of gold" by the ancient Egyptians. Indeed, the name *Nubia* itself is derived from the ancient Egyptian word for gold (*Nub* or *Nubu*) (Ruiz 2001: 80, 140).[142] Present-day Ghana is known as "the Gold Coast" because of the abundance of gold in the region. Gold can also be found in present-day Liberia, Zimbabwe, Algeria, etc. Africa is also a continent where copper, tin, and iron-ore deposits can be found. Therefore, it may have been African copper and tin mines that helped make the Bronze Age in the Middle East possible. Indeed, the middle-east is not a region where such metals are plentiful. These locations coincide with the DNA findings, which traces the original ancestors of all human-beings to Africa.

The Mesopotamian texts confirm that precious metals were highly regarded by the Anunnaki/Elohim:

> An artfully made bright crenellation rising out from the abzu [i.e., deep waters] was erected for Lord Nudimmud [i.e., Enki/Ea]. He built the temple from precious metal, decorated it with lapis lazuli, and covered it abundantly with gold [. . .] the temple of Enki bellows.
> —Enki's Journey to Nibru

In another text, we are informed that the Anunnaki lord, Enlil, apparently preferred copper over silver:[143]

---

[142] This etymology is contested by those who believe that this word refers to a nomadic tribe called the Noubai or Noba. However, the origin of the name of this tribe is uncertain. Therefore, the name Noubai and Noba may originally itself have derived from the word for gold. In this case, the Noubai may denote a tribe of people who dwelled in the land of gold.

[143] Enlil's interest in precious metals is also referred to in the Song of the Hoe text.

> Silver and Strong Copper having carefully had a debate, Strong Copper had the lead over Silver in Enlil's house—Father Enlil be praised!
>
> —The Debate Between Copper and Silver

Similar reports come to us from the Ugaritic texts as well:

> 'Rejoice, Baal! Good news I bring: [. . .] The rocks will yield you much silver, the hills desirable gold. And build a house of silver and of gold, a house of jewels and lapis lazuli!' Valiant Baal rejoiced. He called a caravan into his house, merchandise into the midst of his palace. The rocks yielded him much silver, the hills desirable gold; The quarries brought him choicest gems.
>
> —Baal Cycle

It appears that Earthling servant laborers were used to extract the precious metals from the Earth.[144] In the following account, we are informed of the Earthling king, Gudea,[145] who was assigned the task of constructing a temple to the god Ningirsu (i.e., Nin-jirsu):[146]

> Great things came to the succour of the ruler building the E-ninnu: a copper mountain in Kimas revealed itself to him. He mined its copper onto rafts. To the man in charge of building his master's house, the ruler, gold was brought in dust form from its mountains. For Gudea refined silver was brought down from its mountains. Translucent cornelian from Meluḫa was

---

[144] This is also referred to in the Cursing of Agade text, which reads as follows: "[. . .] to break up its soil like the soil of mountains where precious metals are mined, to splinter it like the lapis lazuli mountain [. . .]"

[145] Ruler of the city of Lagash, 2144-2124 BCE.

[146] Most likely the same god that is more commonly referred to as Ninurta.

spread before him. From the alabaster mountains alabaster was brought down to him.

—The Building of Ningirsu's Temple

Likewise, in the following verse we find a similar situation, in which servants of Lord Enlil were sent out to retrieve the wealth of the land:

> Ores (? [sic]) from Ḫarali, the faraway land, ...[sic][147] storehouses, ..., rock-crystal, gold, silver, ..., the yield of the uplands ......, heavy loads of them, were despatched [sic] by Enlil toward Eres. After the personal presents, the transported goods ......, Ninmaḫ and the minister .... The dust from their march reached high into the sky like rain clouds.
>
> —Enlil and Sud

The precious metals were transported by the Earthling servants of the gods:

> Let the magilum boats of Meluḫa transport gold and silver and bring them to Nibru for Enlil, king of all the lands.
>
> —Enki and the World Order

Likewise, the Old Testament/Tanakh tells us that "gold," "bronze," and "iron," were of great importance to Yahweh (Exodus 25.1–14; Exodus 27.2; Haggai 2.8; 1 Ezekiel 16.17; Kings 7.49–51; Numbers 31.22, 31.50; Genesis 24.35). It is also telling that the book of Genesis specifically mentions that "gold" and "onyx" were located in and/or around one of the rivers that flowed out from the garden of Eden (Genesis 2.10-13).

Archaeological evidence indicates that the smelting of ores originated in the Near East (Cortizas, et al. 2015: 399). As a consequence of the copper mining and smelting, metal pollution first

---

[147] Damaged portions of the original text.

appeared in the Late Neolithic and increased in the Early Bronze Age in Jordan (Cortizas, et al. 2015: 399). Similar evidence was found in the North American state of Michigan, where deposits from copper mining entered into lake Manganese and Copper Falls lake eight thousand years ago (Cortizas, et al. 2015: 399).

Despite these finds, evidence of mining operations in the earlier Paleolithic period—which is when extraterrestrial genesis hypothesis postulates that the first human-beings would have been put to work in the mines—remains elusive at this time. However, this may be because the excavations were not large enough to leave evidence behind hundreds of millennia later, or they simply have not yet been discovered. Another possibility is that the original workers were not used in the mines, but rather were used for some other unknown purpose.

* * *

The question that now arises pertains to the time-line. When the DNA evidence is considered, we are looking at a picture that stretches as far back as 200,000 years ago. However, this conflicts with the extraterrestrial genesis time frame, due to the fact that civilization does not go back nearly that far.

In this case, some might suggest that we should then consider the possibility of the legendary civilization of Atlantis that was reported by Plato (in the Timaeus and Critias). Unfortunately, no substantial evidence of such a lost civilization has ever been found. What Plato called "Atlantis" was most likely a conflation of the island of Crete along with the nearby island of Thera (i.e., Santorini) that was destroyed by a volcanic eruption around 1500 BCE (Feder 2002: 187). Moreover, there are also many problems with the historicity of the Atlantis story. For example, Plato claimed to have received information about Atlantis from the Egyptians; however, there is no record of anything resembling Atlantis in ancient Egypt (Feder 2002: 182); nor does it appear in the records of the other Grecian chroniclers of the era, such as Herodotus and Thucydides (Feder 2002: 182). In recent years, discoveries of underwater symmetrical structures in the Caribbean Sea have renewed the Atlantis debate; however, in the case of the Bimini Road site (i.e., "Atlantis Road"), geologists who have studied the location have concluded that the seemingly artificial formations are nothing more than naturally eroded beach-rock. Indeed, similar type of geometrical features can be found off the coast of Australia (Feder 2002: 200-201). However, a more interesting location was found off the west coast of Cuba in 2001. It was in this location that underwater geometric structures were discovered by marine engineers using sonar-mapping technology (Ballingrud 2002). Unfortunately, at the time of writing, this location has still not been properly studied. Nevertheless, according to Plato, Atlantis was in the geographical vicinity of Athens, which would rule out a Caribbean location.

The remaining prehistoric candidates would then seem to be the legendary continents of Lemuria and Mu. The existence of Lemuria was postulated by the zoologist P. L. Sclater and the biologist Ernst Haeckel, in the nineteenth century. The name *Lemuria* was chosen to help explain the presence of primate mammals, such as lemurs, that are found in Madagascar and India, but not in the surrounding regions, such as Africa and the Middle East. It was believed that a lost continent in the Indian Ocean would have acted as a land bridge for these animals to cross. However, it is now known that the connection between India and Madagascar can be attributed to plate tectonics and continental drift. Indeed, Lemuria is no longer considered a valid scientific hypothesis.

The legend of Mu is based only on the anecdotal testimony of two men: Augustus Le Plongeon and James Churchward. The latter claimed to have received this information from a series of ancient stone tablets that were shown to him by a Hindu priest in the late nineteenth century. Churchward claimed that this text, which was written in an unknown language called *Nacaal*, revealed the story of a large continent in the Pacific Ocean, from which all human races originated. However, the Pacific is the deepest ocean in the world, which is hardly a plausible location for a lost continent—or at least one that disappeared as quickly as is claimed. It was more likely that Churchward fabricated his story in order to help sell the books that he had written on the subject (*The Lost Continent of Mu, Motherland of Man*). Indeed, no evidence of the Nacaal tablets, or even the monastery where they were allegedly kept, have ever been found. What is more, is that there are suspicious differences between Le Plongeon's and Churchward's accounts.

Graham Hancock presents a case in his book *Finger Prints of the Gods*, that the site of a lost civilization may have been the continent of Antarctica. According to Hancock's theory, the reason why a civilization could have existed in such a lifeless location is because of a pole-shift that may have occurred—which he bases on Charles Hapgood's theory of "earth crust displacement." However, not only is there no credible evidence to support this theory but geological studies (i.e., ice core samples) prove that Antarctica has been covered by ice for over 400 thousand years (Hale 2000). This negates Hancock's supposition that a lost civilization existed in 10,000 BCE. (This also rules out Hancock's Piri Reis map evidence as well.)[148] Indeed, even Hancock himself seems to have given up on the Antarctica hypothesis. Furthermore, none of these improbable theories explain how the trail of evidence keeps leading us back to Africa.

The early Mesopotamian definition of the "underworld" seems to have been based on both the concept of a hellish realm, as well as an underground location that was referred to by the Sumerians as the Abzu. The Abzu was thought to be a primeval sea that flowed beneath the Earth. However, we should consider the possibility that there may also be a third definition. It is possible, at least in some instances, that the underworld may also have been a reference to the lower

---

[148] According to Hancock, the Piri Reis map (1513 CE) shows Antarctica without its ice cover, which he believes indicates that the continent could have once been the location of a lost civilization. However, there is reason to believe that what appears to look approximately like an ice-free Antarctica is actually just one of the numerous mistakes that was made by the cartographer. Indeed, the similarity between Antarctican land and the Piri Reis rendering is unreliably crude. For more information on this subject see Gregory Mcintosh's book, *The Piri Reis map of 1513* (University of Georgia Press, 2000).

geographical continent of Africa. It may be possible that such a sub-Saharan African location predates the cities of Egypt, the Indus Valley, and even Sumeria. Perhaps the reason why it has not yet been discovered is because the original settlements were more temporary wood- and reed-constructed work places, rather than the megalithic centers of civilization that developed elsewhere. Nevertheless, it may also be possible that a stone-structured civilization did exist in ancient and/or prehistoric Africa, but simply has not yet been discovered. Perhaps this may even be the location of the legendary city of Nibiru. This distant location may explain why the Mesopotamians were so unsure of Nibiru's identity and why it has not been found in Mesopotamia.

* * *

Present-day contactees report that the NHIs claim that they have the ability to live for hundreds of years. This concurs with the ancient texts, which report the existence of the gods over a similar approximate time-span. In instances where the deity is reported to have existed beyond this time frame, this may actually be a reference to the time-span that the deity was worshiped.

The Mesopotamian accounts report that the creation of the first human-beings occurred in the early years of their own era. The Sumerian gods Enki and Ninti are specifically mentioned as having been involved in the process. However, as was previously noted, the creation of Homo-sapiens could not have occurred in the Sumerian age. However, it must be understood that it was customary practice in ancient Anunnaki culture for the records and mythologies to be revised in order to keep current with the egos of the gods. For example, when Enki's son, Marduk, took over power in Babylon, the myths were revised in order to make Marduk the supreme king of the gods and creator of the world (in the Enuma Elish), as opposed to Enlil, and Anu—both of which had been previously credited with that power. This was done in order to reflect the change in authority and aggrandize the power and the relevance of the newly appointed god. Another example can be found in the Erra and Ishum text, in which

Marduk is also reported to have caused the great Flood. This same type of transference of attributes applies to the gods Assur and Ammon Ra (i.e., Amun-Re) as well (Smith 2002: 10). Therefore, this practice could have occurred in previous generations. In this case, there may have been a time before the ancient era, when previous generations of Anunnaki/NHIs had visited the Earth.

One possible scenario is that a human contactee from Ubaid or Sumeria had been informed of this genetic intervention during a contact event, in much the same way that special information is conveyed to contactees in the present era. When the scribes became aware of this account, they transposed it into a relevant context. Therefore, the Sumerian narratives do not provide a literal first-hand account of what actually happened (which is what Sitchin did not understand). This not only explains the arcane nature of the report but why Sumerian deities, such as Enki, were described as being involved.

What can be extrapolated from the available information, is that the creation of hybrid workers could have initially derived from a need for workers approximately 200,000 years ago. Perhaps after the original extraterrestrial venturers eventually departed, the hybrid worker attendants were left to their own devices. Thousands of years later, an Anunnaki individual, who is referred to in the Mesopotamian records as Enki/Ea, arrived. This may have been when, according to the Eridu Genesis text, "Kingship had descended from heaven." Upon his arrival, an Anunnaki man who either was or who eventually became to be known as "Enki," may have not only established a settlement on Earth—i.e., substation—but made contact with the indigenous people—hybrid beings who his own ancestors had engendered. Therefore, it may have been Enki himself who made the Earthlings aware of their lineage.

In the following Sumerian text, the primitive Earthlings that Enki contacted are referred to as "the Martu nomads":

> Enki presented animals to those who have no city, who have no houses, to the Martu nomads.
> —Enki and the World Order

According to the texts, the colony that Enki established was called Eridu (i.e., Eridug). The remains of this settlement have been found. It is the oldest city ever discovered (circa 5400 BCE).

> When kingship from heaven was lowered, the kingship was in Eridu.
> —The Sumerian King List

According to the Sumerian records, Enki was not some idle unseen imaginary and symbolic figure—at least originally, but rather acted as a hands-on supervisor, who personally oversaw the daily activity of his colony.

> He organized ploughs, yokes and teams. The great prince Enki bestowed the horned oxen that follow ......, he opened up the holy furrows, and made the barley grow on the cultivated fields. [. . .] The lord called the cultivated fields, and bestowed on them mottled barley. Enki made chickpeas, lentils and ...... grow. He heaped up into piles the early, mottled and *innuha* varieties of barley. Enki multiplied the stockpiles and stacks [. . .] He built the sheepfolds, carried out their cleaning, made the cow-pens, bestowed on them the best fat and cream, and brought luxury to the gods' dining places. He made the plain, created for grasses and herbs, achieve prosperity. [. . .] He demarcated borders and fixed boundaries. For the Anuna gods, Enki situated dwellings in cities and disposed agricultural land into fields.
> —Enki and the World Order

Under the guidance of Enki, Eridu became a thriving utopian civilization—along with Edin and the nearby Dilmun civilization.[149] This original peaceful society most likely helped to inspire the biblical concept of the primeval paradise that is reported in the book of Genesis. Indeed, the legend of this ancient paradise of the gods originated in Sumer (Kramer 1963: 148-149).

Archaeologist Juris Zarins believes that he has discovered the location of the Garden of Eden in the Iraqi Persian Gulf region, under what is now the Persian Gulf. Zarins referred not only to the biblical text and archaeological evidence but to geology, hydrology, linguistics, and LANDSAT satellite images from space (Hamblin 1987: 128-135). Zarins believes that the biblical "Gihon" river is what is now known as the Karun River in Iran, and the "Pison" river, is what is now known as the Wadi Batin river system. Therefore, the original Edin is now most likely underwater due to the rise in sea levels that occurred during the final stages of the Flandrian Transgression—perhaps this should also be considered to be a possible location for Nibiru as well. It is now known that Edin was actually a farm type of environment, where food was cultivated for the early Ubaidian and/or Sumerian settlements. Foremost among the settlements was Eridu. The canals that were dug by the hybrids irrigated those gardens.

It is also compelling that visual representations of naked human-beings in the process of serving the gods have been discovered (e.g., Warka vase of Uruk).[150] In these renderings, bald clean-shaven naked men are depicted carrying vases full of food offerings, most likely from the garden of Edinnu, toward the temple of the Elohim goddess, Inanna. Of course, these depictions of the naked servants of the gods are similar to the biblical Adam and Eve account. It appears that the Adam and Eve narrative was a composite of at least two separate

---

149 According to the Epic of Gilgamesh and the Enki and the World Order texts.

150 At the present time, the Warka vase is locked away in the National Museum of Iraq; however, a replica of the artifact is kept at the Pergamon Museum in Berlin Germany.

earlier accounts. The first is the creation of human-beings in the prehistoric era, and the second is the Earthling's lives as naked servants in a utopian civilization of the gods. It should therefore be understood that the presentation of the gifts of the Earth to the gods are no different than the presentation incident that is described in the biblical Cain and Abel account (Genesis 4.3-5).

It may therefore have been in Eridu that Enki disclosed to his Earthling subordinates that they had been created by his kind. It must have been at a later time that the scribes attributed this act of creation to Enki and the other Anunnaki of that age.

* * *

Another subject that must be considered is the Gobekli Tepe site in what is now the country of Turkey. What was discovered were massive t-shaped stone pillars that had been erected and placed into circular formations. The construction was dated back to the early Neolithic era, circa 11,000 years ago—which predates Ubaid and Sumer by approximately 5,000 years. What is also compelling about this site is that it is unlikely that such a permanent construction could have been sustained for an extended period of time on hunting and gathering alone. The people who built Gobekli Tepe must have been among the first to corral wild sheep, cattle, and pigs, as well as domesticate wild grains (Curry 2008).

Some of the megalithic columns at Gobekli Tepe were adorned with the images of anthropomorphic arms. In this case, we must ask the question: who were these figures supposed to represent? Could these have been depictions of the NHI gods? Unfortunately, at the time of writing, the entire site of Gobekli Tepe has not yet been fully excavated. Moreover, unlike Sumer, the site has not yielded any textual records that could provide further elucidation.

Based on the information that is available, I suggest that this site was most likely not dedicated to the NHI gods, but rather to ancestor spirits. I base this theory on the fact that the Gobekli Tepe contains *many* circular temple constructions. Some of these temples were intentionally buried by a subsequent generation, who then built their own temples on top of the older ones. The reason for these multiple constructions may have something to do with the genetic lineages of various clans who inhabited that region. Therefore, each temple may have been dedicated to one specific ancestral lineage and/or patriarchal/matriarchal spirit. It is also telling that human bones have been found. This may indicate that it was used as a funeral site, which accords well with the ancestor spirit theory. Therefore, Gobekli Tepe was most likely a sanctuary temple for the dead. Indeed, this is what the archaeologists who examined the site have deduced as well.

* * *

In the early twentieth century, excavations conducted under the supervision of British archaeologist Sir Leonard Woolley uncovered the lost civilization of the Sumero-Akkadian city-state of Ur. Among the artifacts discovered at the site was a sculpted alabaster stone that is now commonly referred to as the "Disk of Enheduanna."

Enheduanna was the high-priestess (i.e., entu-priestess) of a temple compound called the giparu (i.e., gipar) in Ur. Part of the massive complex and its adjoining buildings were reserved for the Anunnaki deity Nanna. Another section was reserved for his official consort, Ningal; while another section was reserved for the human priestess. Enheduanna was apparently not only the daughter of King Sargon but the "wife" of Nanna (Boomer 2008: 5, 9; Weadock 1975: 101). As the officially appointed priestess, it was Enheduanna's duty to unite with Nanna in the "divine marriage" ritual.

The exact nature of the divine marriage is debated among scholars; although it is generally believed that these were mere symbolic acts. However, the records imply that some temple personnel did actually engage in sexual activities in the giparu (Weadock 1975: 102). Indeed, some of the entu-priestesses are also reported to have bore children (Meador 2000: 61). Therefore, it is possible that there was, at least originally, nothing symbolic about the ritual, and that intercourse with the Anunnaki produced offspring; a race of hybrids who are referred to in the Hebrew texts as the "Nephilim."

However, it is also likely that there were also times when the rituals were symbolic. This most likely occurred in the absence of Nanna, Ningal, and Inanna. In such instances, it seems that idols were used to represent the deities in their place.

The sexual aspect explains why the male priests and kings usually served the female goddesses (e.g., King Enmerkar and Inanna), and the females served the males (Boomer 2008: 6, 8). It seems that there were sections of the giparu that were essentially brothels for the gods. The visiting Anunnaki apparently not only copulated with human-beings but with each other.[151]

During their stay, food and gifts of clothing and gold and silver valuables were offered to the divine beings (Meador 2000: 56; Weadock 1975: 103-104). The literal presence of the Anunnaki explains why images have been found that depict the priests and priestesses standing before these anthropomorphic individuals. Only when the gods were absent were the offerings reduced to mere symbolic rituals that were performed before unseen deities who were represented in the form of idols. In these instances, the psychic NHIs could only be communicated with through dreams, visions, divination, and prayer.

* * *

According to both the Mesopotamian and the Judean/Israelite accounts, the primitive human-beings were unruly and did not always obey orders. The ones who were expelled from the Anunnaki settlements apparently roamed the land in barbarian hordes, which seemed to have caused trouble for their former lords. In the following report, we can see that both the Mesopotamian and the biblical records are nearly identical:

> [. . .] and the country became too wide, the people too numerous. The country was as noisy as a bellowing

---

[151] Song of Inanna and Dumuzi

bull. The God grew restless at their racket, Ellil [Enlil] had to listen to their noise.
—The Atra Hasis

The Lord [Yahweh] saw how great the wickedness of the human race had become on the earth, and that every inclination of the thoughts of the human heart was only evil all the time. The Lord [Yahweh] regretted that he had made human beings on the earth, and his heart was deeply troubled.
—Genesis 6.5-6

According to the Atra Hasis text, the chief deity among the Elohim at that time was the Anunnaki god Enlil, who was supposedly the biological brother of Enki. Just like the biblical account, Lord Enlil ordered the destruction of the heathen Earthlings. Consequently, the hybrids were afflicted with disease, drought, and famine, before it was decided that the they should perish in a catastrophic flood. By doing this, the Anunnaki were allegedly not only ridding themselves of the troublesome savages but may have also been attempting to preserve the integrity of their own genetic line as well. This might explain why primates in the animal world continue to exist to this day, while their more evolved relatives did not. This may have been because the Anunnaki/Elohim wanted to intentionally cut off the link between their genetic pool and the primitive Earthling and animal world, in order to prevent evolutionary regression. Indeed, this may help to explain the disappearance of the Neanderthals.

The texts report that there was at least one hybrid who was permitted to live—which angered Lord Enlil. The Sumerian texts tell us that his original name was Ziusudra (i.e., Zarusthutra). His Akkadian name was Utnapishtim, and in Babylonia he was known as Atra Hasis. The word *atra hasis* literally means "exceedingly wise"—just as Utnapishtim was called the "wise" man of (the Mesopotamian city of) Shurippak. It is therefore possible that the individual who inspired the biblical character Noah was one of the "men of renown" that the book of Genesis tells us were referred to as the Nephilim.

Indeed, in an extra-biblical book that was found among the collection of the Dead Sea Scrolls in the caves at Qumran, the Genesis Apocryphon allegedly reports the early history of Noah and his father (Lamech), who suspected that his wife had given birth to someone else's child. Apparently, young Noah did not share any of his father's genetic traits and he began to wonder if his wife had been impregnated by the so-called "Watchers and the Holy Ones and the Giants." This may be the reason why he was chosen to be spared the sentence of death that was leveled at the more primitive strains.

The designation "Watchers," that is found in the extra-biblical Book of Enoch (Chapters X, XII, XV), refers to the otherworldly fathers of the Nephilim. This term derives from an incident that is reported in the biblical book of Daniel, in which a being, who is described as a "watcher," is reported to have descended from the sky (Daniel 4.13). It is therefore likely that the Watchers and the previously mentioned "starry hosts" are equivalent.

The following report of the great deluge was found in a series of tablets that were unearthed in the royal library of Ashurbanipal in Nineveh. The record describes the exploits of Gilgamesh, who was an Earthling man and king of the city of Erech. In one of the sections of this text, it is reported that Utnapishtim was warned of the impending flood by Enki. Apparently, Enki had taken pity on the Earthling man and did not want to see him suffer and perish, in much the same way that Noah is warned in the biblical account (Genesis 6.9):

> O man of Shurippak, son of Ubar-tutu. Throw down the house, build a ship, forsake wealth, seek after life, hate possessions, save thy life, bring all seed of life into the ship.
> —Enki, The Epic of Gilgamesh

According to the story, that is based on the Sumerian Eridu genesis, Utnapishtim proceeded to construct a large ship that was referred to as the "great house," and loaded it with the seeds of life.

* * *

In order to better appreciate the plausibility of extraterrestrial genesis theory, it is necessary to acknowledge that not only are we *Homosapiens* nearing the scientific threshold of such genetic engineering ability ourselves but reports suggest that a similar type of operation is being carried out by NHIs in our own time as well. What must be considered are the numerous accounts of people who claim to have been taken aboard NHI craft and examined by pale beings with large black eyes who are commonly referred to as the "Grays" (a reference to the color of their skin). According to abductee accounts, the Grays have been carrying out a massive procreational program in an attempt somehow improve and sustain themselves (Jacobs 1992: 104; Mack 1999: 14, 250, 254).

Research into the abduction phenomena was conducted by Harvard University professor and psychiatrist John E. Mack.[152] After working with over two hundred abductees and contactees, Mack presented the findings in his books *Abduction: Human Encounters with Aliens* and *Passport to the Cosmos*. Memories of the abduction incidents were recalled by "experiencers" both with and without the use of relaxation techniques. This is despite the apparent ability of the Grays to cause their subjects to forget their experiences. Indeed, eighty percent were able to recall their abductions without the use of hypnotic regression (Mack 1999: 30).

One purported reason for this operation is because these type of NHIs have become logically practical to such an extreme extent that they have become a mechanized and joyless species (Mack 1999: 264). Therefore, the campaign seems to be an attempt to humanize themselves. However, it must be noted that at least one known abductee "sensed" that the hybrid offspring were going to be used as workers (Jacobs 1992: 314-315). Unfortunately, we do not have a good understanding of the Gray's intentions because of their apparent aversion to revealing too much to their human subjects.

The presence of what appear to be smaller drone-like workers aboard Gray ships may indicate the presence of other artificially

---

[152] johnemackinstitute.org/

generated beings (Mack 1999: 248). Such "manufactured" "proto-beings" may explain the absence of genitalia (Jacobs 1992: 193), and why the smaller more mechanized beings seem to be the servants of the larger ones—which is remarkably similar to the relationship between human-beings and the larger humanoid NHIs in ancient and prehistoric times.

In his book *Intruders,* the UFOlogist Budd Hopkins reports that in many of these cases, the abductees experience a gynecological type of exam, in which sperm and ova are collected:

> Now all of this leads to the unwelcome speculative inference that somewhere, somehow, human beings—or possibly hybrids of some sort—are being produced by a technology obviously—yet not inconceivably—superior to ours. And if that possibility is not enough to induce paranoia in the heartiest, consider this: With our own current technology of genetic engineering expanding day by day, is it not conceivable that an advanced alien technology may already have the ability to remove ova and sperm from human beings, experimentally alter their genetic structure, and then *replant* altered and fertilized ova back into unknowing host females to be carried to term? Ova that can be removed can also be replaced, even by our own present-day medical technology.
> —Budd Hopkins, *Intruders*

Procreational procedures endured by the abductees involve gynecological exams, sperm collection, egg harvesting, and embryo/fetal extraction. Incubation chambers in which fetuses are kept are also seen aboard the ships (Jacobs 1992). In some instances, the abductees are presented with a hybrid baby and informed that they are the parent. In other instances, Gray-Human hybrid children are also seen. In other accounts, hybrids have allegedly even engaged in direct procreative relations with human-beings (Mack 1999: 252-

257). One such case is the Antonio Villas Boas abduction incident[153] (Denzler 2001: 51-53; Jacobs 1992: 39; Ronnevig 2007: 109-110). According to Mr. Boas, he was taken aboard a craft where he claims that he was seduced by a short light-blond haired naked humanoid female with large eyes. After the incident, Boas experienced persisting health problems, which included nausea, headaches, weakness, burning sensation in the eyes, and skin lesions. Scars under his chin where the NHIs had apparently extracted blood were visible for three years after the incident (Denzler 2001: 51). Boas contacted the journalist Jose Martins, who was researching UFOs at the time. Martins arranged for Boas to be evaluated (by Dr. Olavo Fontes) at the National School of Medicine of Brazil. After examination, Boas was diagnosed with radiation exposure sickness.

John E. Mack's investigation into the abduction phenomena was unable to uncover evidence that the experiencers were suffering from mental illness, or were deliberately attempting to be deceitful. Indeed, Mack was impressed with both the consistency and the sincerity of the accounts. It is also telling that some, if not many, of the experiencers were embarrassed, traumatized, and even ashamed when it came to recounting some of the sexual and gynecological aspects of the experience. Another traumatic aspect pertained to experiencers who had to deal with the ridicule and problems at the workplace after going public with their encounters (Mack 1999: 241). Therefore, there does not appear to be any obvious reward or material pay-off for disclosing such information—which, would be the motivation of a hoaxer.

Mack came to believe that investigators were "not dealing with an entirely subjective or internally generated experience," and concluded that the phenomena is indeed real (Mack 1999: 245). He also acknowledged that an extraterrestrial explanation was not something that could be disproved; although, it should be noted that Mack did not claim to provide irrefutable evidence for a physical explanation. Indeed, he was skeptical about the physical nature of the experience, despite the fact that he himself had personally seen skin lesions from

---

[153] 1957, Brazil.

apparent physical operations (Kidnapped by UFOs 1996; Mack 1999: 14, 198). Similarly, he noted that the surgical recovery of alleged material object implants that have been performed by surgeons (e.g., Roger K. Leir) are inconclusive (Mack 1999: 15, 26). Even though Mack attested to the reality of the phenomena, he postulated that the experience is not wholly physical, but rather seems to be occurring outside of the physical level (Mack 1999: 29). In other words, he deduced that the phenomenon was of an inter-dimensional nature.

A separate Harvard University study of the abduction phenomena indicated that "recollections of purported traumatic encounters with space aliens" bear significant correlation with the responses of conventional subjects suffering from post-traumatic physiological stress disorder. Although the report stopped short of claiming to offer incontrovertible proof for the existence of alien abductions, it did acknowledge that reactions indicated that the abduction recollections reflected a genuine "emotional significance" that cannot be dismissed as mere "false memories" (McNally, et al. 2004: 493-496).

These Harvard studies were contradicted years later by another Harvard University researcher, psychologist Susan A. Clancy (author of the book *Abducted: How People Come to Believe They Were Kidnapped by Aliens*.) According to Clancy, people claiming to be abducted by aliens are the victims of false memories that derive from repressed thinking and disingenuous external influences—which contradicts the previous study that she herself had taken part in. However, Clancy admitted that she did not conduct the study because she "was interested in aliens or UFOs," but rather because she was "interested in memory distortion and false memory creation," and "how people could come to believe and then remember things that didn't happen to them" (Abducted 2005). In other words, she started her investigation with the conclusion—the conclusion being that the abductions never happened! Despite her unscientific approach, she did also admit that there is no way to disprove the alien abduction hypothesis, and that people who are reporting these events are not insane. Nevertheless, she contends that such people do have "a tendency to fantasize and to hold unusual beliefs and ideas" (Clancy

2005: 136). Consequently, she faced a backlash from experiencers who were upset at her for her biased study and fallacious diagnosis.

In regard to experiencers who "hold unusual beliefs," there are two explanations for this. The first pertains to the result of someone who has undergone such a profound boundary-breaking experience. In such cases, it is natural for the experiencer's belief system to be effected. Consequently, such an individual may be more likely to adopt beliefs that others might consider to be "unusual." Indeed, this type of transformative experience was repeatedly confirmed in Mack's study. Secondly, it may also be the case that the NHIs are purposely seeking out individuals who they believe are more mentally open to them and less likely to undergo a problematic panic attack when being confronted with the fact that human-beings are not alone in the universe.

Mack's findings were similar to those of Dr. David M. Jacobs.[154] However, unlike Mack, Jacobs found credibility in the physical aspect of the experience (Jacobs 1992: 221). He based part of this interpretation on evidence in the form of scars and colored stains that have been found on clothing that were worn during the abduction (Jacobs 1992: 25, 111). One of these stains was examined using Fourier transfer infrared analysis at Crippen Laboratories (Wilmington, Delaware). Although not enough of the substance was able to be extracted from the fabric to conduct a full chemical analysis, the amount that was able to be collected did indicate that the stain was not a common substance (Jacobs 1992: 242).

Furthermore, in terms of the physical nature of the experience, Jacobs also notes that not only have witnesses seen people being abducted but the experiencers were actually physically missing during the time of the abductions (Jacobs 1992: 302)—which contradicts Mack's non-physical interpretation.

Based on the information provided by the experiencers, it seems that the events can be both physical and non-physical. It also seems that most, although not all, of the physical experiences are more related to the Gray breeding program that the abductees were

---

[154] David M. Jacobs, Ph.D. Former associate professor at Temple University.

subjected to, while the non-physical experiences are more related to contactee events that are of a more profound and spiritual nature. These benevolent types of experiences usually involved more humanoid appearing beings.

Another genuine aspect of the experience is the trauma that is endured by the abductee. Some of the abductees become so frustrated at their inability to stop the repeated abductions that they became suicidal—which further indicates that the phenomena is not based on wishful thinking. Jacobs cites the work of Dr. Ronald Westrum, who identified the mental trauma that abductees endure as "post abduction syndrome." The trauma pertains to the abduction event itself but with the social scorn and ridicule that can sometimes occur as a result. In response to this problem, abduction support groups (e.g., UFO Contact Center International) have formed to help experiencers cope with these issues.

After interviewing sixty people and documenting 300 abduction experiences, both with and without the use of hypnotic regression, Jacobs also concluded that UFOs were involved in the abduction events and the experiences were real (Jacobs 1992: 39). Jacobs believes that the number of people who have been abducted numbers in the thousands; although, he notes, that it may even extend past a million (Jacobs 1992: 306). However, many, if not most, of the abductees may have no memory of their experiences.

Some of the explanations that have been asserted by skeptics to explain these occurrences include: psychosis, mental repression of childhood sexual abuse, hysterical contagion, prewaking and presleeping states, fantasy-prone personalities, the influence of hypnosis, the influence of science fiction, etc. In his book *Secret Life: Firsthand Accounts of UFO Abductions*, Jacobs considers and dispels all of these interpretations (Jacobs 283-304). What the findings indicate is that skeptics who devise such explanations have not done enough honest research into the topic. Furthermore, Jacobs also points out that the charlatans who attempted to perpetrate hoaxes or assert mental delusions were easily found out.

Another common explanation that is used by the skeptics is to dismiss these experiences as "sleep paralysis" (Saler 2007: 131). This is an incident in which someone who is asleep is awakened to find his or herself immobilized. The person will often also sense some otherworldly presence in the room. Of course, skeptical physicalists argue that these are nothing more than bad dreams. The subject of sleep paralysis is another subject that is beyond the scope of this work; although I will mention that I do believe that this has more to do with ultraterrestrials than with extraterrestrials. I base this more spiritual interpretation on my own experience with this phenomenon. Nevertheless, it should be understood that some of the abductions occur in the day-time while the experiencer is awake (Saler 2007: 132). This is only one of the reasons why the sleep paralysis explanation is not plausible and should not be conflated with abduction events.

Based on these accounts, we can now better formulate a picture of what may have happened hundreds of thousands of years ago when a male hominid, perhaps *Heidelbergensis*, was abducted by humanoid NHIs. The specimen would have most likely have been sedated before being taken aboard a craft, where sperm was extracted. After the procedure was completed, the primate would have been returned to the abduction site, with perhaps no memory of what had happened to him. It may even be possible that this archaic human-being of the genus *homo* is the father of us all.

In summary, this possible prehistoric event not only explains the anomalous evolutionary gap between human-beings and the animal world, and not only explains the ancient accounts of genetic intervention, but it explains why the humanoid NHIs look so much like us. In other words, we *Homo-sapiens* are, as the ancient philosopher Plotinus put it, "poised midway between the gods and the beasts."

* * *

It may be possible that the so-called "gods" of the ancient world are the same "tall" and so-called "Nordic" appearing humanoid NHIs that are sometimes referred to—most likely erroneously—in our time as the "Pleiadians." Indeed, present-day contactees report seeing tall blond beings; some of whom are clad in robes (Mack 1999: 194, 223, 233). However, it should also be noted that other reports suggest that the humanoid NHIs are not always fair featured (Mack 1999: 137, 258).

The legends of light-skinned and light-haired god-like beings are not limited to the deities of Nordic mythology. The Hindu god Surya, for example, was said to have had golden hair (Andrews 2000: 195). Likewise, the ancient Greek poet Pindar described Apollo as having golden hair.[155] Aristophanes also described "Phoebus" (i.e., Apollo) as having golden hair.[156] The Muses of Apollo were described as "fair-haired."[157] According to Pindar, the god Bacchus was fair-haired.[158] Hesiod also described "Dionysus" (i.e., Bacchus) as "golden-haired."[159] Anacreon described Cupid as being both "golden" and "fair-haired."[160] However, in regard to Greek mythology, it is true that characters who were not gods, in some instances, were also described as having light features as well. Furthermore, not all of the gods were fair-featured. It must therefore be acknowledged that we cannot place too much credence on these more fanciful classical accounts; most of which developed in a later period. It is therefore necessary to refer to the original Sumerian reports. We know that the gods of Mesopotamia must have bore light features because texts[161] tell us that the Anunnaki lords referred to their hybrid subordinates as "the black-headed ones"; which, of course, indicates that they themselves did not have dark hair.

When the Conquistadors first encountered the Inca in the New World, the Spanish chroniclers reported that they were greeted as gods. This may have been because their taller stature and lighter skin resembled the deity Viracocha (i.e., Kon Tiki). The sixteenth-century

---

[155] Second Pythian Ode

[156] The Birds

[157] First Pythian Ode

[158] The Seventh Isthmian Ode

[159] The Theogony of Hesiod

[160] Ode V: On the Rose; Ode XVII: On a Silver Bowl

[161] E.g., the Lament for Nibru; the Lament for Sumer and Urim

Spanish explorer, author, scientist, and historian Pedro Sarmiento de Gamboa reported in *Viracocha and the Coming of the Incas*, that Viracocha was a white man who dressed in a white robe, and that he had created a race of giants in primeval times. Viracocha allegedly taught the people moral precepts, and when the people did not obey him they suffered from a catastrophic flood. A similar type of situation may have occurred between the Spanish Conquistadors and the Aztecs, who apparently believed that the Conquistadors were related to Quetzalcoatl (Aveni 2001: 15) (which is attested in the Florentine Codex).[162] Quetzalcoatl, or Kukulcan, is sometimes represented not as a humanoid being but rather as a feathered serpent; however, as was previously noted, it may be the case that the serpent image was only a symbol of the humanoid deity (Aveni 2001: 299). Furthermore, it was not only the tall stature and light skin that deceived the natives but also the full beards of the Conquistadors as well, which is an uncommon feature in the South and Meso-Americas at that time.

What is also compelling is that two of those Aztec gods, Tonatiuh and Xochipilli, were depicted with light-colored hair (Milbrath 2013: 49). In this case, what could have possessed the dark-featured indigenous people to attribute such an uncommon feature to their own gods? It is likely that these depictions were influenced by sightings of foreign beings.

In ancient Egypt, it was believed that the pharaohs were the direct descendants of the gods. In other words, at least a few of them were most likely the "mighty" "men of renown"—i.e., the Nephilim—that are reported in the Old Testament/Tanakh. The well-preserved mummy of King Ramses II—who may have been one of those men—was examined by a forensics team (headed by Professor P. F. Ceccaldi) in 1975. What was unusual about the mummy was its light reddish hair, which had been remarkably well preserved. The study revealed that although part of the hair showed signs that it had been treated with henna dye, the root portion indicated that the light reddish pigmentation was not the result of either dye or natural

---

[162] Book XII, chapter 4.

decomposition, but rather was the king's natural hair color. It seems that the servants of the pharaoh had used the dye to cover parts of his hair that was graying. What is also remarkable about the body of Ramses II is that analysis also revealed that the pharaoh was a "leucoderm"—i.e., he was fair-skinned (Brier 1994: 200-203).[163] Ramses II and his immediate ancestors were followers of the god Set (i.e., Seth)—which is why his father and great grandfather were both named Seti. It may be the case that those individuals were direct descendants of that Anunnaki patriarch. However, the mummified corpse of Ramses II is five feet seven inches (1.7 meters), which is about average height for an Egyptian male at that time. Although, if the Anunnaki lords were of so-called "giant" stature then it seems that we should expect that he would have been taller. However, it may have been that Ramses II was several generations removed (if not many more) from his Anunnaki patriarch, which may have effected his height more than his pigmentation due to a maternal side genetic influence. Another possibility is that there are humanoid NHI races that are both short and tall, similar to how some *Homo-sapien* races are taller than others (e.g., the European Dutch and the African Pygmy). This would explain why some humanoid NHIs that are reported in current times are described as being of average human height.

* * *

Due to the genealogical nature of these findings, the argument that might be asserted by some racists who may attempt to use these types of findings to elevate themselves above their fellow human-beings must be addressed. What must be understood is that Caucasians, Aryans, Hyperboreans, or any type of light-complexioned or light-haired people, are not more closely related to the gods, and therefore

---

[163] The final results of the study were published in *La Momie de Ramses II: Contribution Scientifique a l'Egyptologie* (edited by Balout, Roubet, Desroches-Noblecourt).

are not in any way superior to other races. This is because present-day light features evolved naturally.

Light features in human-beings are the result of low levels of the dark pigment *eumelanin,* which is caused by low levels of vitamin D due to low amounts of sunlight. Lighter skin is an evolutionary mutation that allows for more vitamin D to be absorbed. Genetic research indicates that this mutation began to appear around 11,000 years ago, during the Ice Age in Northern Europe (Dobson and Taher 2006). The fact is that the direct lineage of the actual so-called "men of renown" was dispersed among humanity millennia ago.

Apparently, the Elohim saw fit to let all of the present-day human-beings continue to exist. This was ordained because we *Homo-sapiens* are all human-beings.

> Molecular studies of genes have contributed much to the scientific understanding of human origins. Perhaps the most exciting of recent molecular discoveries is that extant human "race" are almost entirely similar genetically. Some of the first evidence came from protein electrophoresis, wherein Caucasoid, Mongoloid, and Negroid populations proved to share identical allelic forms at most surveyed genes.
> —John C. Avise,[164] *The Genetic Gods*

The findings do not pertain to race, but rather to the human-race. It seems that we human-beings all share a common ancestor; an ancestor who not only links every human-being in this world together but connects us with the other forms of life who inhabit this diverse and living universe.

---

[164] John C. Avise, Ph.D., is a professor of ecology and evolutionary biology at the University of California, Irvine.

CHAPTER V

THE TREE OF KNOWLEDGE

All throughout the books of the Bible are references to "secret" matters:

> The secret things belong to the Lord [Yahweh] our God [Elohim]; but the things revealed belong to us and to our children forever, that we may follow all the words of this law.
> —Deuteronomy 29.29

> It is the glory of God [Elohim] to conceal a matter [. . .]
> —Proverbs 25.2

> Oh, how I wish that God [Eloah] would speak, that he would open his lips against you and disclose to you the secrets of wisdom [. . .]
> —Job 11.5-6

Yahweh himself seems to confirm the existence of secrets in the following passage:

> Are you wiser than Daniel? Is no secret hidden from you?

—Yahweh, Ezekiel 28.3

References to secret matters continue in the New Testament, where Jesus makes repeated references to hidden knowledge that was revealed only to his most trusted disciples:

> The knowledge of the secrets of the kingdom of God [Theos] has been given to you, but to others I speak in parables [. . .]
> —Jesus, Luke 8.10

> If you, even you, had only known on this day what would bring you peace—but now it is hidden from your eyes.
> —Jesus, Luke 19.42

> The kingdom of heaven is like a treasure hidden in a field [. . .]
> —Jesus, Matthew 13:44

> [. . .] the knowledge of the secrets of the kingdom of heaven has been given to you, but not to them.
> —Jesus, Matthew 13:11

> For whatever is hidden is meant to be disclosed, and whatever is concealed is meant to be brought out into the open.
> —Jesus, Mark 4:22

Because the secret matters have either been relegated to the status of the forbidden, or assumed to exist outside the range of human comprehension, they have never been fully disclosed and understood. However, clues left to us in the ancient records, including the Bible itself, have made it possible to uncover these hidden matters for the first time in the present age.

The first secret of the Bible refers to the "tree of knowledge of good and evil."

According to the biblical version of events, after the Elohim transported the first human-beings to the Garden in Eden they instructed their young servants what they were and what they were not allowed to do:

> And the Lord God [Yahweh Elohim] commanded the man, "You are free to eat from any tree in the garden; but you must not eat of the tree of knowledge of good and evil, for when you eat of it you would certainly die.
> —Genesis 2.16-17

This so-called "tree" was the first symbol in the garden. The second symbol was the serpent:

> "You will not certainly die," The serpent said to the woman, "For God [Elohim] knows that when you eat of it your eyes will be opened, and you will be like God [Elohim], knowing good and evil.
> —Genesis 3.4-5

And the serpent showed her the "fruit," and she saw that it was "pleasant" to the eyes," and that it was a "tree to be desired." Eve partook from the tree and also "gave to her husband," and the "eyes of both of them were opened," and they became aware of their naked bodies. The experience eventually compelled them to sew leaves together to cover themselves.

The first clue to the identity of the symbol is that it was an experience that was "to be desired." After Eve partook of the "fruit," she went on to share it with Adam. Afterward, they covered up their genitalia—which indicates that the so-called "tree" refers to sexuality.

As for the title "of good and evil," this is a reference to the biblical sense of these terms, where good was equated with obedience, and evil was equated with disobedience.

The reference to intercourse with a so-called "serpent" can be explained by understanding that just like the Tree of Knowledge of Good and Evil, the serpent in the Genesis story was a symbolic representation. The character was not, of course, a literal talking snake, but rather was an individual among the Anunnaki/Elohim who was associated with this symbol (more on this subject later).

There are several reasons why the Elohim did not want their servants engaging in sexual intercourse. The first pertains to the practical concerns that human-beings procreating on their own created for them. We are told that the hybrids multiplied upon the land, and that the so-called "noise" of mankind became too much for the gods to bear. Not only were the hybrids reproducing among themselves but they were interbreeding with the Anunnaki/Nephilim as well—who, in the following verse, are referred to as the "sons of God":

> When human beings began to increase in number on the earth and daughters were born to them, the sons of God [Elohim] saw that the daughters of humans were beautiful, and they married any of them they chose.
> —Genesis 6.1-2

Another concern was that the offspring of these unions were becoming rivals who posed a threat to the authority of the Elohim. If the hybrids were not being disobedient barbarians, the later more developed progeny—some of whom were the Nephilim—were purportedly posing a challenge to the sovereignty of the Elohim by attempting to equal them in power. This is what the biblical tower of Babel story refers to.

Another likely reason why the Elohim did not want the hybrids to engage in unsanctioned sexual intercourse is because this experience distracted them from their duties as servants. We find many examples in the biblical record of Yahweh and his servants denouncing "whores," "fornicators," and "harlots." Such sexually active persons were to be considered "shameful," "disgusting," and "sinful." Of course, this was a policy that helped to ensure a more effective worker, warrior, and worshiper class; which the Elohim needed to be ready and willing to carry out their assigned duties. Indeed, the Mesopotamian texts report that the reason for the creation of the hybrids in the first place was to ease the workload of the gods, so that they themselves could have more time to pursue a leisurely life.

In the following passage from the Epic of Gilgamesh, we find an account in which a man named Enkidu embarked on a journey of internal resurrection. One of the ways that he was able to do this was from an act of sexual intercourse:

> The lass freed her breasts, bared her bosom, and he possessed her ripeness [. . .] She treated him, the savage, to a woman's task [. . .] Now he had vision, broader understanding [. . .] The harlot says to him, to Enkidu: "Thou art knowing, Enkidu; Thou art become like a god!"
> —The Epic of Gilgamesh

Of course, becoming "knowing," "like a god," was exactly what the Elohim did not want their servants to attain for themselves.

In the following passage from the gnostic Christian Nag Hammadi codices, we find a reference to the Tree of Knowledge. The report indicates that the experience was not negative or harmful, but rather was related to a healthy life and personal growth:

> "But the tree which they call 'of the knowledge of good and evil,' this is the Thought of Light, on whose account the commandment was given not to taste of it, that is, 'do not hear them,' since the commandment was directed against him, so that he might not look up towards his perfection nor know his nakedness in relation to his perfection. But I have brought you to eat of it."
> —Jesus, Nag Hammadi Codices, The Secret Book of John

The reference to "them" in this pericope refers to the Anunnaki rulers, i.e., the Elohim, who the gnostics referred to as the "Archons" and the "Demiurge" (more on this subject later). In the following passage, a faction within this group are referred to the as the "authorities" (i.e., the Archons):

> Then the authorities came to their Adam. And when they saw his female counterpart speaking with him, they became agitated with great agitation; and they became enamored of her. They said to one another, "Come let us sow our seed in her," and they pursued her.
> —The Nag Hammadi Codices, The Hypostis of the Archons

Among the authorities who became enamored of Adam's female counterpart was an adversary in the council of the Elohim who was associated with the symbol of the serpent.

Apparently, the ruling Elohim (as opposed to the rebellious Archons) did not have the best interest of their hybrid subordinates in mind when they restricted the Tree of Knowledge from them. Even the biblical source confirms that Adam and Eve were not told the truth by the Elohim. For instance, they were told that if they partook of the tree that they would "surely die"; and yet, after they partook of the fruit they did not die. Judeo-Christian apologists explain that this is a reference not to physical death, but rather to spiritual death. However, as we proceed further along in this investigation, it will become evident that this interpretation cannot be justified.

The gnostic Christians were aware that sexuality was related to the "light" of salvation, which is referred to in the following passage as the "immortal food" of the "bridal-chamber":

> She reclined in the bride-chamber. She ate of the banquet for which she had hungered. She partook of the immortal food. She found what she had sought after. She received rest from her labors, while the light

that shines forth upon her does not sink. To it belongs the glory and the power and the revelation, for ever and ever. Amen.
—The Nag Hammadi Codices, The Authoritative Teachings

Come into the bridal-chamber! Be illuminated in mind!
—The Nag Hammadi Codices, The Teachings of Silvanus

Other passages from the gnostic texts indicate that Jesus himself was open-minded when it came to the fruit of the Tree of Knowledge. Not only did he encourage his disciples to be unashamed of their nakedness[165] but was also seen kissing Mary Magdalene.[166]

The type of teachings that were being espoused by many in the gnostic community can also be found in other cultures and religions as well—such as the Tantric practices of Hindu India. In Tantra, it is believed that there are energy points in the body called "chakras." The ultimate experience of the Tantric practitioner is the *Kundalini* experience, which is the blissful rush of cosmic energy known as the "serpent power." This is the same type of sensual experience that the Buddha may have engaged in when he partook in the "Diamond Glory Partaking of All Desires." Likewise, in the Guhyasamaja Tantra, we are told that the Buddha also had an encounter with the Dakini woman, who he entered into *Samadhi* bliss with.

In ancient Greece, the mystical union was represented in the *theia mania* of the god Dionysus and the *unio mystica* of the mystery rites. This is the type of experience that is referred to in the book of Genesis, when the serpent told Eve that if she partook of the Tree of Knowledge that her "eyes would be opened," and that she would "be like Elohim."

---

[165] Gospel of Thomas
[166] Gospel of Philip

CHAPTER VI

THE TREE OF LIFE

Like the Tree of Knowledge of Good and Evil, the identity of the Tree of Life is veiled behind the guise of symbolic representation. According to the biblical account, the first man and woman were banished from the garden before they had the chance to discover what this second type of mystery experience was.

The first mention of this so-called "tree" appears in the following passage:

> And then Yahweh Elohim said, "The man has now become like one of us, knowing good and evil. He must not be allowed to reach out his hand and take also from the tree of life and eat, and live forever.
> —Genesis 3.22

This symbol is not mentioned again in the Old Testament until the book of Proverbs, where we read the following:

> Wisdom is a tree of life for those who take firm hold of it. Those who cling to it are blessed.
> —Proverbs 3.18

> The fruit of the righteous is a tree of life. He who is wise wins souls.
> —Proverbs 11.30

In the final book of the New Testament, it is reported that the tree was offered as a reward by God to those who proved themselves worthy:

> Every person who has ears should listen to what the Spirit says to the churches. To those who win the victory I will give the right to eat the fruit from the tree of life, which is in the garden of God [Theos].
> —Revelation 2.7

> [. . .] On this side of the river and on that was the tree of life, bearing twelve kinds of fruits, yielding its fruit every month. The leaves of the tree were for the healing of the nations.
> —Revelation 22.2

> If anyone takes away any words from this book of prophecy, God [Theos] will take away his portion of the tree of life and the holy city that are described in this book.
> —Revelation 22.19

What is especially significant about these passages is that they all describe the Tree of Life in positive terms. In other words, the fruit of this tree was planted in "paradise," and was considered to be "happy" and "righteous." Therefore, we know that just like the Tree of Knowledge it was an object, and/or experience, that was not harmful. It is therefore likely that the only reason why this experience was denied to the hybrids is because the Elohim did not want their servants to be treated as equals.

We also know that, just like the Tree of Knowledge, it was something that was in the midst of the garden itself, and was something that the hybrids could have access to, and possibly partake of behind the back of the Elohim.[167]

In order to solve the mystery of the second type of "tree," it is necessary to refer back to the original Mesopotamian sources. One of the most well known of the records from that time and location is the Epic of Gilgamesh. This text was discovered in the ruins of Nineveh[168] in 1872. The clay tablets were dated to around the seventh century BCE—although, even older versions were later found, making it one of the oldest known works of literature. Like most ancient mythological lore, the characters, situations, and symbols reported within were most likely inspired by actual events, people, places, and objects. Indeed, we know that the character of Gilgamesh was based on an actual person because his name appears on the Sumerian King list, along with other known historical figures of that location and age.

Gilgamesh was the King of the city of Uruk (the biblical Erech). He was reported to have been one of the direct hybrid human offspring of the gods. It is therefore possible that he was one of the Nephilim that the book of Genesis tells us were so influential in those times.

According to the account, the king set out on a quest to obtain the legendary Plant of Immortality, which he believed would resurrect his dead companion Enkidu (whose name means creature of Enki). It was a divine object that he believed would also preserve and rejuvenate his own life as well.

> This plant of all plants is unique: By it a man can regain the breath of life!
> —The Epic of Gilgamesh

However, the Plant of Life was not easy to obtain, partly because the gods had hidden it from humankind.

---

[167] This negates Sitchin's interpretation of the Tree of Life, who believed that this object referred to life spans (Sitchin 1990: 189).

[168] Nineveh is the ancient capital of Assyria.

According to the Enki and Ninhursaga text, the tree and the plant were similar:

> His minister Isimud had the answer for him. "My master, the 'tree' plant," he said to him, cut it off for him and Enki ate it.
> —Enki and Ninhursaga

In the Epic of Gilgamesh, we are told that after a frustrating period of searching Gilgamesh eventually came across a man named Utnapishtim (the Akkadian name of Noah), whose name literally means "He found Life" (Dalley 1989: 2, 330). Gilgamesh consulted him since he was a wise man who was favored by the gods:

> Utnapishtim said to him, to Gilgamesh: Thou hast come hither, toiling and straining. What shall I give thee to take back to your land? Let me disclose, Gilgamesh, a closely guarded hidden matter—a secret of the gods I will tell thee: A plant there is, like a prickly berry bush is its root. Its thorns are like a briervine's; thine hands the thorns will prick. (But) if with thine own hands the plant you could obtain, Rejuvenation you will find.
> —The Epic of Gilgamesh

Because the identity of the Plant of Life was a "hidden matter" of the gods, Utnapishtim tells Gilgamesh not to attempt to solve this forbidden mystery, and instead accept his fate as a subordinate mortal:

> Gilgamesh, whither rovest thou? The life thou pursuest thou shalt not find. When the gods created mankind, Death for mankind they set aside, Life in their own hands retaining.
> —The Epic of Gilgamesh

Despite this warning, Gilgamesh continues his quest, which eventually leads him to the elusive object. However, just as he is about to take it, it is stolen away by a snake. But before this happens we are given a brief description of the Plant of Life:

> When he saw the plant of rich rose color and ambrosial shimmering in the water like a prism of the sunlight [. . .]
> —The Epic of Gilgamesh

What is particularly significant about this report, is that the translation describes the plant as being "ambrosial," which is a term that is derived from the original Greek word *ambrotos*—which means "immortal." Ambrosia is known in Greek mythology as the nectar of the immortal gods.[169] Likewise, in the scriptures of the ancient Hindus, there are references to a sublime "ambrosia" that was ingested by the gods. The following Vedic Hindu passage tells us that the partaking of the immortal "ambrosia" of life was a privileged experience that was reserved for the gods:

> [. . .] promise them that by drinking the ambrosia they shall become immortal. But I shall see to it that they have no share of the water of life, but theirs shall be labor only.
> —The Churning of the Ocean[170]

The water of life—which is most likely the symbolic representation of the ambrosia that is made from the Tree of Life—was not to be imbibed by the common person. This elixir was referred to in the ancient Vedic scriptures as "Soma."

---

[169] e.g., Hesiod's Theogony.

[170] The Churning of the Ocean myth (i.e., Sumudra Manthan or Sagar Manthan) appears in the Bhagavata Purana, the Mahabharata, and the Vishnu Purana.

In the following passage, a mortal man boasts of his sublime experience after having tasted the forbidden nectar of the gods:

> We have drunk Soma and become immortal; we have attained the light, the Gods discovered [. . .] Absorbed into the heart, be sweet [. . .] O Soma, lengthen out our days for living.
> —Rig Veda, Book VIII, Hymn XLVIII. Soma

The word "Soma" appears throughout the ancient Hindu scriptures in relation to the mystical experience, although it is never explicitly described. The records only tell us that it was made from an object that was pressed with stones and mixed with milk. The exact ingredients of this substance were known only to the gods and the initiated elite.

In the Rig Veda, we find references to immortality that are related to Soma:

> O Soma, of thy juice for wisdom, and all Deities for strength [. . .] So flow thou on as bright celestial juice, flow to the vast, immortal dwelling-place.
> —Rig Veda, Book IV, Hymn CIX. Soma Pavamana

> Soma hath risen in us, exceeding mighty, and we are come where men prolong existence.
> —Rig Veda, Book VIII, Hymn XLVIII. Soma

Like the Mesopotamian Plant of Life, Soma had the ability to not only restore life but to bestow wisdom as well. To attain this higher and divine perception was the ultimate aim of the mystic Yogis, Rishis, and Holy God men of India.

> Flow on, Sage Soma, with thy stream to give us mental power and strength,
> —Rig Veda, Book IV, Hymn C. Soma Pavamana

> Thou, Soma, art preeminent for wisdom;
> —Rig Veda, Book I, Hymn XCI. Soma

> He becomes the one ocean, he becomes the sole seer! This, Your Majesty, is the world of *brahman*, So did Yajhavalkya instruct him. This is his highest attainment! This is his highest bliss! On just a fraction of this bliss do other creatures live.
> —Brhadaranyaka Upanishad

According to the ancient Vedic texts, the Rishis (the Seers) ritualistically partook of Soma in their religious ceremonies. The mystery drink was used to raise themselves up from out of the inferior world of man and up into the supernatural realm of the gods.

> Soma advances to the special place of Gods.
> —Rig Veda, Book IX, Hymn LXXXVI. Soma Pavamana

> So, God, for service of the Gods flow onward, flow, drink of Gods, for ample food, O Soma.
> —Rig Veda, Book IX, Hymn XCVII. Soma Pavamana

Soma imbued the partaker with exceptional "power":

> Flow on thy way to win us strength, to speed the sage who praises thee: Soma, bestow heroic power.
> —Rig Veda, Book IV, Hymn XLIII. Soma Pavamana

This heroic power awakened the mind and rejuvenated the heart:

> Send us a good and happy mind, send energy and mental power [. . .] O Soma, rest thy powers that influence the heart.
> —Rig Veda, Book X, Hymn XXV. Soma

According to clues in the texts, the special ingredient in Soma derived from a special type of plant:

> Of all the many Plants whose King is, Soma, Plants of hundred forms, Thou art the Plant most excellent, prompt to the wish, sweet to the heart.
> —Rig Veda, Book X, Hymn XCVII. Praise of Herbs

> And, Soma, let it be thy wish that we may live and may not die: Praise-loving Lord of plants art thou.
> —Rig Veda, Book I, Hymn XCI. Soma

Soma was a sacrament that was considered to be so especially sacred that its secrets had to be guarded by an elite priestly class known as the Brahmins.

> One thinks that he has drunk the Soma when they press the plant. But the Soma that the Brahmins know—no one ever eats that. Hidden by those charged with veiling you, protected by those who live on high, O Soma, you stand listening to the pressing stones. No earthling eats you. When they drink you who are a god, then you are filled up again.
> —Rig Veda, Book X, Hymn LXXXV. Surya's Bridal (The Marriage of Surya)

In the following verse, we are told of the relationship between the gods and Soma—although, in this pericope it is referred to not as Soma, but rather as "Homa":

> Bright Immortals robbed in sunlight sailed across the liquid sky. And their gleaming cloud-borne chariots rested on the turrets high [. . .] Ida, adja, homa offerings pleased the "Shinning Ones" on high [. . .]
> —The Mahabharata. Book III (Rajasuya), Section II (Feast and Sacrifice)

The immortal Soma/Homa (i.e., "Haoma") elixir also appears in the records of the Zoroastrians of Persia:

> All men became then of one voice in praise lifted loud to Ahura Mazda and his angels. Saoshyant with his assistants slaughter ceremonially an ox, from whose fat and the white Haoma drink of immortality is prepared, which is given to all men, who become thereby immortal forever.
> —The Bundahish (Book of Creation)

"Haoma" was imbibed by Zoroastrian priests in order to liberate their minds from the veil of repression and deception that was cast by the devilish god, Angra Mainyu—whose epithet was the "Demon of the Lie" (Campbell 1964: 192). This fiendish deity prevented humankind from empowering themselves, and thus, from realizing the reality of their oppression by attempting to eliminate the "White Haoma Tree" from the Earth; however, he was thwarted by the benefactor of humankind: Ahura Mazda, the good god of light.

> Praise to Haoma. Good is Haoma, and the well-endowed, exact and righteous in its nature, and good inherently, and healing, beautiful of form, and good in deed, and most successful in its working, golden-hearted, with bending sprouts. As it is the best for drinking, so through its sacred stimulus is it the most nutritious for the soul. The first blessing I beseech of thee, O Haoma, thou that drivest death afar! I beseech of thee for heaven, the best life of the saints, the radiant, all-glorious. The second blessing I beseech of thee, O Haoma, thou that drivest death afar! This body's health before that blessed life is attained. The third blessing I beseech of thee, O Haoma, thou that drivest death afar, that I may stand victorious on earth,

conquering in battles, overwhelming the assaults of hate, and conquering the lie [. . .]
—The Avesta, Yasna

The passage indicates that like the biblical Tree of Life, the partaking of Haoma was a "righteous" experience that was used for "healing." It was also a substance that was used to "drivest death afar"; which is a reference to the ability of the object to endow its user with some type of life-extending ability. The substance is also described as being "nutritious for the soul," which signifies that it bore a spiritual property as well.

Ahura Mazda brought for Haoma the star spangled spiritual girdle, that is, the Mazdaysian religion [of Zoroaster].
—The Avesta, Yasna

The following passage indicates that Haoma was a substance that was made from a divine "plant" called "*Hadhanaepata*":

[. . .] for the worship of the Creator Ahura Mazda, the resplendent, the glorius, and for that of the Bountiful Immortals, I desire to approach this Haoma with my praise [. . .] this plant Hadhanaepata.
—The Avesta, Yasna

We are told in the Zoroastrian Bundahishn text (i.e., The Knowledge of Zand), that Haoma was a liquid that was distilled from the "Gaokerena Tree"—a tree that, according to eminent mythologist Joseph Campbell, was the "tree of life" (Campbell 1949: 151).

In one version of the Epic of Gilgamesh, we are told of not only a "Plant of Immortality" but of a sacred *Hulluppu* tree. Just like the Plant of Life, this tree was also guarded by a serpent. In this version of the story, Gilgamesh's friend, Enkidu, retrieves the tree but falls into the underworld of the dead—which may be a reference to the spiritual unworthiness of Enkidu to receive such a divine sacrament.

The Sumerian Hullupu Tree may also be related to the Babylonian/Assyrian Mesu Tree. In the Mesopotamian Erra and Ishum text, there is a report of a "Mesu tree" that "reach by its roots the bottom of the underworld, and by its top the heaven of Anu." This was, of course, not a literal description of a large tree, but rather an allegorical representation that was intended to describe the spiritual properties of the divine object. The literal meaning of the Mesu Tree is considered uncertain, although some scholars suspect that it may be the Rose-wood tree, which is based on the supposition that its name may be a reference to the word *Elmesu,* which could be translated as "amber." It is indeed possible that the light brownish orange color of amber may be a reference to rose-wood; however, the Mesopotamian language scholar Stephanie Dalley acknowledges that the word and the tree itself contain "wider mythological connotations" (Dalley 1989: 313-314).

In the following Mesopotamian text, we find another example of a tree that is connected to both heaven and to hell. A special kind of tree that bestows "lordship," and is referred to as "the flesh of the gods":

> Now then warrior Erra, as concerns that deed you said you would do, "Where is the wood, flesh of the gods, suitable for the lord of the universe. The sacred tree, splendid stripling, perfect for lordship, whose roots thrust down a hundred leagues through the waters of the vast ocean to the depths of hell, whose crown brushed Anu's heaven on high?
> —Marduk, Erra and Ishum

The existence of a sacred tree of the gods is also reported in the Old Scandinavian mythologies, such as the *Edda* text as well (800-1100 CE). In his quest to obtain wisdom, the god Wodan (i.e., Odin) hung on the symbolic *Yaggsdrasil* "World Tree" for nine days. During that time, he entered into a trance and obtained wisdom and rebirth.

The story is also reminiscent of the story of the Buddha, who attained Enlightenment while sitting under the *Bodhi*—from the Sanskrit word for Enlightenment—tree. It was during that experience that he broke free from the mental veil of delusion and achieved a state of god-like awareness.

> In Buddhist legend [. . .] the whole sense of the teaching is that one should penetrate that guarded gate and discover that tree—the *Bodhi*-tree, the tree of the "Waking to Omniscience," which is the very tree beneath which the Buddha sat when he opened to mankind the way of release from those same two conditions [desire and fear] [...]
> —Joseph Campbell,[171] *The Mythic Image*

Likewise, in the following Buddhist verse we find a reference to the "nectar of immortality" that bestows "life" and awakens "consciousness":

> As a blind man feels when he finds a pearl in a dustbin, so am I amazed by the miracle of awakening rising in my consciousness. It is the nectar of immortality that delivers us from death, the treasure that lifts us above poverty into the wealth of giving to life, the tree that gives shade to us when we roam about scorched by life [...]
> —Shatideva, The Bodhicharyavatara

In the Buddhist Lotus Sutra text, the tree is likened to the "*Dharma* (meaning: righteous natural law) Flower," which is compared to the lotus, the symbol of Enlightenment:

> The Storehouse of the Dharma Flower Sutra is deep, solid, and far reaching. No one could reach it except that now, the Buddha, in teaching and transforming the

---

[171] Joseph Campbell, MA, was a twentieth-century American mythologist and professor at Sarah Lawrence College.

Bodhisattvas and bringing them to accomplishment, demonstrates it for their sakes.

—The Lotus Sutra

The symbol of the lotus, which is found throughout both the Near and Far East, is an object that is related to both the Tree of Life, the gods, and enlightenment.

> The Tree of Life in various stylized forms is known as a distinctive feature of palace architecture and ornamentation. It is found, for instance, in the proto-Ionic capitols from columns in palaces in Jerusalem, Samaria, Hazor, and Ramat Rahel between Jerusalem and Bethlehem, and on an ivory relief of the royal couch in the palace at Ras Shamra, where it is combined with the Egyptian motif of a flourishing lotus.
>
> —John Gray,[172] *Near Eastern Mythology*

Likewise, Plato makes a reference to a group known as the "lotus eaters" in his classic work, The Republic.[173] The lotus was not only a symbol of health and beauty but of wisdom. Indeed, Gilgamesh was not only seeking eternal life but "wisdom" as well. This accords with the previously cited biblical verse, in which the Tree of Life is associated with wisdom (Proverbs 3.18).

In the Mahayana Buddhist tradition, the Buddha Avalokiteshvara—the Lord who looks down in pity—was called the "Lotus bearer" (Campbell 1949: 127). To him the mantra *Om mani padme hum* ("Hail to the Jewel in the Lotus") is chanted. In the role of shaman, the Buddha received the sacrament of the lotus while sitting under the influence of the Bodhi Tree.

---

[172] John Gray, Ph.D., was a twentieth-century Scottish professor of Hebrew and Semitic languages at Aberdeen University.

[173] Book VIII

> Luminous is this mind, brightly shining, and it is free of the attachments that visit it.
>
> —Anguttara Nikaya

It may be from this experience that Buddhism derived the psychological nature of its creed.

> The Buddha is no more the one who is in the world conceivable in space and time. His consciousness is not that of an ordinary mind which must be regulated according to the senses and logic [. . .] The Buddha of the Gandavyuha lives in a spiritual world which has its own rules.
>
> —Daisetz Teitaro Suzuki,[174] *On Indian Mahayana Buddhism*

As the Enlightened Buddha looked back upon the world through the eyes of a god (Elohim), he could no longer see value in the figures and concepts that he once viewed as authoritative and absolute:

> I consider the position of kings as that of dust motes in a sunbeam. I see the treasures of gold and gems as broken tiles. I look upon the finest silken robes as tattered rags. I see the myriad of worlds of the universe as small seeds and the great Indian ocean as drops of mud that soils one's feet. I perceive the teachings of the world to be illusions of magicians. I look upon the judgment of right and wrong as the serpentine dance of dragons, and the rise and fall of beliefs as the traces left over by the four seasons.
>
> —The Buddha, The Sutra of 42 Chapters, Seeing the Illusions of the World. (The Eye of Wisdom)

---

[174] D. T. Suzuki was a nineteenth- and twentieth-century Japanese Buddhist scholar.

The dissolution of the artificial world and the authority of its rulers is exactly what the Elohim in the Garden of Eden story did not want their servants to attain for themselves. This is one of the reasons why the first man and woman had to be denied access to the Tree of Life, which would have contradicted the purpose of their creation in the first place. Indeed, the gnostic Christians were also aware of this fact:

> And this tree is to the north of Paradise, so that it might arouse the souls from the torpor of the demons, in order that they might approach the tree of life and eat of its fruit and so condemn the authorities and their angels.
> —Nag Hammadi Codices, On the Origin of the World

Here we find another description of the Tree that is described as having the ability to "arouse" "souls," which is another reference to the spiritual nature of the Tree of Life. The fact that it also had the ability to impel the user to condemn the authorities, refers to the ability of the Tree to raise the consciousness of the individual from out of the mundane subordinate level.

> [. . .] where the Soma flows over, there the mind is born.
> —Svetavatara Upanishad, Second Adhyaya

In the following extra-biblical text, we find another revealing report that derives from a devout servant of the Elohim:

> Then they [the Anunnaki] took wives, each choosing for himself; whom they began to approach, and with whom they cohabitated, teaching them sorcery, incantations, and the dividing of roots and trees. And the woman conceiving brought forth giants [. . .]
> —The Book of Enoch

In this account, the rebellious Annunaki are reported to have shared both the Tree of Knowledge as well as the Tree of Life with human females. The so-called "trees" that were associated with the Old Testament concept of "divination" and so-called "sorcery" indicate that the Tree of Life produced some type of mystical state that could be classified as sorcery—that is, by the uninitiated.

* * *

It is clearly evident that the Tree of Life was both a literal object as well as a preternatural experience that can be related to higher states of consciousness, wisdom, and eternal (i.e., spiritual) life. Amazingly, such an extraordinary object does exist. It is a substance that can be classified as a psychoactive stimulant, psychotropic, or psychedelic.

The nomenclature *psychedelic* is constructed from the word *psyche,* which is the Greek term for "soul" (In more recent times, the term has been altered by physicalist in order to refer to the mind); and *delic*, from the Greek word *delos,* which means to "manifest," or to "clarify." Together it is a term that represents a substance that brings about the eternal world of the spirit/soul, euphoria, and higher states of consciousness. These stimulants are also sometimes classified under the alternate nomenclature: *entheogen*. The word "entheogen" is a term that was coined in 1979 by ethno-botanists and mythology scholars from the Greek root-words for "god within." An entheogen can be defined as any psychedelic substance that is used in a religious, spiritual, or shamanic context. This is what the Tree of Life was. Indeed, the use of hallucinogens to attain the supernatural realm of the gods by prophets and shamans stretches back over the course of millennia (Richards 2016: 7; Schultes 1976: 155; Schultes, et al. 1979: 9, 26). Archaeological evidence in the form of paintings, rock carvings, amulets, ceramic artifacts, and stone figurines confirm the presence and the use of psychoactive plants in both hemispheres of the world (Schultes 1998: 1).

* * *

There are several possible candidates for the Tree of Life. This is most likely because there was more than one type of so-called "tree."

The first candidate is the acacia tree, which contains the psychoactive compound DMT (i.e., *dimethyltryptamine*). It is therefore interesting that references to the acacia appear in the Egyptian texts, where it is reported that it was a "holy" object that was kept in a temple at Heliopolis.

> Thereupon the keepers of the doors who were in the [temple of][175] the holy Acacia Tree started up at the voice of Horus. And one sent forth a cry of lamentation, and Heaven gave the order that Horus was to be healed.
> —The Narrative of Isis

The acacia wood may have been burned as incense that was inhaled by those who were allowed to enter into the temple area:

> Thou canst snuff at will the odours of the holy Acacia of Anu (An, or Heliopolis).
> —Book of Breathings

In the following passage, it is reported that the rebellious god Seth partook of the acacia tree without the consent of the ruling gods:

> He has committed sacrilege against the splendid space of *ius-a*. As with the acacia tree, in which death and life are (decided).[176] He has thought of eating the *mAfd.t* [sic] before the faces of Mut and Bastis.
> —Seth's Misdeeds in the Sites of Egypt

---

[175] Text insert in brackets from: Budge, Ernest A. Wallis. *Legends of the Gods: The Egyptian Texts* (Cosimo Classics, 1912).

[176] Text insert in parentheses from an English translation: (www.reshafim.org.il/ad/egypt/texts/victory_over_seth.htm) of an earlier work: Schott, Siegfried. *Bücher und Sprüche gegen den Gott Seth, Urkunden des ägyptischen Altertums, sechste Abteilung*. Heft 1, 1929.

The acacia tree is associated with the god Osiris,[177] as well as Unas and Horus;[178] the latter of which is said to have been born from within (or under) an acacia tree.[179]

> Unas is Horus, who came from the acacia.
> —The Pyramid Text of Unas (Spells for the Spirit's Rebirth)

This divine tree[180] also appears in early Hindu Indian seals, where it was also associated with the gods (James 1966: 23-24). It is also possible that the acacia may have some relation to the *ished* tree. Although the identity of this legendary ancient Egyptian tree is not known for certain, it does appear to be associated with the life-giving powers of the Sun and the gods—specifically Ra (i.e., Re) (James 1966: 41). Moreover, it is significant that the ished tree was also associated with eternal life (Breasted 1906: 129).

Benny Shanon, a professor of cognitive psychology at the Hebrew University of Jerusalem, postulated that Moses was under the influence of a psychotropic substance when he was communing with God (Wood 2008).[181] It is Shanon's belief that the substance was derived from the acacia. Indeed, the acacia tree (i.e., *shittah*, or *shittam* wood) is mentioned throughout the Old Testament/Tanakh (Exodus 26.15; Isaiah 41.19). (The MAO inhibitor necessary for

---

[177] Osiris and Isis Text

[178] Pyramid Texts

[179] According to various internet sources, another ancient Egyptian deity that may have been associated with the acacia tree was the goddess Iusaaset (i.e., Saosis). However, I have been unable to verify this.

[180] Along with the acacia, depictions of what may be the pipal tree are also extant. The Indian pipal tree (i.e., the Ashvattha or *Ficus religiosa*) is thought to be endowed with supernatural properties and is associated with the god Vishnu. The pipal tree can also be used for medicinal purposes; however, it does not contain psychotropic properties.

[181] Benny Shanon. "Biblical Entheogens: A Speculative Hypothesis." (*Time and Mind*. Vol. 1, No. 1, Jan 2008): 51-74.

DMT activation may have been the *peganum harmala* plant.) Shanon was inspired to pursue this thesis after personally experiencing "visions that had spiritual-religious connotations," while under the influence of the DMT-bearing *ayahuasca* plant decoction in a Brazil rain-forest.

Although DMT is structurally similar to other psychotropic substances, the effect itself is so uniquely potent in high doses that the user will usually sense his or her psyche being propelled from out his or her body and transported into an otherworldly environment. It is in this alternate modality that the experiencer will often see, hear, or feel the presence of autonomous beings who appear in various fantastical forms. This other realm bears similar characteristics to what is traditionally referred to as the spirit world. Indeed, the similarities between the DMT experience and out-of-body near-death events are significant (Strassman 2001: 154-155, 221, 226). For example, it is not uncommon for the experiencer to hear a buzzing or a ringing sound shortly before the out of body experience begins, as well as a sensation of the psyche departing and entering through the top of the head, and encounters with beings. Both DMT and NDE experiencers also report that when they open their eyes when they are still in their bodies, it looks as if this other dimensional environment is "superimposed" over our own.[182]

Between 1990 and 1995, DMT experiments involving 60 participants were conducted under the supervision of Rick Strassman[183] at the research center at the University of New Mexico's School of Medicine. The experiments confirmed the mystical nature of the experience, which compelled Strassman to refer to DMT as the

---

[182] These examples were taken from: Raymond A. Moody Jr.'s book *Life After Life: The Investigation of a Phenomenon—Survival of Bodily Death.* (Harper One, 1975), and Rick Strassman's book, *DMT: The Spirit Molecule: A Doctor's Revolutionary Research into the Biology of Near-death and Mystical Experiences* (Park Street Press, 2001).

[183] Rick Strassman MD, clinical associate professor of psychiatry at the University of New Mexico School of Medicine.

"spirit molecule."[184] It can therefore be proposed that the closest method that is presently available to repeatedly replicate what may be related to the near-death experience is through the use of DMT.

According to Strassman, "contact with beings predominated people's sessions" (Strassman 2001: 184, 249, 334). These autonomous entities are sometimes described not only as creatures but as angels, aliens, and gods. Moreover, most of the participants in the experiments were convinced that what they had experienced was not a mere drug hallucination, and that DMT did not create this other world but rather revealed it (Strassman 2014: 165-167). Indeed, many believed that they had received profound insights into the afterlife; which can be described as a parallel universe that exists along-side our own.

> DMT provides regular, repeated, and reliable access to "other" channels. The other planes of existence are always there. In fact, they are right here, transmitting all the time! But we cannot perceive them because we are not designed to do so; our hard-wiring keeps us tuned in to Channel Normal. It takes only a second or two—the few heartbeats the spirit molecule requires to make its way to the brain—to change the channel, to open our mind to these other planes of existence. How might this happen? I claim little understanding of the physics underlying theories of parallel universes and dark matter. What I do know, however, causes me to consider them as possibles places where DMT might lead us, once we have rushed past the personal.
> —Rick Strassman, *DMT: The Spirit Molecule*

These experiences were so genuine, so sublime, and so powerful that it compelled Strassman to conclude that religion may have derived

---

[184] *DMT: The Spirit Molecule: A Doctor's Revolutionary Research into the Biology of Near-death and Mystical Experiences* (Park Street Press, 2001), and the documentary movie, *DMT: The Spirit Molecule* (2010).

from ancient drug use.[185] Indeed, one user reported feeling during the event as if he himself had become a god (Strassman 2014: 173).[186] However, the DMT experience is not only brief (usually ten to thirty minutes) but where the experiencer arrives at in this other realm seems to be entirely random—or at least based on some sort of karmic trajectory. I suspect that if DMT was at least one of the identities of the Tree of Life, it was used in only a minority of possible cases. Indeed, most of the reports of theophany in the Hebrew Bible occur not in some other world, but in this one.

The second candidate for the Tree of Life is the boswellia tree. The hardened sappy resin from this tree, which is commonly referred to as frankincense, contains a psychoactive compound (*incensole acetate*). The smoke from Boswellia resin has the ability to produce sensations of exaltation and has been used for religious purposes for centuries (Moussaieff, et al. 2008: 3024). The amber color of Frankincense can be related to the amber color of the previously mentioned mythological Mesu Tree. It is also compelling that Boswellia trees are most common in southern Arabia, which is a location that is within the geographical vicinity of the earliest settlements that are referred to in the book of Genesis. However, if the tree was plentiful in the region this would conflict with the biblical account that reports that human-beings were cut off from access to the Tree of Life. If this account is accurate, this indicates that the "tree" was only accessible in the garden, which would eliminate both the boswellia tree and the acacia tree as the Tree of Life.

This same reasoning may also apply to the "Blue Lotus" (*nymphaea caerulea*). The sacred water lily contains a psychoactive alkaloid (*apomorphine*) that produces states of euphoria and

---

[185] This theory is continued in Strassman's subsequent book: *DMT and the Soul of Prophecy* (Park Street Press, 2012).

[186] In his book *True Hallucinations*, author Terence Mckenna reports that while under the influence of the mushroom entheogen his brother experienced a profound revelation concerning the link between other dimensions of space and time, ancient Egypt, and "acacia tryptamine cults."

enhanced perception. Besides the previously mentioned relation between the lotus and the Buddha, the image of this divine flower was also depicted in the funerary art of both the ancient Egyptians and the Mayans (Emboden 1978: 395-407; Prance and Nesbitt 2012: 203). The association with the water lily with the after-life and with the gods is a reference to the mystical nature of its effect. This effect is referred to in the Egyptian Book of the Dead,[187] where it is referred to as an object that bestowed eternal life in the realm of the gods.

> It is essentially a magical shamanic transformation. The water lily was initially the favorite of Ra, and a product or emanation from his being. Ani wished to have the power to transform himself into the sacred blue water lily so that his body might have new birth and ascend daily into heaven. Another version of this transformation allowed Ani to transform himself into Ptah (creator god). Importantly, the accompanying vignette is a human head springing from the open flower of Nymphaea caerulea growing in a pool of water. The text of this is attributed to "Osiris Ani" who says, "I am the holy water lily that comes forth from the light which belongs to the nostrils of Ra, and which belongs to the head of Hathor. I am the pure water lily that came forth from the field of Ra." Later versions of the same text petition the water lily with requests for visions and soul flight. Such supplications suggest the power of the water lily and are important stylistic clues to the chemical nature of the flower which might be used to provide such transcendent experiences.
> —William A. Emboden,[188] *The Sacred Narcotic Lily of the Nile*

---

[187] In the Transformation into a Water Lily Flower chapter.

[188] William A. Emboden, Ph.D., is an emeritus professor of biology at California State University, Northridge.

Indeed, the sacred water lily was prominently displayed in the throne-rooms of the Egyptian gods (Figure 12).

(Fig. 12)

It has been suggested by some scholars that Soma may have been made from ephedra (from the plant *ephedra sinica*). The evidence exists in the form of organic remains that have been uncovered at a Bactrian site that seems to have been a temple complex (Kellens 1999: 384; Parpola 1995: 371). Ephedra is a substance that stimulates the nervous system. It most notably increases the heart-rate and blood pressure. In China, ephedra has been used as a medicine for centuries. In more recent times it has been used in the occident for weight-loss purposes, because of the effect that it has on the metabolism. Among the numerous dangerous side-effects of ephedra are heart-attacks, stroke, and death—which is why this substance has been banned by the American Food and Drug Administration. Moreover, the effects of ephedra do not concur with how the effects of Soma are described in the ancient texts. I therefore contend that ephedra was not the sacred Soma ambrosia. What is more interesting about the Bactrian (i.e., Togolok-21 and Gonur-1) site, however, is the cannabis and poppy remains that were found inside ritualistic vessels (Parpola 1995: 371).

Another possible candidate for the Tree of Life was investigated by the amateur mycologist R. Gordon Wasson, in the 1960s. After extensive travels in Asia, Wasson published his findings that pertained to the identity and the history of Soma.

> In a word, my belief is that Soma is the Divine Mushroom of Immortality, and that in the early days of our culture, before we made use of reading and writing, when the *Rgveda* [sic] was being composed, the prestige of this miraculous mushroom ran by word of mouth far and wide throughout Eurasia, well beyond the regions where it grew and was worshiped.
> —R. Gordon Wasson, *The Divine Mushroom of Immortality*

A similar libation was made by the shamans of Siberia, who mixed the mushroom together with liquids (Schultes 1976: 24).

Although mushrooms, such as the *amanita muscaria* and the *stropharia cubensis*, are not indigenous to the hot and arid regions of the Near East, the conditions may have been favorable in Eden: the artificial garden environment that was located in the biblical "land between two rivers." It is therefore a possibility that the mushrooms were transported to and cultivated in the garden by the Anunnaki/Elohim. This explains why human-beings in the biblical garden of Eden would have been cut off from immediate access to this substance. Furthermore, unlike DMT, the psychoactive mushroom effect takes place in this world, not in another. This more accurately reflects what is reported in the Old Testament/Tanakh. (The subject of the vision that takes place in another world that is documented in the book of Enoch and Revelation will be examined later. What will be shown is that these "visions" do not apply.)

* * *

Entheogens, such as Soma, should not be confused with narcotics and intoxicants. Whereas those type of stimulants shut the individual off by producing sensations of mindless abandon, the entheogen rouses the psyche from the limitations of the common state. Indeed, the partaking of Soma was a sacred rite that raised the bio-electric frequency of the partaker to that of the astral gods.

> All other toxicants go hand in hand with the Rapine of the bloody spear, but Haoma's stirring power goes hand in hand with friendship.
> —The Avesta, Yasna

Whether the user undergoes a deeply profound spiritual experience while under its influence or not, such diverse experiences are, nevertheless, insights into the spiritual/mental state of the user. The effect of the entheogen only intensifies whatever internal condition that the individual is already experiencing; and, of course, the amount ingested is a significant factor as well.

The relationship between entheogens and the spiritual experience was confirmed by experiments that were conducted by Stanislov Grof,[189] who examined the link between the entheogen and the psychology of religion.[190] This finding was corroborated in a 2002 study (published in 2006) conducted at Johns Hopkins University, which also confirmed the relation between the *psilocybin*-bearing mushroom and what the participants referred to as one of the most profound spiritual experiences of their lives (Edwards 2008). Both of these studies validate the findings of the earlier Marsh Chapel

---

[189] Stanislov Grof M.D., Ph.D., is a psychiatrist and one of the founders of transpersonal psychology. He is a professor of psychology at the California Institute of Integral Studies in the Department of Philosophy, Cosmology, and Consciousness in San Francisco, CA, and at the Wisdom University in Oakland, CA.

[190] Grof has subsequently documented his discoveries in his books: e.g., *The Adventure of Self-Discovery* (State University of New York Press, 1988); *LSD Psychotherapy* (M.A.P.S., 2001); *Realms of the Human Unconscious* (Condor Books, 1975).

Experiment, which was affiliated with Timothy Leary's[191] historic Harvard Psilocybin Project.

In summary: a majority of the partakers were not experiencing mere recreational folly that is commonly associated with drug use, but rather something far more spiritual and profound.

* * *

The following image is from a Babylonian clay relief plaque (1000 BCE-1700 BCE). In the hand of the male is a mushroom-shaped object (Figure 13).

(Fig. 13)

The following image is from a royal seal of a Hittite king. (1250-1220 BCE). On each side of the seal impression are mushroom-shaped objects (Figure 14).

---

191 Timothy Leary, Ph.D., was an American psychologist and philosopher.

(Fig. 14)

There are several possible reasons why the mushroom may have been identified as a "tree." The primary reason is that the true identity of the object needed to be protected in order to preserve the sacred secret. The secret purportedly began when the Elohim needed to preserve their own interests by keeping their servants away from this key to discovery, euphoria, and freedom. The second reason evolved over a course of time as human-beings themselves began to discover the mystery for themselves, and came to realize that it was an experience that not everyone would not have the capacity to partake in and to comprehend. Due to the connotation that psychedelics have with degenerate drug use and so-called "hallucinations," the general public would only dismiss the experience, and the visions received while under its influence would be relegated to the status of mere delusional mirages—despite the fact that such seemingly meaningless pseudo-hallucinations are still related to the projections of the individual's underlying psyche state, which is significant in itself.

Part of this reaction is due to the anti-mysticism that developed as a result of the prevalence of Judeo-Christianity, which views such activities as having sinful affiliations that are related to the occult, divination, and sorcery.

> The virtues of society are the vices of a saint.
> —Ralph Waldo Emerson, *Circles*

The sanctity of the secret had to be protected from the immature by obscuring it behind the veil of symbolic representation. This policy of secrecy is what the Buddha referred to in the following verse:

> At that time, the Buddha further told the Bodhisattva, Mahasattva Medicine King [. . .] the [Dharma Flower] sutra is the treasury of the Buddha's secrets and essentials. It must not be distributed or falsely presented to people. That which the Buddha's, the World Honored Ones, have guarded from the distant past until now, has never been explicitly taught.
> —The Lotus Sutra

Some might disagree with these findings—especially the Buddhist laity, based on the fact that the Buddha taught that intoxicating stimulants clouded the mind, and were nothing but harmful distractions; however, it must be understood that the entheogen is not an "intoxicant." While it is true that for the "purpose of training" it is necessary for the neophyte to avoid alcohol etc., the psychedelic experience has nothing in common with escapist abandon. Indeed, the epiphany that was induced by the Tree of Life may even, in some cases, lead to the Buddhist state of "Nirvana"—which means "extinction," or "blowing out," of the lower mental veil of distractive delusion. Indeed, this is the type of experience that was reported by modern entheogen advocates, such as Timothy Leary and Terence Mckenna.

* * *

In South and Meso-America, the native shamans partake in a decoction called *ayahuasca* (i.e., "yage"), which translates as spirit vine. Besides this DMT-based substance, the indigenous people also ingest the sacred *teonanacatl* mushroom (i.e., the "flesh of the gods") (Schultes 1976: 58). Teonanacatl was sometimes mixed with a honey-flavored beverage, similar to the Asian Soma (Aguilar-Moreno 2006: 360-361). Indeed, hundreds of sculptures from Guatemala (circa 500-

1000 BCE) have been discovered that depict the sacramental image of the stone mushroom figure (Figure 15).

(Fig. 15)

It is also known that hallucinogens, such as the psychoactive mushroom and morning glory seeds, were sacred to the Aztec religion (Schultes, et al. 1979: 26-27). Indeed, Mushroom cults were prevalent in the ancient and the early Common Era in both South and Meso-America (Schultes 1976: 59; Schultes, et al. 1979: 162).[192]

Similarly, the psychoactive datura flower was not only used in sacred ceremonies at the Temple of the Sun (Schultes 1976: 145) but is found in other cultures and religions throughout the world, including the Middle East (Schultes 1976: 9, 52; 1979: 28).

> DHATURA and DUTRA (*datura metel*) are the common names in India for an important Old World species of Datura. The narcotic properties of this purple-flowered member of the deadly nightshade family, Solanaceae, have been known and valued in

---

[192] Mushroom and peyote use by the indigenous people is attested in the sixteenth-century *Florentine Codex (General History of the Things of New Spain)* (Book IX; chapter 8; Book X, chapter 29).

India since prehistory. The plant has a long history in other countries as well. Some writers have credited it with being responsible for the intoxicating smoke associated with the Oracle of Delphi. Early Chinese writings report an [sic] hallucinogen that has been identified with this species. And it is undoubtedly the plant that Avicenna, the Arabian physician, mentioned under the name jouzmathel in the 11th century. Its use as an aphrodisiac in the East Indies was recorded in 1578. The plant was held sacred in China, where people believed that when Buddha preached, heaven sprinkled the plant with dew.

—Richard Evans Schultes,[193] *Hallucinogenic Plants*

These different types of entheogens were used as a sacred tool for the purpose of healing, spiritual discovery, wisdom, and communication with supernatural beings (Schultes 1976: 102). It can therefore be concluded that there was more than only one type of Tree of Life.

* * *

In order to better appreciate the veracity and significance of this experience, it is necessary to understand the relationship between the brain and the electrical signals that exist in the form of charged subatomic particles. The alteration that is induced by the entheogen has a profound effect upon this bio-chemical activity. It is also apparent that this experience expands past physical boundaries, via emanating fields that interact with the surrounding environment, and perhaps even beyond via non-local quantum activity. This type of activity may be related to the psychic experience. Indeed, users have reported experiencing telepathic experiences while under the influence of the entheogen (McKenna 1993b: 40, 125, 133). It is

---

[193] Richard Evans Schultes, Ph.D., was a twentieth-century American biologist and professor at Harvard University. He is considered to be the father of ethnobotany.

likely that this type of experience—which is exemplified in the philosophical relationship between the *nous*, the *psyche*, and the *anima mundi*—is related to Teilhard de Chardin's concept of a sphere of consciousness that he referred to as the "noosphere."

While the brain exists in the three physical dimensions of space (and time), it is apparent that the psyche simultaneously exists on a separate level that is not understood by mainstream science at this time (Radin 2006: 250). In their benchmark book *The Invisible Landscape*, brothers Terence and Dennis McKenna present the theory that the psyche/mind is like a hologram that exists as a fifth dimensional "overstructure" (McKenna 1975: 51, 101). In the "modular wave hierarchy" paradigm that the Mckenna's posit, the entheogen actuates a shift in "harmonic vibrations" and "waveform phenomenon" that stimulates quantum mechanisms in the brain. This shift in frequency most likely explains why some users are able to see light in the ultraviolet spectrum (McKenna 1975: 95). What is also compelling about Mckenna's holographic mind theory, is how well it relates to Karl Pribram's[194] holonomic brain theory, which also describes the brain as a holographic storage network that is effected by vibrations, waves, and frequencies.

These theories are pointing toward a quantum mechanism regarding the brain's higher functioning.[195] Indeed, it is apparent that quantum activity is effecting neurons in the brain (Penrose 1989: 400-401). (More on this subject later.)

> [. . .] the divine process operates in time dimensions which are far beyond our routine, secular, space-time limits. Wave vibrations, energy dance, cellular

---

[194] Karl H. Pribram, M.D., Ph.D. (Hon.), was a board-certified neurosurgeon, professor at Georgetown and George Mason Universities, emeritus professor of psychology and psychiatry at Stanford University, distinguished professor at Radford University, professor of neurophysiology and physiological psychology and researcher at the Yerkes Laboratory of Primate Biology at Yale University.

[195] For more information on this topic see *Irreducible Mind: Toward a Psychology for the 21st century*, by Edward Kelly, et al. (Rowman & Littlefield, 2007).

transactions. Our science describes this logically. Our brains may be capable of dealing with these processes experientially.

So here we are. The great process has placed in our hands a key to this direct visionary world. Is it hard for us to accept that the key might be an organic molecule and not a new myth or a new world?

—Timothy Leary, *The Politics of Ecstasy*

Rick Strassman suggests that the center of activity for psychedelic/mystical phenomena might be occurring in the brain's pineal gland (Strassman 2001: 58-84). The idea that the pineal serves a spiritual function is not new. It was also postulated by the seventeenth-century philosopher Rene Descartes, who regarded this mysterious object in the center of the brain as "the principal seat of the soul."[196] However, recent scientific findings indicate that the pineal appears to be an endocrine gland that releases melatonin, which plays a part in circadian sleep patterns, and may also be indirectly related to the immune system (Macchi and Bruce 2004: 177). This is a sufficient explanation for the existence of the pineal. Nevertheless, it is also true that this gland is not fully understood at this time (Macchi and Bruce 2004: 177).

It is compelling that DMT has been found in the pineal gland of rats (Barker, et al. 2013: 1690-1700), and, as was previously postulated, DMT may be acting as some type of transducing interface agent between wavelength states. However, as far as I have been able to tell, significant evidence for DMT production in the human pineal gland remains unproven at this time. Although, it must also be noted that the pineal does contain the unique enzymes that are necessary to convert high levels of serotonin into tryptamine, which is a precursor to the formation of DMT. Strassman posutlates that DMT could be created by the pineal when triggered under specific conditions (Strassman 2001: 69). This is a compelling theory that deserves further scientific research.

---

[196] In a letter titled: To Meyssonnier, 29 January 1640.

What is also notable is that the pineal gland is photo-receptive to light and dark. McKenna theorized that the photosensitive nature of the pineal may be playing a part in a symbiotic process that is based on the properties of light.

> Tryptamine hallucinogens certainly fill the head with light. Serotonin, their near relative in brain chemistry, is transduced to melatonin by a light-mediated reaction. In other words, light actually enters through the eyes and follows a part of the visual pathway that branches off and goes to the pineal gland, where photons work a chemical change on serotonin and turn it into melatonin. These compounds are near relatives of the tryptamines, which are psychoactive compounds occurring in psychedelic mushrooms. All this is going on in the pineal, and it's all light-driven chemistry.
> —Terence McKenna, *Trialogues at the Edge of the West*

The interface between the brain and mind, and mind and the world of spirit, might therefore be found in the bio-electric energy of light. If this is true, then one of the properties of light may have something to do with it being a carrier of information. Therefore, the ultimate discovery that will likely one day reconcile the physical world of science with the supernatural world of the mystics, may be discovered in the science of light.

> The time has come to seek new levels of reality. Your ego and the (name) game are about to cease. You are about to set face with the Clear Light.
> —The Bardo Thodol (The Tibetan Book of the Dead)

> Win thou the light, win heavenly light, and, Soma, all felicities; and make us better than we are.
> —Rig Veda, Book IV, Hymn IV, Soma Pavamana

It can therefore be proposed that the expansion of sense experience into the realms of light energy is actuated in the numinous epiphany of the psychedelic experience.

* * *

The ecstatic epiphany that is induced by the entheogen may be related to what the psychologist Abraham H. Maslow[197] referred to as "peak experiences" that were necessary for the achievement of fulfillment and self-realization. This empowering epiphany is apparently what the Elohim did not want the hybrid servants to attain for themselves.

> The millions are awake enough for physical labor. But only one in a million is awake enough for effective intellectual exertions, only one in a hundred million to a poetic or divine life [. . .] To be awake is to be alive.
> —Henry David Thoreau, *Walden*

These peak experiences were not only achieved by the use of the Tree of Life but by the Tree of Knowledge as well. Likewise, in the following biblical passage, it is reported that one of the problems that also aroused the anger of Yahweh was the intoxicating stimulant of alcohol—which gave men "energy," and made them "mighty":

> Woe to those who are mighty in drinking wine, and to the men with vital energy for mixing intoxicating liquor [. . .]
> —Isaiah 5.22

The ancient records indicate that alcohol was also present in early Mesopotamian society. The Babylonian Code of Hammurabi includes regulations related to the drinking of alcohol. Indeed, the drinking of alcohol was an activity that the records tell us that the gods

---

[197] *Religions, Values, and Peak Experiences* (Penguin, 1964)

themselves partook in. The Anunnaki goddess Ninkasi (daughter of Enki), was even referred to as the goddess of beer.

Therefore, not only were the hybrids using alcohol to make themselves "mighty" and fill themselves with "vital energy," but the gods and the ancient kings of the Earth were also using the Tree of Life to invigorate themselves as well.

* * *

If it still seems unlikely that the Tree of Life is what we would refer to in the present day as a "drug," then first consider that the ban on both premarital sex—i.e., the Tree of Knowledge—and on so-called "drugs"—i.e., the Tree of Life—is still adhered to by many of the followers of this tradition to this very day. Indeed, the ban on the Tree of Knowledge was inadvertently enforced by the Spanish ecclesiastical Conquistadors, who imposed Judeo-Christian regulations on the indigenous people regarding what they viewed as proper sexual attitudes—which mainly pertained to restrictions.[198] Likewise, Spanish missionaries also inflicted a similar type of subordinate morality upon the populace by prohibiting the native Mexicans from partaking in the use of the psychedelic mushroom, teocanacatl (Schultes 1998: 2; Schultes, et al. 1979: 156) and peyote (Schultes 1976: 116).

It is a fact that ancient people used narcotics. Evidence for this is found in the following image that was found in King Sargon II's palace at Khorsabad, which shows what is either a king or one of the Anunnaki gods holding an opium poppy plant (Figure 16).

---

[198] This view is exemplified in the writings of the fourth- and fifth-century Judeo-Christian theologian, Augustine of Hippo. In his book Confessions, he equates sexuality with sinful desire that tempts Christians away from God.

(Fig. 16)

Consider also the following Hindu text, in which the word for Soma is actually substituted with the word "drug":

> Long-hair holds fire, holds the drug,[199] holds sky and earth. Long-hair reveals everything, so that everyone can see the sun. Long-hair declares the light.
> —Rig Veda, Book 10, Hymn CXXXVI, Kesins (The Long Haired Ascetic)

The one who "holds the drug" is the one who holds the "fire." This experience is also represented in the ancient Grecian legend of Prometheus; the god who gave so-called "fire" to mankind. As a result of this act of rebellious treason against the dominant institutional god, Prometheus was punished by Zeus, who sought to

---

[199] In an older nineteenth-century English edition (by Ralph T.H. Griffith, 1896) the word drug is inexplicably translated as "moisture." However, applying coherence theory to the interpretation implies that the most accurate contextual translation is drug. It is likely that Griffith may have felt compelled to offer a more conservative translation considering the time in which his work was published.

maintain power over humankind by keeping mortal man without the light of the metaphorical "fire."

* * *

The ancient Indo-Iranian/Proto-Indo-European tribes (formerly referred to as the Aryans) not only brought with them the traditions of their culture into the Indus Valley but perhaps even the Tree of Life as well.

A form of the word "ambrosia" that the Indo-Europeans used is etymologically related to the Indo-Hindu term *amrita* (Abraham 1913: 63)—which is how it appears in the ancient Sanskrit and Punjabi languages. The Indo-European root-word for death is *mer-to*. When it is preceded by the letter *a*, it means "non-death"—i.e., immortal (Merriam Webster Dictionary Online s.v. Amrita). Although the full Sanskrit word is found in the Hindu texts, it also sometimes appears in the abridged form of *rita, amrut, amrit,* or *amrta*. The Rig Veda tells us that amrita (which is also sometimes referred to in the texts as "nectar") was related to "Soma."

> [. . .] well-made food, meath blent with Soma juice [. . .] thence Amṛta is produced [. . .] The kindred Four have been sent downward from the heavens: dropping with oil they bring Amṛta and sacred gifts.
> —Rig Veda, Book IV, Hymn LXXIV, Soma Pavamana

> From the ocean rose the honeyed wave, together with Soma, it acquired the properties of amrita.
> —The Rig Veda, Book IV, Hymn LVIII. Ghṛta

In the Japanese and Chinese solar sect of Buddhism, the Buddha is also referred to as the *Amitayus*—which means "forever enduring" (Campbell 1964: 263). The eternal nature of the spirit is associated with the spiritual consciousness that is induced by the entheogen.

In the traditional Buddhist votive chant *Namu Amida Butsu*— which means "Glory to the Amida Buddha," there is a reference to the immortal effect that is bestowed to the seeker who sits under the branches of the sacred Bodhi tree. Likewise, in the following Hindu text we read:

> He is the beginning, producing the causes which unite (the soul with the body), and, being above the three kinds of time (past, present, future), [. . .] He is beyond all the forms of the tree and of time, he is the other, from whom this world moves round, when one has known him who brings good and removes evil, the lord of bliss, as dwelling within the self, the immortal, the support of all.
> —Svetasvatara Upanishad, Sixth Adhyaya

The passage also relates to the words of the Hindu demi-god hero Krishna, when he said (in the Bhagavad Gita) that "Although I am unborn and my transcendental body never deteriorates [. . .] I still appear in every millennium in my original transcendental form."

> [. . .] with Soma bringing forth delight [. . .] place me in that deathless, undecaying world, wherein the light of heaven is set, and everlasting lustre shines [. . .] Make me immortal in that realm where they move even as they list. In the third sphere of inmost heaven where lucid worlds are full of light.
> —Rig Veda, Book IV, Hymn CXIII. Soma Pavamana.

Likewise, in a collection of Kabbalahist commentaries on the Torah (that is dated to the Medieval age), we find reoccurring references to immortality, the Tree of Life, and spirituality:

> [. . .] blessed be he [. . .] to whom he gave the Torah of truth, the Tree of Life. Whoever takes hold of this achieves life in this world and in the world to come. Now the Tree of Life extends from above downward, and it is the Sun which illumines all.

—The Zohar[200]

It would seem that life in the spiritual state is not as subject to the effects of biological decay as it is in the lower material wavelength. Perhaps this is the key to understanding the nature of immortality as it relates to both spirituality and the Tree of Life. Indeed, this is the world that Jesus referred to as the eternal abode of the heavenly Father; which is the realm of "everlasting life" (John 6.47).

> The Spirit gives life; the flesh counts for nothing. The words I have spoken to you—they are full of the Spirit and life.
> —Jesus, John 6.63

This concept accords with the findings of Doctor Raymond A. Moody Jr.,[201] and is presented in his book *Life After Life*. According to interviews with over fifty people who have had brief experiences crossing over into the "after-life," the spirit world is not as subject to the same limited parameters of time and space as it is in our own lower level (Moody 1975: 107, 116-117).[202] Likewise, some psychedelic experiencers reported losing all sense of time, as if they had entered into an environment that exists outside of our own temporal regimentations (Richards 2016: xxi, 10; Strassman 2001: 149, 226). It is as if time becomes relative to the internal state of the experiencer. Likewise, this same feeling is reported in some mystical experiences as well (Richards 2016: 46). It is also notable how well this concept relates with the Hindu concept of *kala*. Near-death experiencers will often report the gnostic Christian belief that the

---

[200] The Zohar, Vol. 5, trans. Harry Sperling, Maurice Simon, and Paul Levertoff (The Soncino Press, 1934): 203.

[201] Raymond A. Moody Jr. Ph.D., M.A., is a psychologist, physician, philosopher, and author.

[202] Non-linear time is also corroborated in Eben Alexander's OBE/NDE account (Alexander 2012: 40, 70, 131).

transcendent world of spirit is superior to the lower world of temporal matter (Moody 1975: 77, 91).

It is therefore likely that the key to understanding immortality may be found in the world of spirit, which is not only a state of spiritual existence but spiritual consciousness:

> Everyone already has the lamp of mind, but it is necessary to light it so that it shines; then this is immortality.
> —Secret of the Golden Flower, The Opening of the Mysteries of the Golden Flower

* * *

In regard to the immortality of the energy of the "spirit," according to the theory of the transmigration of souls (i.e., *metempsychosis*), individuals who cannot adapt and/or sustain themselves in the hyperspacial environment are compelled to reinvest their energy back down into the lower plane of matter in order to re-undergo the evolutionary process—that is, it seems, if such individuals do not choose by their own free will to remain in a dysfunctional environment. In these instances, this might be what is occurring during so-called "ghost" "haunting" situations. This is most likely where the Catholic concept of an intermediate "purgatory" and the Tibetan Buddhist "bardo" state derive from. If soul-transfer and survival theory is true, then it may be possible that we human-beings have the ability to live on in a different vibratory wavelength of the space-time continuum for as long as we are able to sustain ourselves, and do not choose by our own free will to reinvest our energy back down into the lower level of temporal matter. (The scientific veracity of this theory will be examined further ahead.) Although, whether this experience is based on free will or not, of course, remains unknown at this time—I suggest that it is likely. When this happens, it seems that the individuated energy form—i.e., psyche/soul—receives the genetic coding of its biological progenitor hosts, as well as a new consciousness, and therefore essentially begins anew, while simultaneously retaining the same monad psyche/soul that derives from the Infinite Energy Circuit—which is in accordance with conservation of energy laws.

In his book *The Emperor's New Mind*, Roger Penrose[203] agrees that there seems to be a transcendent, "non-material," and "divine or mysterious" element of "consciousness" that has not yet been revealed to scientists (Penrose 1989: 405. 444-446). Indeed, physicalist neuro-scientists can explain basic brain and sense perception functioning, and yet cannot explain the existence of consciousness (Kaku 2014: 41-42; Penrose 1989: 405; Sheldrake 2013: 117). However, it is more likely that it is not consciousness that is itself this transcendent element that Penrose speaks of, but rather the psyche/soul; which may be the underlying source of consciousness.

> Characteristics of mental states, especially *qualitative* characteristics, appear utterly unlike physical characteristics of brains. If minds and brains exhibit dramatically different characteristics, how could minds be brains?
> —John Heil,[204] "Mental Causation," *The Cambridge Dictionary of Philosophy*

In the following passage, the physicalist position is referred to as "central state materialism":[205]

> Some central state materialists also have held that the mind is the brain. However, if the mind were the brain, every change in the brain would be a change in the

---

203 Roger Penrose, Ph.D., is the emeritus Rouse Ball Professor of Mathematics at the University of Oxford.

204 John Heil, Ph.D., is a professor of philosophy at the University of Washington in St. Louis.

205 Central state materialism refers to the theory that mental states are closely associated with the central nervous system.

mind; and that seems false: not every little brain change amounts to change of mind. [. . .] So, against central state materialism and the identity theory, it is claimed that mental states cannot be states of our central nervous system.
—Brian P. McLaughlin,[206] "Philosophy of Mind," *The Cambridge Dictionary of Philosophy*

The reason why physicalist scientists have not been able to comprehend this phenomenon is because the psyche cannot be reduced to a physical mechanism or to a mental definition. Indeed, the original ancient Greek definition of the *psyche* did not pertain to the mind, but rather to the soul—which also pertains to the ancient Greek philosopher's concepts of the *ousia* and the *hypostasis*. It may therefore be the case that these philosophers, who developed the concept of the psyche, were most likely aware of something that modern-day neuro-scientists are not.

The psyche may be an entity that attaches onto its biological host sometime between conception and birth. When this happens, the monad entity receives physical state sense perception and is able to inhabit a form in this vibrational level, or so-called "membrane" (more on this subject later). After the biological host expires, it seems that the psyche detaches, along with its associated anthropomorphic spirit form—which can be defined as an astral doppelganger (or so-called "ghost in the machine"),[207] and shifts into a parallel energy state—albeit in new form, after having been effected by its previous attachment. This is due to a separate but connected type of symbiosis that is occurring between these two energy states. Therefore, reincarnation can be defined as a cyclical renewal process that circumvents the physical laws of disorder and decay. The

---

[206] Brian P. McLuaghlin, Ph.D., is a professor of philosophy at Rutgers University.

[207] This term refers to Oxford philosopher Gilbert Ryle's description of Cartesian dualism, which pertains to the separation of psyche and body that was proposed by Rene Descartes.

preexistence of the psyche, which is effected by its experiences, helps to explain why identical twins can have different personalities.

The scientific investigation into the subject of reincarnation was conducted by Jim B. Tucker, a professor of psychiatry and neurobehavioral sciences at the University of Virginia. Dr. Tucker studied hundreds of cases of individuals who claim to remember people, places, and events from a previous life.[208] [209] Most of those who were examined were children. This is due to the apparent fact that the inherent mental state of a child is less affected, and therefore is more able to recall impressions that have become infused into the energy of the underlying constitution his or her's psyche.

> In attempting to understand this, we should keep in mind that some physicists now consider consciousness to be an entity separate from the brain and one with important functions in the universe. [. . .] and if consciousness is indeed a fundamental part of the universe—if Stanford physicist Andrei Linde is correct when he says that a consistent theory of everything that ignores consciousness is unimaginable—then the world is a far more complex and wondrous place than what the physical world shows us in every-day life.
> —Jim B. Tucker, *Life Before Life*

It is interesting that in some of the cases that were studied, the children reported suffering from a violent death (Tucker 2005: 12, 93, 214). It seems that the person was so traumatized by their death that they were unable to continue on in that spiritual environment and felt compelled to essentially wipe the slate clean by starting over.

---

[208] More than 2,500 cases are stored at the Division of Personality Studies at the University of Virginia (Tucker 2005: xiv).

[209] J.B. Tucker. *Life Before Life: A Scientific Investigation of Children's Memories of Previous Lives* (St. Martin's Press, 2005).

Dr. Tucker's research continued the earlier work of Ian Stevenson (1918–2007).[210] It is especially compelling that the children were aware of people, places, events, and things that were not available to them, and no other plausible explanations were found for such knowledge. Indeed, both Tucker and Stevenson's findings attest to the validity of soul-transfer and survival hypothesis.

Critics argue that most of these reincarnation cases were found in regions where the belief in reincarnation is prevalent. This suggests that the children were influenced by cultural factors. However, it can be countered that a child in a region where reincarnation is not as known, indeed even socially unaccepted, may be more inclined to suppress and dismiss the experience until such memories are forgotten. Indeed, people who have undergone out-of-body/near-death events report that they feel pressure not to report the experiences because of the concern of the possibility of being ridiculed and perceived as insane. Some conclude that it is better to remain silent rather than endure the cynical ridicule and scorn that is often inflicted by the critical physicalists (Moody 1975: 12, 78-80).

* * *

Despite information that suggests that immortality is related to the world of the spirit, we should consider that it may also relate to the emotional feeling of internal rebirth as well.

> This is the King Soma, the food of the gods, and the gods eat it [. . .] When someone eats that food [. . .] from him one comes into being again.

---

[210] Professor of psychiatry at the University of Virginia School of Medicine and founder and director of the university's Division of Perceptual Studies. Dr. Stevenson's most notable works are: *Twenty Cases Suggestive of Reincarnation* (University of Virginia Press, 1966), *Reincarnation and Biology: A Contribution to the Etiology of Birthmarks and Birth Defects* (Praeger, 1997), and *European Cases of the Reincarnation Type* (McFarland, 2003).

—Chandogya (Khandogya) Upanishad, Adhyaya V, chapter 10

Immortality is participation in the eternal now of the divine ground; survival is persistence in one of the forms of time. Immortality is the result of total deliverance.

—Aldous Huxley,[211] *The Perennial Philosophy*

Perhaps when the mind and spirit remain revitalized there is no demise and therefore no need for reincarnation. In this case, the partakers become rejuvenated by the energy of the "vital-force" that is bestowed by Soma. This may also relate to the nature of "immortality."

> To Indra flow these running drops, these Somas frolicsome in mood. Exhilarating, finding light; Driving off foes, bestowing room upon the presser, willingly bringing their praiser vital-force.
> —Rig Veda, Book IV, Hymn XXI, Soma Pavamana

In the following Buddhist verse, we find a direct reference to the one who is awakened and who "lives forever," which is a reference to someone who has taken the fruit from the Tree of Life and now sees as the Elohim do:

> Wakefulness is the way of life. The fool sleeps as if already dead, but the master is awake and he lives forever.
> —The Buddha, The Dhammapada

Indeed, the relationship between esoteric Buddhism, higher states of consciousness, and the psychedelic experience is undeniable.

---

[211] Aldous Huxley was a twentieth-century writer and philosopher.

The awakening of the senses is the most basic aspect of the psychedelic experience [. . .] This is the Zen moment of satori, the nature mystic's high, the sudden centering of consciousness on the sense organ, the real-eye-zation that this is it!
—Timothy Leary, *The Politics of Ecstasy*

* * *

If the "immortality" that is bestowed by the Tree of Life is not literally physical but rather spiritual and/or emotional, how then can the unnatural life-spans of the Annunaki/Elohim gods be explained? For example, the Mesopotamian records seem to indicate that the gods lived for millennia. This incredible time-span would seem to indicate that the immortality that is bestowed by the Tree of Life was literally physical. However, it must be understood that the ancient texts were not entirely literal accounts. Therefore, the millennia-old life-spans of the gods may have actually been based on the time-span that the deity was worshiped, rather than the actual life time of the deity.

It also seems that the anomalous life-spans of the Anunnaki/Elohim may have had more to do with their extraterrestrial genetics, rather than with the Tree of Life. What should also be considered are the reports that indicate that the NHIs possess methods in which they use to extend their life-spans. One alleged example of this is what is being reported by the Raelian[212] UFO cult, who believe—based on information that they say was imparted to them by NHIs themselves—that the NHIs can achieve immortality by the transference of one's "being" from one genetic clone to another. In other words, the NHIs have allegedly achieved the ability to

---

212 The Raelians were founded by a man named Claude Vorhilhon, who claims to have witnessed a flying-saucer near Clermont Ferrand France, from which an extraterrestrial being emerged. The NHI apparently disclosed to Vorhilhon that human-beings were created by extraterrestrials, and that they one day would like to re-establish relations with humankind, just as they did in ancient times.

reincarnate without losing consciousness. Furthermore, contactee George Van Tassel claimed that the NHIs had shown him technology that could be used to rejuvenate cells in the human body. Van Tassel also claimed to have met a humanoid NHI who looked to be in his mid-twenties, and yet claimed that he was over seven-hundred Earth years of age.[213] Based on the information that was imparted to him about this technology, Van Tassel began building a wooden domed structured device that he referred to as the *Integratron*. However, Van Tassel passed away in 1978 before the project was completed.

Another possibility that has been suggested by other "ancient astronaut" theorists, is that the NHIs were traveling back and forth between their own distant planet and our own. According to the theory of relativity, time would slow down for the traveler at the speed of light. However, this is unlikely due to the apparent ability of the NHIs to be able to transcend the limitations of known physics. Indeed, this is what contactees are reporting (Mack 1999: 58). This not only explains why so-called "UFOs" are seen making high-speed turns that are impossible to do without its pilot being killed by g-forces, and this not only explains why sonic booms are never reported (Davis 2014: 97, 144; Kean 2010: 124), but it also explains how they can travel extreme distances without taking millennia to do so.

Another way that transcendent travel might be achieved is through what physicists refer to as "wormholes," which are essentially tunnels through the fabric of space and time. If this is true, then the NHIs are able to somehow manipulate the space-time in order to significantly shorten travel distances—which might be related to the concept of "topological folding" (Sheldrake 2013: 320).

> Is [traveling faster than the speed of light] possible? To answer this question, we have to push the very boundaries of modern quantum physics. Ultimately, things called "wormholes" may provide a shortcut through the vastness of space and time. And beings

---

[213] www.theblackvault.com/casefiles/kvos-webster-reports-the-extraordinary-equation-of-george-van-tassel-1964/#

made of pure energy rather than matter would have a decisive advantage in passing through them.
—Michio Kaku, *The Future of the Mind*

Another possibility is that the NHIs are able to enter into a transcendent state in which time and space as we know it does not exist—which is something else that is being reported by contactees (Mack 1999: 59). We can also refer to virtual particle physics in regard to these matters. Virtual particles are theoretical particles (that are predicted by quantum field theory calculations) that are able to violate the laws of known physics, such as travel faster than the speed of light. If advanced intelligences were able to shift into this type of virtual modality, perhaps they too would be able to violate the laws of known physics. Indeed, the cosmologist Paul Davies postulates that if extraterrestrials do exist, the most realistic way for them to traverse such distances would be to do so outside of their biological forms (Kaku 2014: 311). For a civilization that may be millions of years ahead of ours, this may be possible. Therefore, the argument made by the skeptical critics, namely, that interplanetary travel is impossible, is based on limited thinking. Perhaps this other massless virtual state is related to what is more commonly referred to as "hyperspace" by the theoretical physicists (and Terence McKenna), and the "spirit world" in the religious tradition. Even in some other transcendental energy state, where the effects of time may not decay matter at the same rate as it does in the lower vibratory level, it may be possible that they were able to continue to influence the affairs of man through the use of mediums, who are referred to in the biblical record as the "prophets." Therefore, it can be concluded that the mystics were using the Tree of Life to raise their vibrational frequency to that of the inter-dimensional Elohim, in order for communion to take place.

* * *

According to contactees, the NHIs are interdimensional beings (i.e., ultraterrestrials) who exist in a higher vibrational wavelength. Therefore, in order for contact to occur, the experiencer must be able to raise his or her vibratory rate up to their level (Mack 1999: 88, 216).

> The UFO intelligences, they [contactees] say, come from higher dimensions all around us which function on different vibratory levels, just as there are various radio frequencies operating simultaneously in our environment. We can attune ourselves to these higher dimensions in much the same manner as a radio receiver tunes into the frequencies of broad-casting stations. Different entities travel on various frequencies, according to their vibratory rate.
>
> —Brad Steiger,[214] *The Fellowship*

Likewise, according to Van Tassel, the NHIs could only be communicated with if one was in "attunement" with their "wavelength" or "vibration" (Van Tassel 1952: 16). In this case, when UFOs appear to vanish they may actually only be shifting their frequency into a wavelength that is no longer visible to observers in the lower spectrum. According to Van Tassel, these beings not only personally visited him but they demonstrated their ability to shift in frequency by disappearing and re-manifesting their physical bodies at will—which he says was witnessed by nineteen people who were present on his California property.[215] According to Van Tassel, the NHIs refer to Earthling human-beings as "mortals," and refer to the majority of people on this planet as "materialists" (Van Tassel 1952: 31, 34).

Other modern-day experiencers also report that during the contact event it feels as if their bodies are vibrating at a higher level (Mack 1999: 65-66, 72, 78). Experiencers also report altered states of

---

[214] Brad Steiger is a prominent American independent researcher and prolific writer on subjects pertaining to the paranormal.

[215] "The Extraordinary Equation of George Van Tassel." *KVOS Channel 12 Films. Center for Pacific Northwest Studies, Western Washington University.* June 18, 1964. Film. You Tube video, posted by Western Washington University.

consciousness while they are in contact with these beings (Mack 1999: 56, 17-18, 55-59). During the event, the NHIs usually communicate telepathically. Some of the experiencers also report undergoing a profound spiritual awakening both during and after these encounters (Mack 1999: 81, 277). It is as if the experiencers become initiated into a great mystery that exists behind the scenes of our everyday world.

> The cosmos that is revealed by this opening of consciousness, far from being an empty place of dead matter and energy, appears to be filled with beings, creatures, spirits, intelligences, gods—the names vary according to the apparent world view of the observer of function and behavior of the entity at hand—that have through the millennia been intimately involved with human existence. In some instances, it would appear, certain of these entities may even cross over the divide that we created in order to keep unseen realities and mysteries apart, ideologically speaking, from the material world.
> —John E. Mack, *Passport to the Cosmos*

What is also significant about these encounters with NHIs is their relationship with the psychedelic experience.

> This ecstasy provides preparations for the sacred flight that enables man to experience mediation between his mortal existence and the supernatural forces—an activity establishing direct contact through this plant of the gods.
> —Richard E. Schultes, Albert Hofmann, Christian Ratsch. *Plant of the Gods*

The shamanic practice of employing the entheogen to raise the vibratory frequency of the psyche to that of the "realm of light" that is inhabited by the celestial gods is referred to in the following ancient Hindu texts:

> We have drunk the Soma; we have become immortal; we have gone to the light; we have found the gods.
> —The Rig Veda, Book VIII, Hymn XLVIII, Soma (We Have Drunk the Soma)

> Soma flows bright and pure between the earth and heaven [. . .] the Holy One, filling the firmament stationed amid the worlds, Knowing. the realm of light, hath come to us [. . .]
> —The Rig Veda, Book IV, Hymn LXXXVI. Soma Pavamana.

> On flows the stream of Soma who beholds mankind: by everlasting Law he calls the Gods from heaven.
> —Rig Veda, Book IV, Hymn LXXX. Soma Pavamana.

The realm of light is not just the light of higher states of being but of higher dimensions of space and time. It is likely that these other dimensions are referred to as "spaces" in the following verse:

> All through the three Soma days, he flies to the six broad spaces and the one great one.
> —The Rig Veda, Book 10, Hymn XIV, (Yama and the Fathers)

These other dimensions are also referred to as "worlds":

> This Soma for the gods effused, entering all their various worlds. Resplendent is this deity, immortal in his dwelling place, Foe-slayer, feaster, best of gods.
> —The Rig Veda, Book IV, Hymn XXVIII. Soma Pavamana

> It is this person—the one that consists of perception among the vital functions (*prana*), the one that is the inner light within the heart. He travels across both worlds, being common to both [. . .] Now, this person has just two places—this world and the other world.
> —Upanishads, Brhadaranyaka Upanishad

In this sublime hyper-spacial energy state, the celestial traveler becomes, or at least seems to become, immortal:

> Make me immortal in that realm, wherein is movement glad and free, in the third sky, third heaven of heavens, where are the lucid worlds of light.
> —The Rig Veda, Book IX, Hymn CXIII. Soma Pavamana

* * *

For some experiencers, the contact event is not physical. In these cases, the person may feel their spirit form lift out of their body and up into an awaiting craft (Jacobs 1992: 50, 250; Mack 1999: 150). Some also report traveling through a tunnel. Similar tunnels are also seen in some DMT experiences (Strassman 2001: 147, 224-226), as well as out-of-body experiences (i.e., OBE) during near-death experiences (i.e., NDE).

> Near-death experiences have been reported throughout history in many diverse cultures. Although cultural expectations and parameters of the close brush with death may influence the content of some NDEs, the core near-death phenomenology has not varied through the centuries and around the globe. That invariance may reflect universal psychological defenses, neurophysiological processes, or actual experience of a transcendent or mystical domain. Research into these alternative explanations has been hampered by the

spontaneous and unpredictable occurrence of NDEs, and has provided only indirect evidence supporting the psychological, neurophysiological, and transcendent interpretations. The challenge of complex consciousness, including thought processes, perceptions, and memory formation at a time of severely compromised brain function, suggests the need to expand our models of consciousness and its relation to brain. The prevailing notion that the mind can be reduced to neural processes is a philosophical assumption, and there is nothing inherently unscientific about exploring data challenging that assumption and supporting others. Developments in postclassical physics over the past century offer empirical support for a new scientific conceptualization of the interface between mind and brain. Regardless of the causes or interpretations of NDEs, however, they are consistently associated with profound and long-lasting aftereffects on experiencers, and may have important implications for non-experiencers as well.

—Bruce Greyson,[216] Western Scientific Approaches to Near-Death Experiences[217]

Three significant aspects of the NDE are as follows: (1) enhanced mentation when the physical body is unconscious, or in some way impaired. (2) An out of body experience, in which the subject is able to view his or her body from a higher vantage point. (3) Awareness of remote events outside of the immediate location. Compelling aspects of these events concern patients who were able to accurately recount everything that transpired in the room when the patient was technically deceased (Moody 1975: 93). Another significant aspect

---

[216] Bruce Greyson, M.D., is a professor of psychiatry and neurobehavioral sciences at the University of Virginia and editor of the Journal of Near Death Studies.

[217] Greyson, Bruce. "Western Scientific Approaches to Near-Death Experiences." (*Humanities*, Vol. 4, 2015): 775-787.

pertains not only to reports of the sightings of deceased family members but family members who are not known by the subject to have died (i.e., "Peak in Darien" experiences) (Greyson 2010: 159-169).

Enrico Facco[218] and Christian Agrillo[219] explain in their report ("Near Death Experiences Between Science and Prejudice." In *Frontiers in Human Neuroscience*) why physicalist interpretations of NDEs, who assert that such reports are nothing more than neurobiological disturbances (e.g., "temporal lobe dysfunction" and "neurotransmitter imbalances," etc.) are not plausible (Facco and Agrillo 2012: 2-3). Indeed, NDE experiences have been concluded to be real (Kelly, et al. 1999-2000: 518).

Critical physicalists argue that so-called "supernatural" types of experiences can be replicated by manually stimulating specific areas of the brain. They then use these experiments to argue that what appears to be supernatural phenomena is nothing more than biological perturbations. For example, the neurologists Olaf Blanke was able to induce out-of-body experiences through the use of electrodes attached to the brain. Similarly, by stimulating the temporoparietal area of the brain in another subject, he was able to make the subject become aware of a shadowy malevolent spirit form in the room (Kaku 2014: 268). However, these experiments do not disprove the traditional supernatural explanation. This is because what is actually more likely occurring is that the brain is not fabricating the phenomena *ex nihilo*, but rather it is being made more receptive to it—in much the same way that the entheogen works. Furthermore, these experiments do not account for instances of supernatural experiences in which a scientist is not present to manually stimulate areas of the brain. Therefore, these experiments cannot be used as evidence to support a physicalist interpretation. It can therefore be held that not all paranormal cases

---

[218] Enrico Facco, Ph.D., is an associate professor in the Department of Neurosciences at the University of Padova.

[219] Christian Agrillo, Ph.D., Department of General Psychology, University of Padova.

can be explained as mere mistakes, hoaxes, or natural fluctuations in the brain (Almeder 1992: 129, 159, 194).

After careful examination of the evidence, Robert Almeder[220] concludes in his book *Death & Personal Survival: The Evidence for Life After Death*, that post-mortem survival cannot always be ruled out, and that although further study is needed, he notes that the "dualist" (in which the mind can be separated from the body) interpretation of OBE is "more rationally warranted than any alternative interpretation presently available" (Almeder 1992: 198).

We find similar NDE/OBE accounts in the following ancient Egyptian text, which also recalls the "Doors of Perception" that the philosopher Aldous Huxley referred to in his classic work:

> The hitherto closed door is thrust open and the radiance in my heart hath made it enduring. I can walk in my new immortal body and go to the domain of the starry gods. Now I can speak in accents to which they listen, and my language is that of the star Sirius.
> —The Book of the Coming Forth by Day (The Egyptian Book of the Dead)

* * *

In his book *True Hallucinations*, Terence Mckenna examines the link between "hyperspacial" phenomena, telepathy, UFOs, and the psychedelic experience. He reported that while under the effects of the psychoactive mushroom *stropharia cubensis*, he received the overwhelming impression of a highly-evolved benevolent presence "from outer space or from another dimension," that was monitoring events on Earth. McKenna believed that the presence of what he referred to as the "Other" was using the mushroom as a contact agent. He concluded from his experiments that the entheogen is a "transdimensional doorway" that was linked to some important

---

[220] Robert Almeder, Ph.D., is a professor emeritus of philosophy at Georgia State University.

"Gnostic Truth" (Mckenna 1993b: 41-42). He referred to the omniscient telepathic voice that spoke to him as to the divine "Logos."[221]

> The ontological mode of the higher dimensions into which humanity is being propelled is being anticipated by the singularity that we call the wholly Other or the alien. The alien is teaching something through its reinforcement schedule: It is preparing us to confront the God facet of ourselves that our explorations into the nature of life and matter are about to reveal.
> —Terence Mckenna, *True Hallucinations*

In one of his DMT sessions, McKenna also experienced an out-of-body experience, in which he ascended into the sky. While in this state, he witnessed the presence of flying saucers above Earth's atmosphere (Mckenna 1993b: 60).

Other experiencers report that when the contact event occurs in hyperspace, the type of beings that are seen are sometimes luminous and translucent (Mack 1999: 65, 67). In some cases, tall humanoid celestial beings wearing robes are also witnessed (Mack 1999: 72, 137, 152).

Artistic depictions of the Tree of Life alongside the winged-disk have been uncovered that clearly show a relation between these two objects. As was previously noted, the symbol of the mushroom and the winged-disk were depicted on the royal seals of the Hittite kings. In the following relief from the Neo-Hittite period, we find an example of a king who is not holding the mushroom, but rather the lotus during the contact event (Figure 17).

---

[221] The word *Logos* has multiple meanings. In ancient Greece, it was a word that was associated with sophic discourse, reason, order, and knowledge. In Christianity, it commonly refers to the word of God.

(Fig. 17)

The king used the psychoactive effects of the lotus to raise the vibrational frequency of his psyche to that of the celestial gods, who came down from out of the heavens in their flying disk to meet with their earthly protegees. The two beings on either side of the disk are the equivalent of the mythological cherubim/seraphim creatures that are depicted in the Old Testament/Tanakh.

Likewise, in the following Assyrian sculpture, a goddess holding the sacred lotus flower is depicted. Although it has been badly damaged, a winged-disk is visible above her head (Figure 18).

(Fig. 18)

The following neo-Elamite seal (Figure 19) not only depicts a stylized lotus but also shows the god who rides in the flying disk, partaking of the Tree of Life:

(Fig. 19)

In the following relief (Figure 20) from the northwest palace of Ashurbanipal (or Ashurnasirpal) in Assyria, the Tree of Life is represented as a flower—possibly the lotus.

(Fig. 20)

The following image, from a Neo-Assyrian ivory panel from the eighth century BCE (figure 21), portrays a winged disk that appears from out of the sky to the beholder of the divine flower:

(Fig. 21)

Likewise, in the extra-biblical Ascension of Isaiah text (in the Secondary Apocrypha), there is a mention of a "door" to an "alien world" that was "granted" "to a man" in order for the prophet to reach the "Most High in the highest world."

* * *

References can be found in the New Testament Gospels not only to "sacred secrets" and "hidden" matters (Mark 4.22) but something that is referred to as the "keys to the kingdom":

> I will give you the keys of the kingdom of heaven; whatever you bind on earth will be bound in heaven, and whatever you loose on earth will be loosed in heaven.
> —Jesus, Matthew 16.19

Here we are told that the "keys to the kingdom" relate to a connection that is made between the lower earthly world and the higher realm of the heavenly Father. Not long after this statement, Jesus took Peter, James, and John up to a remote location where he underwent the "transfiguration." During that event, his face and garments were transformed into light as he rose in vibrational frequency of that of the "Father." It was also during that event that he was visited by spirit beings, as well as a luminous so-called "cloud" that descended from out of the sky; from which the voice of the Father of Jesus emanated. We know from the previously mentioned findings that these so-called clouds were an ancient way of describing aerial phenomena that was associated with the otherworldly Elohim. The transfiguration of flesh into the light of spirit and the meeting with supernatural beings, which were all events preceded by references to secret hidden matters and a so-called key that is used to access the realm of the heavenly Father, all bear the tell-tale qualities of the Tree of Life.

> The ability to turn on the senses, to escape the conditioned mind, to throb in harmony with the energies radiating on the sense organs, the skillful control of one's senses, has for thousands of years been the mark of a sage, a holy man, a radiant teacher.
> —Timothy Leary, *The Politics of Ecstasy*

Indeed, principles related to spirit, wisdom, and rebirth, are found all throughout the accounts that are reported in the gospels. Jesus himself fits the archetypal role of not only the hero but of the shaman:

> In a word: the first work of the hero is to retreat from the world scene of secondary effects to those causal zones of the psyche where the difficulties really reside, and there to clarify the difficulties, eradicate them in his own case (i.e., give battle to the nursery demons of his local culture) and break through to the undistorted, direct experience and assimilation of what C.G. Jung has called "the archetypal images." This is the process known to Hindu and Buddhist philosophy as Viveka, "discrimination."
> —Joseph Campbell, *The Hero with a Thousand Faces*

This is the type of experience that Jesus underwent in the wilderness when he confronted the temptations.

* * *

Another aspect of the psychedelic experience that must also be addressed pertains to what has been referred to as the "Dark Night of the Soul." This didactic ordeal can be described as a spiritual state that is devoid of positive life-force energy. The role of the explorer is to learn from this experience and use it to make his or herself wiser and more spiritually aware. In the end, the individual is rewarded by attaining the divine apotheosis; which pertains to the awareness and mastery of what the mythologist Joseph Campbell referred to as "the Two Worlds."

The psychedelic experience reveals the state of one's own spiritual condition. The lesson to be learned is how the energy of our thoughts, actions, and emotions, relate to the spiritual life and the spiritual world.

> But the man who comes back through the Door in the Wall will never be quite the same as the man who went out. He will be wiser but less cocksure, happier but less self-satisfied, humbler in acknowledging his own ignorance, yet better equipped to understand the relationship of words to things, of systematic reasoning to the unfathomable Mystery which it tries, forever vainly to comprehend.
> —Aldous Huxley, *The Doors of Perception*

It is also interesting how well this principle concurs with the type of panspsychism that is espoused by the philosophic Idealists, who posit that reality is fundamentally a mental and subjective construction. It is apparent that mentalistic causation is more prevalent in the spiritual level.

> A man [. . .] goes with his action to that very place to which his mind and character cling. Reaching the end of his action, of whatever he has done in this world—from that world he returns back to this world, back to action.
> —Brihadaranyaka Upanishad, Book IV, Fourth Brahmana

If one is securely aligned with the joyous, peaceful, respectful, strong, and loving forces of the energy of life, the experience will be inspirational and blissful. If one is not, the result may be an unpleasant lesson in the sacred importance of maintaining a healthy mental, emotional, and spiritual state of well-being.

The ancient texts are full of references to a hellish nether-world in which those who cling to negative and destructive energy congregate. Although this state does not have to be quite as "eternal" as traditional Judeo-Christian doctrine asserts, it does take a considerable amount of conflict, penance, and karmic reformation in order to reverse such negatively charged habits and repercussions.

It is this lesson that gives birth to the phrase, "we are what we think"—which was a maxim that was professed by the Buddha after returning from his own trial, in which he too faced a barbaric antagonist (i.e., Mara), who attempted to distract him away from the goal of attaining internal mastery.

* * *

Themes of death, discovery, and rebirth, are common all throughout the ancient records. These are also the same type of archetypal stages that are found in the ancient Grecian rites at Eleusis. During those ceremonial gatherings, a mystery drink (*kykeon*) was ingested that induced visions and profound revelations (*epopteia*) (Richards 2016: 49, 154). Part of the experience for some may have also involved a descent into the netherworld (*katabasis*), from which the initiate would eventually emerge a new and wiser person:

> Eleusis is a shrine common to the whole earth, and of all the divine things that exist among men, it is both the most terrible and the most luminous.
>
> —Aelius Aristides[222] [223]

Although what the ambrosia at Eleusis exactly was is debated among scholars, some (e.g., Albert Hoffman; Richard E. Schultes; R. Gordon Wasson) believe—based on hymns and artwork featuring the goddess Demeter—that it may have derived from barley that was infused with ergot fungus (*Claviceps*) that bore psychoactive alkaloids (Schultes and Hoffman 1979: 26, 102). However, a stele relief from the same period (circa 470-460 BCE) depicts persons (possibly Persephone and Demeter) who seem to be engaged in what may be the *epopteia* sacrament. Moreover, both of the figures are shown holding what appear to be mushrooms (Figure 22).[224]

---

[222] Aelius Aristedes was a second-century CE Greek orator and author.

[223] It is likely that this quote derives from Aristedes work Orationes; unfortunately, I have not been able to locate an English translation of the original text. The above cited passage appears in the book: *The Mysteries: Papers from Eranos Yearbooks*, edited by Joseph Campbell (Bollingen Series XXX, 2, Princeton University Press, 1955 [reprint 1990]: 21).

[224] It is commonly thought that the mushrooms that are depicted in this work are flowers. Indeed, this work (which is on display at the Louvre in Paris) is known as the "Master of the Exaltation of the flower" (*L'Exaltation de la Fleur*). However, it can be contended that the image does not depict flowers but rather mushrooms, and that the standard interpretation is in need of correction.

(Fig. 22)

By the time of Socrates and Plato, Eleusis was a major festival. The *epopteia* may have played such a profound influence upon the consciousness of the Greeks that it caused them to question everything that they knew. This process most likely inspired new ways of thinking that may have even contributed to the emergence of Greek philosophy.

> Every philosophy is tinged with the coloring of some secret imaginative background, which never emerges explicitly into its chains of reasoning.
> —Alfred North Whitehead, [225] *Science and the Modern World*

Indeed, some of the earliest philosophers, such as the Pythagoreans, held secretive teachings among themselves. Nevertheless, it is unknown for certain if the Eleusian mysteries played a significant philosophical role because the initiates were sworn to secrecy;

---

[225] Alfred North Whitehead was an influential nineteenth- and twentieth-century British mathematician, logician, and philosopher.

although it is true that Socrates (according to Plato) did refer to the importance of "initiation" into the "mysteries."[226]

> [. . .] we philosophers following in the train of Zeus, others in company with other gods; and then we beheld the beatific vision and were initiated into a mystery which may be truly called most blessed, celebrated by us in our state of innocence [. . .].
> —Socrates, Phaedrus

> [. . .] he who enters the next world uninitiated and unenlightened shall lie in the mire, but he who arrives there purified and enlightened shall dwell among the gods.
> —Socrates, Phaodo

Likewise, the ancient essayist Plutarch related philosophical illumination to the Eleusinian mystics (Kerenyi 1967: 91).

Just like the secret of the Tree of Life, the Eleusisian mysteries were concealed from the common public.

> The later pupils of Plato, the Neoplatonists, attribute to him a secret teaching, to which he admitted only those who were worthy, and then strictly under the "seal of silence." His teaching was considered secret in the same sense as the Mystery wisdom.
> —Rudolf Steiner,[227] *Christianity and Occult Mysteries of Antiquity*

The Neo-Platonic concept of the "Divine Mind" (i.e., *nous*) not only relates to the Hindu concept of "*samadhi*" but to the modern-day concept of "higher states of consciousness." This elevated state of

---

[226] In both Meno and Phaedrus.

[227] Rudolf Steiner, Ph.D., was a nineteenth- and twentieth-century Austrian philosopher and proponent of "spiritual science."

awareness can be associated with what Stanislov Grof referred to as the "holotropic" mind, and what nineteenth-century psychiatrist Richard Maurice Bucke referred to as "cosmic consciousness."

> This state of cosmic consciousness that under favorable circumstances may be attained with hallucinogens is related to the spontaneous religious ecstasy known as the *unio mystica* or, in the experience of Eastern religious life, as samadhi or satori. In both of these states, a reality is experienced that is illuminated by that transcendental reality in which creation and ego, sender and receiver, are One.
> —Richard Evans Schultes, Albert Hofmann, Christian Ratsch. *Plants of the Gods*

* * *

The texts indicate that the divine spirit of Soma inhabited different types of plants. In the following Hindu text, we are told that there was more than just one type of Soma:

> One thinks that he has drunk the Soma when they press the plant. But the Soma that the Brahmins know—no one ever eats that.
> —Rig Veda, Book X, Hymn LXXXV. Surya's Bridal (The Marriage of Surya)

The Soma that the Brahmin priests ingested and was brown-colored and "golden-hued." This may indicate that it was made from the psychoactive mushroom. In the following passage, a "red" color is mentioned:

> Sing a praise-song to Soma brown of hue, of independent might. The Red, who reaches up to heaven.
> —Rig Veda, Book IX, Hymn XL Soma Pavamana.

This could be a reference to the red color of the psychoactive *amanita muscaria* mushroom cap.

However, the Soma that the common people partook of may have been pressed from another type of "plant." A plant that was green:

> Engendering the Sun in floods, engendering heaven's lights, green-hued, robed in the waters and the milk, according to primeval plan this Soma, with his stream, effused flows purely on, a God for Gods.
> —Rig Veda, Book IX, Hymn XLII. Soma Pavamana.

A clue to the identity of this other substance is found in the following passages, where it is described as a plant and a healing herb:

> O all ye various Herbs whose King is Soma, that overspread the earth, urged onward by Brhaspati, combine your virtue in this Plant.
> —Rig Veda, Book X, Hymn XCVII. Praise of Herbs

> Herbs rich in Soma, rich in steeds, in nourishments, in strengthening power, all these have I provided here, that this man may be whole again. The healing virtues of the Plants stream forth like cattle from the stall, Plants that shall win me store of wealth, and save thy vital breath, O man.
> —Rig Veda, Book X, Hymn XCVII. Praise of Herbs

> Herbs rich in Soma, rich in steeds, rich in nourishments [. . .] The healing virtues of Plants steam forth like cattle from the stall [. . .]
> —Rig Veda, Book X, Hymn XCVII. Praise of Herbs.

The second type of Soma was a type of "herb" that bore leaves:

> Broad-leaved plant sent by the gods to bring happiness
> and the power to triumph [. . .]
> —Rig Veda, Book X, Hymn CXLV. Sapatnibadhanam
> (Against Rival Wives)

It is a description that corresponds with the following biblical verse:

> [. . .] And the leaves of the tree [of life] are for the healing of the nations.
> —Revelations 22.2

Marijuana is the plant that is associated with herbs, happiness, and healing.

The Hindu texts also report that when this substance was mixed into beverage form it was referred to as *bhang*, which was the ambrosia that was imbibed by the god Shiva. Just like Soma was made from the psychoactive mushroom, bhang was a potion that was made from the leaves of the *cannabis sativa* plant (Schultes 1973: 63). In Hindu Indian mythology, this plant was considered to be a gift to humanity, vouchsafed by the god themselves (Schultes 1973: 61).

The history of the marijuana plant is extensive. This plant was also referred to as *cana* in Sanskrit. In Arabia, it was called *kannab*; in Assyria, it was *kunubu* (or *qunnabu*); in Persia, it was referred to as *kenab*, and in ancient Chaldea it was known as *kanbun* (Benet 1975: 40). The Greco-Latin word is *cannabis*.

Cannabis was burned as incense in the ancient temples of the Anunnaki gods in both Babylon and Assyria (Benet 1975: 40; Schultes 1973: 63). Indeed, a reference to cannabis was discovered in a Neo-Assyrian clay tablet fragment from the library of Ashurbanipal (650 BCE).[228] Other evidence indicates that the Assyrians also used cannabis for medicinal purposes (Brown 1998: 3). Hemp fabric that was made from the *cannabis sativa* plant was found in an Egyptian tomb that was dated to over three millennia ago (Schultes 1973: 59). High level traces of cannabis were also found in the lungs of an

---

228 British Museum. Library of Ashurbanipal. Registration number K.4345.

Egyptian mummy (Brown 1998: 3; Parsche and Nerlich 1995: 380-384). Cannabis is also mentioned in the Egyptian Ebers papyrus from the sixteenth-century BCE (Brown 1998: 3).[229]

The ancient Greek historian Herodotus reported that the Scythians intoxicated themselves with hemp.[230] This report has been confirmed by archaeological evidence in the form of unearthed copper vessels from that region that contained the remains of burned cannabis (Brown 1998: 5). Likewise, the Greek philosopher Democritus, who was also known as the "laughing philosopher," apparently spiked his wine with hemp (Ratsch 2001: 93).

> Upon eating hemp, the euphoric, ecstatic and hallucinatory aspects may have introduced man to an otherworldly plane from which emerged religious beliefs, perhaps even the concept of deity. The plant became accepted as a special gift of the gods, a sacred medium for communion with the spiritual world, and as such it has remained in some cultures to the present.
> —Richard Evans Schultes, *Man and Marijuana*

The cannabis plant was not only used ritualistically and by the ancient wealthy elite but the hemp material that is derived from the plant was used for practical reasons in some parts of the world. Hemp could not only be made into fabric but could be burned as oil.

> The history of cannabis use dates to ancient times. Hemp fabrics from the late 8th century B.C. have been found in Turkey. Specimens have turned up in an Egyptian site nearly 4,000 years of age. In ancient Thebes, the plant was made into a drink with opium like effects. The Scythians, who threw cannabis seeds and leaves on hot stones in steam baths to produce an

---

[229] The translation is from the hieroglyphic symbol *shemshemet*, which denotes an object that is related to both fiber and medicine.

[230] *The Histories*

intoxicating smoke, grew the plant along the Volga 3,000 years ago. Chinese tradition puts the use of the plant back 4,800 years. Indian medical writing, compiled before 1000 B.C., reports therapeutic uses of cannabis. That the early Hindus appreciated its intoxicating properties is attested by such names as "heavenly guide" and "soother of grief." The Chinese referred to cannabis as "liberator of sin" and "delight giver." The Greek physician Galen wrote, about A.D. 160, that general use of hemp in cakes produced narcotic effects. [. . .] Objects connected with the use of cannabis were found in frozen tombs of the ancient Scythians, in the Altai Mountains on the border between Russia and Outer Mongolia.

—Richard Evans Schultes, *Hallucinogenic Plants*

If the inhabitants of the ancient Near East, including Egypt, had access to cannabis, this would indicate that their neighbors in Israel and Judah were equally aware of this plant. Sula Benet[231] postulated that the biblical "calamus," (i.e., "*Keneh bosem*") (Exodus 30.23; Song of Songs 4.14; Isaiah 43.24; Jeremiah 6.20; Ezekiel 27.19) was cannabis (Benet 1975: 40-41). She contended that the word calamus was a mistranslation that was inserted into the Septuagint Bible. However, there is reason to believe that this is not an accurate interpretation. This is because the context in which the word calamus is presented is much more befitting of "sweet cane" (Jeremiah 6.20; Song of Songs 4.14), which can also be identified as sweet flag grass (i.e., *acorus calamus*). Cannabis, on the other hand, is not sweet. However, a more compelling reference to cannabis appears in the book of Ezekiel, where the merchant product *pannag* is referred to (Ezekiel 27.17). Raphael Mechoulam[232] postulates that the word *pannag* may have originally derived from the Hindu word for

---

231 Sula Benet, Ph.D., was a twentieth-century anthropologist.

232 Raphael Mechoulam, Ph.D., is a professor of medicinal chemistry at Hebrew University of Jerusalem.

cannabis, due to the fact that the letters P and B are interchangeable in Hebrew. Therefore, *bannag* could have derived from *bhanga*, or *bhang*. It is very possible that bannag was brought in to Israel and Judah by merchants from the east (Brown 1998: 3). Indeed, in 1993 human remains from the fourth century CE were unearthed near Jerusalem. Next to the body of a young woman and her baby, burned cannabis was found (Brown 1998: 4).

Although the use of cannabis/marijuana is censured by traditional Judeo-Christians, the Rastafarian Christian sect freely enjoys the use of it. The belief in the benevolent value of the Tree of Life seems to have been shared by some in the gnostic Christian community as well:

> And this is in the north side of Paradise in order to raise up the souls from the stupor of the demons, so that they might come to the Tree of Life and eat its fruit and condemn the authorities and their angels.
> —Nag Hammadi Codices, On the Origin of the World

The ultimate goal of the gnostic Christian was not just simply to follow according to blind faith, but rather to follow in the Christ's foot-steps by realizing the intrinsic divinity that lies within all of the descendants of the Elohim. Partaking of the Keys to the Kingdom was a way in which to do this.

* * *

All throughout the historical record, signs of the effect that the Tree of Life has had in the world can be found. However, many of these signs have been overlooked since the significance and meaning of this substance has been misunderstood, unrecognized, and even purposely concealed. Some of this arcanum can be found in the writings of the alchemists.

Although the alchemists publicly professed only to the practice of a type of metallurgy that was based on chemistry—that being the insoluble quest to turn base-metals into precious metals, such as gold (i.e., the exoteric tradition)—their writings and activities were also infused with mysterious references to mystical subjects and symbols (i.e., the esoteric tradition). One of the mystery subjects that was involved in the "Great Work" was the "Elixir of Life," which was the ambrosia of rejuvenation and immortality that was associated with an object that was referred to as the "Philosopher's Stone." According to the alchemical texts, the Philosopher's Stone could not be explicitly defined.

> There are many names given to it, and yet it is called by one only, while, if need be, it is concealed. It is also a stone and not a stone, spirit, soul, and body.
> —*Turba Philosophorum*

The seventeenth-century alchemist Elias Ashmole believed that the Philosopher's Stone had the ability to grant the user spiritual and cognitive powers. Likewise, the scientist and philosopher Robert Boyle, believed that the stone could make it possible to summon and communicate with angels (Principe 1998: 187). It is therefore likely that, at least in some instances, the philosopher's stone and the alchemical Elixir of Life were occidental terms for the Tree of Life and Soma. This explains why some alchemists traced the history of the Philosopher's Stone to the Garden of Eden (Patai 1995: 19).

The extensive history of alchemy is complex. This is partly because the subject meant different things to different people. What must be understood is that alchemy did not always refer to either metallurgy or chemistry, and that most of the alchemists were using their public work as a sort of protective cover story. This practice allowed them a certain amount of impunity from the persecution that would otherwise have been directed at them from the officials of the Judeo-Christian Church; a powerful institution—especially during that age—that not only had the ability to put an end to their endeavors but also charge them with the practice of black magic and put an end to their lives as well. Indeed, it is not coincidence that the rise of

Hermetic[233] alchemy coincided with the age in which Judeo-Christendom was at the height of its power. The mystics responded to this threat by developing a way of infusing just enough esoteric information into what they referred to as the "Great Work" so that only the perceptive of mind would be able to comprehend its true meaning.

> This art must ever secret be, the cause whereof is this, as ye may see; if one evil[234] man had thereof all his will, all Christian peace he might easily spill [. . .]
> —Thomas Norton,[235] *The Ordinal of Alchemy*

For the mystical alchemists—as opposed to the material alchemist—the ultimate aim of transmuting base-metals into gold was not literally a physical operation, but rather an internal process within the psyche of the individual. The aim of the alchemist was to redeem "vulgar" base material of the self, and therefore achieve the illumined golden state of the *chrysopoeia*. In this idealized state, the practitioner attains a state of harmony with the world soul (*anima mundi*)—which was one of the ultimate aims of the great work (*magnum opus*). For the esoteric alchemists, this is what the "transmutation" actually was. The rest of the procedures and additional ancillary symbols were either added to throw off the uninitiated, or invented by unwitting practitioners who were taking the practice of turning metals into gold literally. The pseudo-alchemists who made such literal claims were most likely either charlatans who were endeavoring to defraud wealthy investors, or they naively believed that material transmutation was literally possible.

---

[233] Hermeticism is an esoteric philosophy that is focused on the writings of the mystic Hermes Trismegistus, who may have lived in between fourth century BCE, to second and third centuries CE.

[234] This term pertains to rebellious disobedience, which is the original Yahwehist definition of evil.

[235] Thomas Norton was a fifteenth- and sixteenth-century English alchemist.

The first known alchemical texts from Egypt—e.g., the Leyden Papyri (circa 250-300 CE) and the Stockholm Papyrus (i.e., the Papyrus Graecus Holmiensis) (circa 300 CE)—did not actually profess the ability to turn metal into gold, but rather offered techniques in which to cause metal to *appear* to look like gold (Holmyard 1990: 41). In other words, the first alchemists were looking for ways to refine metals into purer and more attractive forms by applying coloring techniques through the use of various forms of acidic compounds and tinctures—such as the *aqua regia*, which was a substance that was composed of sulphuric acid, nitric acid, and other base compounds (Forbes 1970: 33, 170). Like the Freemasons who came after them, a guild of smelters eventually formed into a secretive brotherhood as a higher meaning became attached to the work. They were able to use the symbols of their material craft to encrypt mystical knowledge.

The following passage is taken from the writings of the progressive Judeo-Roman-Christian theologian, Albertus Magnus (twelfth century CE):

> The first rule is that the alchemist should be silent and discreet and not reveal to a living soul his secret; not for any reason whatsoever, being strong in the conviction that should others learn the secret it would be surely be divulged; and once the secret has been divulged, the alchemist would be called an imposter, and set on the road to ruin [. . .]
> —Albertus Magnus, *Libellus De Alchemia*

Why would the secret lead to the alchemist being called an "imposter" unless they were not actually involved in literally turning led into gold? Indeed, the Renaissance scientist and occultist Paracelsus stated that his belief was that alchemy did not pertain to the making of gold and silver, but rather to the study of medicines (Holmyard 1957: 170). (The alchemical process of collecting and extracting essential plant substances into tinctures that he helped to pioneer is referred to as "spagyrics.")

The esoteric alchemists also believed in the concept of a "Universal Mind" that is diffused into nature. It was their intention to uncover this universal energy and use it for the advancement of the "Art."

> That which is above, is like that which is below.
> —*The Emerald Tablet*

This special relationship of the *physika mystika* extended through both the organic and the non-organic worlds—including plants.

> The operations of the craftsman were carried out to the accompaniment of religious or magical practices, and supposed connexions [*sic*] were seen between metals, minerals, plants, planets, the Sun and gods.
> —Eric John Holmyard,[236] *Alchemy*

The following passage was written by Zosimos, a gnostic Egyptian/Greek alchemist who lived in Upper Egypt in around 300 CE:

> [. . .] the splitting off of the spirit from the body, and the fixation of the spirit on the body are not operations with natures alien one from the other, but, like the hard bodies of metals and the moist fluids of plants, are One Things of One Nature, acting upon itself.
> —Zosimos of Panopolis, *Ars Nostra*

By drawing a connection between metals, plants and the spirit, the alchemists may have been offering a clue to the mystery. Indeed, this notion was alluded to by the fifteenth-century German alchemist and Benedictine monk, Basil Valentine:

---

[236] E. J. Holmyard was a twentieth-century scholar and instructor at Clifton College in England.

> All herbs, trees, and roots, and all metals and minerals, receive their growth and nutriment from the spirit of the earth, which is the spirit of life.
> —Basil Valentine, The *Twelve Keys*

The first of the "Twelve Keys" that Valentine proposed is the drinking from the "golden fountain." The effect of which allowed the "king" to ascend into the heavenly "regions."

> The king travels through six regions in the heavenly firmament, and in the seventh he fixes his abode. There the royal palace is fixed with golden tapestry. If you understand my meaning, this key will open the first lock, and push back the first bolt; but if you do not; no spectacles or natural eyesight will enable you to understand what follows. But Lucious Papirius has instructed me not to say any more about this key.
> —Basil Valentine, The *Twelve Keys*

It is likely that the "key" that opens the doors of perception is referred to in the gospels as the "keys to the kingdom."

In the following passage from the same text, Valentine discloses the color of the philosopher's "stone":

> [. . .] being composed of white and red. It is a stone, and no stone; therein Nature alone operates.
> —Basil Valentine, The *Twelve Keys*

It may not be coincidence that white and red are the two colors of the *amanita muscaria* mushroom. Indeed, this is not an artificial substance that can be fabricated in an alchemist's lab, but rather is a substance that can be found in "Nature."

The following passage was written by an Arabic alchemist:

> This prime matter which is proper for the form of the Elixir is taken from a single tree which grows in the

> lands of the West [. . .] And this tree grows on the surface of the ocean as plants grow on the surface of the earth. This is the tree of which whosoever eats, man and *jinn* (spirits) obey him; it is also the tree of which Adam (peace be upon him!) was forbidden to eat [. . .]
> —Abu'l Qasim, *Kitab al-'ilm*

The Tree of Life and its relationship to the ocean is also a reoccurring theme that is found in the ancient texts of India. According to Hindu mythology, the gods used the serpent (King Vasuki) to wrap around and turn the "milky ocean" of Soma. Several versions of the Churning of the Ocean (*Samudramathana*) myth exist (also called the "Churning of the Ocean of Milk"—which is a reference to one of the ingredients in Soma). According to some versions, Soma was lost after the time of the flood. The angels and demons then worked together—which is a reference to its spiritual qualities—in order to churn the ocean and stir the *amrita* up to the surface:

> From the ocean rose the honeyed wave, together with Soma, it acquired the properties of amrita.
> —The Rig Veda, Book IV, Hymn LVIII. Ghṛta

It is highly likely that Soma is analogous with both the Grecian ambrosia of the gods and the alchemical Elixir of Life.

In the Middle Eastern lands of what once was Mesopotamia, the Arabian alchemists were also devising methods in their laboratories to refine the Tree of Life:

> According to an anonymous seventeenth-century book entitled The *Sophic Hydrolith*, the Philosopher's Stone—or the ancient, secret, incomprehensible, heavenly, blessed, and triune universal stone of the sages, is made from a kind of mineral by grinding it to powder, resolving it onto its three elements, and recombining these elements into a solid state of the fusibility of wax.

—Eric John Holmyard, *Alchemy*

It is evident that what was being created had nothing to do with metallurgy, but rather with the turning of the *cannabis sativa* plant into hashish.

One of the ways that hashish was made was by a mechanical process called sieving. After the marijuana plant was dried out and broken up into a powder (*kief*), it was sieved though various screen meshes until only the most potent THC (*tetrahydrocannabinol*) remained. The powder could then be pressed and condensed into a single solid or soft pasty waxy state using a heating process. However, the alchemical method most likely involved using a chemical separation method that used a solvent (most likely ethanol) as a dissolving agent. The solvent was then evaporated until only the potent THC waxy resin remained. This process brings to mind the alchemical maxim: "dissolve and combine" (i.e., *solve et coagula*).

One of these pharmacological alchemists was Abu-Bakr-Muhammad-ibn-Zakariyya-alRazi (or simply "Rhazes"). This tenth-century Persian alchemist classified plants by their structure and wrote of the similarities between metals as well. He also wrote of a mysterious liquid called the *Aqua Vitae* (the "Water of Life"), which may be the same substance that was known to the European alchemists as the Elixir of Life—which is what the Tree of Life was referred to when it was dissolved and combined with liquids.

One of the documents that was referred to by the Chinese alchemists is The Secret of the Golden Flower. This is a Taoist/Buddhist text that is said to be centuries old, but was only committed to paper over two hundred years ago (Cleary 1991: 2-3).

> The golden flower is the same thing as the gold pill.
> The transmutations of spiritual illumination are all guided by mind.
> —The Secret of the Golden Flower, The Celestial Mind

In the esoteric alchemical tradition, gold represents the radiant illumination of Enlightenment.

> The highest secrets of alchemy are the water of vitality, the fire of spirit, and the earth of attention. The water of vitality is the energy of the primal real unity. The fire of spirit is illumination. The earth of attention is the chamber of the center, the celestial mind.
> —The Secret of the Golden Flower, The Origin Spirit and the Conscious Spirit

Cannabis has a long history in China, where the hemp fabric was not only used for practical purposes but its effects were associated with oracles, love magic, and shamanism (Ratsch 2001: 21). Shennong's[237] ancient pharmacopeia identifies a type of cannabis, that is referred to as *ta-ma*, as an essential ingredient of the elixir of immortality (Ratsch 2001: 22). Indeed, the contents of some of the "pills" and "elixirs" that were being used by the Chinese Taoist alchemists were made from the *cannabis sativa* plant.

In the nineteenth century of the Common Era, it was no longer the alchemists who were investigating psychoactive substances, but rather the chemists (Schultes, et al. 1979: 196). What chemists, pharmacists, and psychopharmacologists learned to do was to isolate the active psychoactive compound from the plant. For example, in the 1890s German chemists isolated mescaline from peyote. In 1946, Oswaldo Goncales isolated DMT from the South American plant *mimosa tenuiflora*. A similar discovery was published in 1955 by chemists: Fish, Johnson, and Horning, regarding the cohoba/yopo tree (i.e., *piptadenia peregrina*) (Fish, et al. 1955: 5892-5895; Strassman 2001: 44).

> The next step to be explored by scientists is: What constituents—which of the substances in those plants—actually produce the effects that have led to their use in religious rites and magic? What the chemist

---

[237] Shennong was a legendary ancient Chinese character. He is considered to be part human part god.

> is looking for is the active principle, the quintessence or *quinta essentia*, as Paracelsus called the active compounds in plant drugs.
> —Richard Evans Schultes, Albert Hofmann, Christian Ratsch. *Plants of the Gods*

The quintessence of the psychoactive mushroom is the *psilocybin* and *psilocin*. Indeed, the term "quintessence" (i.e., *quinta essencia*) is found in the writings of the alchemists. Although some definitions relate it to an "element," it can also be described as the purest substance drawn out of any natural body, and a medicine that is made from the essential "spirit," or "chief force" of a substance (Furnivall: 30). Therefore, the purpose of distilling the psychoactive plant down to its most basic form was not only a way in which to conceal the truth from the Judeo-Christian authorities but to produce a better tasting and more potent alternative.

According to the fifteenth-century alchemical manuscript titled *The Book of Quintessence or the Fifth Being*,[238] not only was the quintessence a secret matter that was related to "Heaven," "gold," the "Water of Life," and the restoration of a "lost" "nature" (Furnivall: 2-3), but it could be associated with "wine." It is evident that in this instance, the author was not referring to the Tree of Life, but rather to another type of stimulant: alcohol. It must therefore be acknowledged that the alchemical process can pertain to a number of subjects.

Another modern-age example of an alchemist is Albert Hofmann, who is the discoverer of LSD (*lysergic acid diethylamide*). Not only was Hofmann able to isolate the compounds psilocybin and psilocin from the psychoactive mushroom but he was able to crystallize these compounds into a hardened form (Schultes, et al. 1979: 22). Perhaps this is the type of solid form that the alchemists referred to as the "philosopher's stone."

Despite all of the precautions taken by the original alchemists, the secrecy did not stop Pope John XXII from issuing the Papal Bull (*Spondent Pariter*) in 1317 CE, which banned the practice of

---

[238] Edited from the Sloane Manuscript (no. 73).

alchemy. The theologian Clement of Alexandria also publicly condemned the practice, which he declared to be the work of the devil. Among the accusations that Clement proclaimed was that the alchemists knew the secret powers of "plants" and the "science of the stars" (Thompson 2003: 11). Indeed, church officials used passages in the Old Testament to justify condemnation of activities that they could associate with so-called "sorcery" and "divination."

* * *

Among the extra-biblical narratives that emerged in the European medieval age was the story of the Holy Grail. This artifact was thought to have once contained the blood of Christ, and was the cup that Jesus himself supposedly drank from during the Last Supper. This sacred vessel was also said to be imbued with supernatural powers that healed wounds, empowered kings, and produced food.

It was the twelfth-century French writer Chretien de Troyes, who wrote the earliest known version of the Grail legend in his story *Perceval, the Story of the Grail*. It is also believed that an earlier Celtic version of this legend exists; although this cannot be substantiated (Barber 2004: 237, 242-245). Indeed, folktales, such as the magic cauldron of Bran the Blessed, bear little significant resemblance to the Grail romances.

What then was the Grail, and did it even literally exist?

The etymology of the word itself is disputed among scholars. The word most likely derives from the Roman word *garalis*, which was a sauce receptacle. Likewise, the French word *gradale* is a word that is defined as a broad dish that was used to carry food. These descriptions accord with how it is described by Chretien in the original account. It is therefore most likely that the original Grail was more of a deep platter, rather than a cup (Barber 2004: 94-95). If the Grail authors had wanted to indicate that the vessel was a cup, they would have most likely used the word *calice*, (i.e., chalice) (Barber 2004: 97). Indeed, the original Grail was not identified as the holy cup of the Crucifixion and the Eucharist, but rather appeared only briefly as a shallow container that was made of gold and ornamented by precious stones. This is a description of a royal medieval vessel, not a first millennium Judean cup.

It was not until Robert de Boron's version of the legend that the Grail became a holy relic that could be traced back to the time of Jesus. It was around that same period that others submitted their own Grail stories that also veered away and expanded upon the original account. In these other more fanciful and pious narratives, the Grail was not only described feeding and healing people but also flying through the air on its own accord.

According to the original account, Perceval eventually learned that the "Fisher King," who was the guardian of the Grail, was the son of the king who was served by the Grail:

> "[. . .] he is served a single host which is brought to him in the Grail. It comforts and sustains his life—the Grail is such a holy thing. And he, who is so spiritual that he needs no more in his life than the host that comes in the Grail has lived there for twelve years [. . .]."
> —Chretien de Troyes, *Perceval*

Here we see that it is not the Grail itself that was supernatural, but rather the so-called "host" that was *inside* of it!

This brings up the question: What does the author mean by "host"? As was previously explicated: the word host derives from the Hebrew *sabat/sabaoth/tsebaoth*, which literally means troops or army. However, this definition is not only not appropriate for the standard astral interpretation but also for the context in which it is used in the Grail legends. Another possibility is that it may apply to the later era Judeo-Christian interpretation, which defines it as angels. The biblical information that is used to justify this interpretation can be found in passages that describe the angels of God performing a military function (e.g., 2 Kings 19.35; Revelation 12.7). In this case, the host can essentially be defined as a holy and mighty spiritual entity that acts under the authority and guidance of God. However, it is also telling that the symbol of the Grail knights was the turtle-dove, which was the Judeo-Christian symbol of the Holy Spirit (Barber 2004: 180). This motif derives from the biblical baptism of Jesus scene, in which the spirit of God was seen descending like a "dove" (Matthew 3.16). When all of the data is compiled and considered, it is reasonable to conclude that the term that appears as "host" in the Grail legends was intended to denote the Holy Spirit, or the "Holy Ghost," of God. (The identity of the Holy Ghost will be examined further ahead.)

The Grail story was expanded further by the German author, Wolfram von Eschenbach, in the thirteenth century. Wolfram claimed that he himself had been initiated into the deeper mystery by a man named Kyot, who claimed to have discovered the secret of the Grail hidden in an old Arabic manuscript that he had recovered in Toledo Spain. The manuscript itself was purportedly written by an eccentric astrologer/astronomer named Flegetanius, who claimed that he had learned the secret of the Grail while examining the stars. According to this account, when Flegetanius was observing the sky one evening, he witnessed what he referred to as a "host" leave the Grail on Earth, before ascending back to the stars. It is also true that Chretien also claimed to have received his knowledge of the Grail from an earlier mysterious source as well. Whether Chretien's source was the same as Wolfram's is unknown—most scholars would argue that it is unlikely.

What is also notable about Wolfram's account, is that Flegetanius was said to be a direct descendant of King Solomon. This reference to Israel and the Old Testament tradition may be another clue to this mystery. Indeed, Spain was a destination for displaced Jews during the medieval age.

Wolfram also wrote about "the hidden tidings concerning the Grail." He claimed that those who "are summoned to the Grail" "make the blissful journey to that place." Indeed, the host that was inside the Grail was said to be a gateway to the spiritual world that induced transcendent visions (Barber 2004: 112, 147).

> The Grail represents the fulfillment of the highest spiritual potentialities of human consciousness.

—Joseph Campbell, *The Power of the Myth*

In Wolfram's *Parsival*, there is also a reference to a mystery symbol that he referred to as a "stone":

> I will tell you of their [the templar knights][239] food: they live by a stone whose nature is most pure. If you know nothing of it, it shall be named to you here: it is called lapsit exillis. By that stone's power the phoenix burns away, turning to ashes, yet those ashes bring it back to life. Thus the phoenix sheds its moulting plumage and thereafter gives off so much bright radiance that it becomes as beautiful as before. Moreover, never was a man in such pain but from that day he beholds the stone, he cannot die in the week that follows immediately after. [. . .] That stone is called the Grail.
>
> —Wolfram von Eschenbach, *Parzival*

The *lapsit exillis* was not a literal stone, but rather a "stone from the heavens." It is therefore likely that that enigmatic object was none other than the Philosopher's Stone of the medieval alchemists.

The alchemical transmutation process explains the question that has perplexed those who have attempted to solve this mystery; namely, what is the relationship between the Grail and the phoenix, and what is the relationship between the stone and immortality? (Barber 2004: 182-183). The legend of the phoenix bird that burns to death before arising anew from its own ashes derives from ancient Greek mythology. This archetypal rebirth process is not unlike the Buddhist state of nirvana, which is the "extinction" of the old self and the world of delusion, and rebirth into a higher state. Allusions to spirituality and the phoenix explains the Grail's link to immortality. Indeed, according to Wolfram, the stone had the ability to make one

---

[239] The Templar knights were an elite military Roman Catholic order that existed during the Crusades.

become young again. This is a reference to the spiritual rejuvenation that is induced by the Philosopher's Stone.

It is therefore evident that the legend of the Grail pertained to much more than simply a physical relic. It is apparent that what was actually inside this vessel—at least according to Wolfram von Eschenbach—were the Keys to the Kingdom. Although, it is also true that in one of continuations of Chretien's original account, titled *Perlesvaus* (i.e., *The High History of the Holy Grail*), are allusions to the "secret things of the sacrament," and "the secrets of the Savior," as well.

It was the substance that was inside the Grail that awakened mortal men and transformed them into glorious Christ-like kings—that is, for those of them who qualified for such an experience. However, those who were impure of mind and spirit were banned from this experience. This may have been to prevent what the modern-day counter-culture commonly refers to as a "bad trip" (i.e., "the dark night of the soul") from adversely effecting the environment, and thus interfering with the will of God.

If we are able to essentially read between the lines, it is possible to see how the following passage may be a reference to the Tree of Life that was in the Garden of Eden:

> Upon a green achmardi[240] she carried the perfection of Paradise, both root and branch. This was a thing that was called the Grail, earth's perfection's transcendence.
> —Wolfram von Eschenbach, *Parzival*

Likewise, according to Wolfram, "the Grail was bliss's fruit." By referencing trees, fruit, and a perfected state of paradise, the author may have been referencing the Tree of Life in the Garden of Eden.

Wolfram also reported a dynasty of rulers who served the Grail, and who ruled through its power. What is especially compelling about this Grail dynasty is how similar it is to the ancient Grecian philosopher kings, who had been initiated into the Eleusinian

---

[240] Arabian or Syrian silk.

Mysteries, as well as something along the lines of an esoteric mushroom cult. Through the use of the host within the Grail, the elite knights were able to access a higher supernatural state. It was through them, like the shamans and the prophets throughout the history of the world, that a higher power was supposedly able to directly intervene in the affairs of man.

A revealing report comes to us from the controversial twentieth-century occultist and psychotherapist Israel Regardie, who, in a book that he wrote on the subject of the *Tree of Life*, disclosed the secret:

> The Eucharist too is implicit, and the chalice is used as the communion cup, the hallowed contents of which—thaumaturgic and iridescent; the sacramental wine, in short—must be dedicated and consecrated to the service of the Most High. The oblation to be consumed with the Eucharistic wine is, by this interpretation, the secret essence of both the intoxicated magician and the supreme God whom he has invoked. In this method also is imputed to a very large degree the alchemical technique, inasmuch as it concerns for the most part the production of the potable Gold, the Stone of the Philosophers, and the Elixir of Life which is Amrita, the dew of immortality.
>
> —Israel Regardie, *The Tree of Life*

Whether Wolfram's interpretation of the Grail was the same as the original Chretien version is debatable. What is known for certain is that Chretien died before he was able to finish his version. Therefore, the original author's intention will never be known. Although, it must be acknowledged that Chretien did attribute a higher spiritual significance to the Grail as well.

However, if the Grail was the Tree of Life, why then would Wolfram describe it as doing such impossible and uncharacteristic things as providing food? One possible explanation is that such instances were not intended to be literal but rather symbolic descriptions. Indeed, symbolic language appears throughout

Wolfram's writings (Barber 2004: 180, 182, 184). This may have been a deliberate strategy on his part to obscure the secret behind the facade of allegorical ambiguity. The purpose may have been to not only preserve the mystery but his own life as well. Indeed, Wolfram von Eschenbach lived in a time in which heretics were regularly sentenced to death by the church. It is also possible that the author was deliberately attempting to obscure the controversial message by employing the same type of protective diversion technique that was practiced by the alchemists. In these instances, if an accusation was made by a hostile inquisitor, an exoteric example could be cited by the alchemists to exonerate themselves. For example, the alchemists were able to explain that they were only attempting to turn metals into gold—which would be of interest to those in power, including the heads of the church. If a similar charge had been leveled at Wolfram, he could cite such overtly fictitious instances in his own work that have much in common with the mythological horn of plenty (i.e., the cornucopia), in order to deny accusations pertaining to so-called "necromancy."

* * *

The message of the Grail is not restricted to the Arthurian romances. We find similar arcanum occurring in Francois Rabelais's *Gargantua and Pantagruel*. In this story from the same age, we are told of a "herb" that was called the "Pantagruelion," that bestowed "energy" of a "transcendent nature" that was related to both "trees" and "prophets":

> It is likewise called Pantagruelion, because of the notable and singular qualities, virtues, and properties thereof; for as Pantagruel hath been the idea, pattern prototype and exemplar of all jovial perfection and accomplishment; so in this Pantagreulion have I found so much efficacy and energy, so much completeness and excellency, so much exquisiteness and variety, and so many admirable effects and operations of a

transcendent nature that is the worth and virtue thereof had been known, when those trees, by the relation of the prophet, made election of a wooden king, to rule and govern over them, it without all doubt would have carried away from all the rest the plurality of votes and suffrages.
—Francois Rabelais, *Pantagruel*

CHAPTER VII

THE EXTRA-SENSORY REALITY

In regard to evidence for the type of extra-sensory experiences that have been referred to throughout this work, the scientific studies have been conducted into parapsychology and all of its related subjects, such as telepathy, precognition, clairvoyance, etc., must be considered. Skeptics argue that such experiments are corrupted with methodological imperfections. While this may be true in some cases, others cannot be dismissed.

The word "skeptic" is a term that connotes scientifically-minded humanists who maintain rational doubt. However, in many cases such an objective designation cannot be applied. This is because many are actually committed deniers of the very possibility of such things. Consequently, no amount of empirical experience or even credible scientific evidence will alter their rigid purview. Therefore, a more accurate designation for such individuals who relegate reality solely to the physical (i.e., material) level is *physicalist*.

Having never had such experiences themselves, the physicalists conflate subjects related to mysticism with primitive world superstitions, erroneous pre-Enlightenment religious dogma, and modern-day charlatans. Because of such misunderstandings, some have even made it their mission to devise ways of discrediting the subject altogether.

One of the techniques that the physicalists employ is to set the qualification level for scientific verification for this particular subject of research so high that it cannot be met, which is then used as so-called "proof" that the activity does not exist. Indeed, the standard for research in the field of parapsychology and its related branches are so demanding that it exceeds that of other subjects of study (Denzler 2001: 108-109; Mack 1999: 35; Sheldrake 2013: 165). Moreover, findings that are incompatible with the physicalist stance have also been unjustly disregarded (Facco and Agrillo 2012: 1; Radin 2006: 278-280, 285-290; Sheldrake 2013: 166). Indeed, there is a difference between the way that science is described in the rhetoric that is used by these skeptics and the way that it is actually defined and employed in standard scientific practice (Lagrange 2007: 190). It should also be understood that the accusations that have been directed at parapsychology researchers and their studies by the physicalists implies a level of incompetence never before seen in the history of scientific examination.

In his book *Entangled Minds*, Dean Radin[241] reveals that most of the common physicalist complaints about parapsychology experiments are not credible. Likewise, in his book *The Sense of Being Stared At*, Rupert Sheldrake[242] also affirms that physicalist complaints of cheating, sensory cues, hand-scoring errors, implicit learning, peripheral vision, etc., are, for the most part, not credible.

---

[241] Dean Radin, Ph.D., is an adjunct faculty member in the Department of Psychology at Sonoma State University, co-editor of The Journal of Science and Healing, member of the distinguished consulting faculty at Saybrook Graduate School, and senior scientist at the Institute of Noetic Sciences. He has conducted research into psi phenomena at the Princeton University, University of Edinburg, University of Nevada, Interval Research Corporation, and SRI International.

[242] Richard Sheldrake, Ph.D., is a former research fellow of the Royal Society and former director of studies in biochemistry and cell biology at Clare College, Cambridge University. He is the former director of the Perrott-Warrick Project on unexplained human abilities, funded by Trinity College, Cambridge. He is a fellow of the Institute of Noetic Sciences and a visiting professor at the Graduate Institute in Connecticut.

Some critical physicalists even reject philosophy, which they see as nothing more than grandiloquent musings that have no bearing in the real world. It is therefore ironic that some of the great philosophers have understood that the mentality of an observer can affect what is being observed (e.g., ontological idealism), which coincides with scientific studies that confirm the effect that the mind has on perception (e.g., observation theory) (Radin 2006: 218-220, 224, 251-252). Indeed, related issues that pertain to the subjective interpretation problem is a factor that mainstream scientists are aware of (Davies and Gribbin 1992: 21-22; Zukav 1979: 30-31).

> What we observe is not only nature itself, but nature exposed to our line of questioning.
> —Werner Heisenburg,[243] *Physics and Philosophy*

It must be understood that critical physicalists are both consciously and subconsciously influenced by their own experience, and in some instances, lack of experience.

The truth is that numerous studies that have been conducted under highly-controlled laboratory studies point toward the reality of psi phenomena (Radin 2006: 82, 275; Schwartz 2015: 252-259; Sheldrake 2013: 89, 230, 236).[244] [245] [246] Not only have rigorous

---

[243] Werner Heisenburg, Ph.D., was a Nobel-Prize-winning twentieth-century German physicist.

[244] Examples of experiments include: (1) reception of information from deceased individuals by mediums under conditions that eliminated conventional explanations (Beischel et. al. 2015: 136-141). (2) A University of Toronto study that demonstrated telepathic phenomena (Eisenberg and Donderi 1979: 19-42). (3) The various Princeton Engineering Anomalies Research studies that demonstrated the effects of mind intention (www.princeton.edu/~pear/publications. Html).

[245] See also the following studies conducted by Rupert Sheldrake: "The sense of being stared at: Experiments in schools." *Journal of the Society of Psychical Research.* Vol. 62, 1998: 311-323. "The 'sense of being stared at' confirmed by simple experiments." *Biology Forum*, Vol. 92. 1999: 53-76. "Experiments on the sense of

studies been conducted by qualified researchers at reputable institutions[247] and published in peer-reviewed scientific journals,[248] but many similar findings have been repeatedly observed (Radin 2006: 115, 153, 2013: 85, 87, 134, 151, 162-163, 188; Rhine and Pratt 1957: 60-62; Sheldrake 2013: 165-166, 266-267; Targ 2012: 158-159). For example, numerous studies in mind and matter interactions have produced correlative anomalous results that cannot be explained by conventional determinism (Achterberg, et al. 2005: 965-971; Barry, 1968: 237-243; Benor 260; Nash 1984: 145-152; Radin 1992: 1-40, 2006:186-192; Teddler 1980). The odds of chance in the experiments that demonstrated psi in these studies ranged from twenty to one, to seventy-five million to one, to 100 billion to one, to ten million billion billion billion to one! (Radin 2006: 118,162; Sheldrake 2013: 163, 267)

Nevertheless, critical physicalists declare that they will only believe what can be repeatedly demonstrated at will in a laboratory

---

being stared at: The elimination of possible artefacts." *Journal of the Society for Psychical Research*, Vol. 65, 2001: 122-137.

[246] Other studies can be found at the following internet link: deanradin.com/evidence/evidence.htm

[247] The Princeton University PEAR laboratory, the Boundary Institute, Institut für Grenzgebiete der Psychologie und Psychohygiene, the Institute of Noetic Sciences, the Koestler Unit of Parapsychology, the International Consciousness Research Laboratories, the University of Edinburgh, the Society of Psychical Research, the Rhine Research Center, and the Division of Perceptual Studies at the University of Virginia.

[248] e.g., European Journal of Parapsychology, International Journal of Parapsychology, Journal of the American Society for Psychical Research, The Journal of Parapsychology, The Journal of Scientific Exploration, The Journal of the Society for the Psychical Research, etc. I anticipate that critics will complain that these are small "fringe" publications. However, this is not a legitimate complaint due to the fact that most of the more popular journals, which are presided over by the physicalists themselves, mostly refuse to even consider subjects related to parapsychology. Moreover, most of these smaller specialty journals are presided over by groups of credentialed scholars.

setting under regulations that they deem to be appropriate—that is, unless such a study actually does prove the existence of paranormal phenomena; in which case, a flaw in the methodology must be found, or at least assumed; or they conduct their own examination, in which the intention is to discredit the previous study. Of course, this intention can play a part in skewing the perspective toward a predetermined direction—which, ironically, is what experiments in mind intention have proven (e.g., the Princeton PEAR studies, etc.).

Physicalists often complain about bias, and yet overlook such impropriety when it occurs in their own outlook. This fallacy manifests in the interpretation of the results of studies by these skeptical extremists (Bem and Honorton 1994: 16). This problem is related to perspective factors such as "confirmation bias," which occurs when evidence supporting one's belief is seen as plausible, while evidence that contradicts one's beliefs is interpreted as implausible (Radin 2006: 102). This problem is also related to what is referred to as the "experimenter expectancy effect" (Radin 2006: 285-287), as well as "inattentional blindness," which occurs when attention is placed on what the observer *chooses* to be aware of (Radin 2006: 43-44). Unfortunately, many of the skeptical physicalists are under the erroneous impression that these factors only apply to parapsychologists and not to themselves.

> I shall not commit the fashionable stupidity of regarding everything I cannot explain as a fraud.
> —Carl Jung, The Psychological Foundations of Belief in Spirits

The critical physicalists support their argument by pointing to instances of fraud in parapsychology experiments (e.g., the Uri Geller and James Randi group incidents). However, despite these straw-man cases, not only has fraud and mistakes occurred in mainstream scientific research as well (Sheldrake 2013: 9) but parapsychology is actually less susceptible to these types of problems because of heightened scrutiny.

It must also be understood that there are many mainstream scientific studies that are not only selectively reported but have not been able to be replicated (Handwerk 2015; Ioannidis 2011). For example, fewer than half of attempts made to replicate psychology studies that were published in some of the world's leading scientific journals were able to be repeated (Handwerk 2015); and yet, psychology is not labeled a "pseudoscience"; nor are the journals that these studies published in considered to be "fringe." This is because as long as psychology is not related to the supernatural, then, even if it fails the physicalist own standard, it is still considered permissible. Likewise, in his book *Rigor Mortis: How Sloppy Science Creates Worthless Cures, Crushes Hope, and Wastes Billions*, investigative science journalist Richard Harris reveals that most biomedical studies were also unable to be replicated.

In some parapsychology studies, rigorous experiments that were presided over by skeptical scientists themselves, to make sure that no chicanery or missteps occurred, still yielded positive results that were above the statistical odds of chance (Sheldrake 2013: 165). Indeed, some of the experiments that were conducted by the physicalists actually validated psi phenomena! (Sheldrake 2013: 83-87, 164) However, in these instances, the physicalists did not feel compelled to publish these unwanted results (Sheldrake 2013: 86). In the case of negative results, when the physicalists were asked to show their data so that it could be examined by the parapsychologists, they were not always able to fully comply (Sheldrake 2013: 85-87). Indeed, other purported negative results were not always substantiated with published data (Sheldrake 2013: 68).

In regard to conflicting data derived from other studies that is used by physicalists to disparage the subject, what needs to be understood is that psi activity is natural and often spontaneous, and therefore cannot always be forced in a laboratory setting. Furthermore, there is also a subjective nature to this experience that the critical physicalists seem to be completely unaware of. This is due to the natural and verifiable fact that some test subjects are more open to subtle energies than others (Eisenberg and Donderi 1979: 42; Storm, et al. 2010: 1). For example, one study showed that people who meditated had greater results than those who did not (Sheldrake 2013: 325); while individuals who harbor a preexisting bias toward this subject are more likely to score negative results (Rhine and Pratt 1957: 46-50). Therefore, as long as different laboratories are using different participants, diverse data will result. This diversity should not be used to indicate that psi does not exist.

Nevertheless, critical physicalists demand precise repeatability; like a falling apple from a tree, which has been used since the time of Newton to demonstrate the effect of gravity. However, human beings are not inanimate objects, and therefore cannot be held to the same type of narrow criterion. Indeed, because psi phenomena, the brain, consciousness, bio-electric activity, and quantum mechanics are related, it should be understood that just like quantum activity is unpredictable, so is psi. It must therefore be understood that just because something is unpredictable does not mean that it does not exist.

> Like every other physicist, I prefer sharp, precise, and unequivocal predictions. But I and many others have come to realize that although some fundamental features of the universe are suited for such precise mathematical predictions, others are not—or, at the very least, it's logically possible that there may be features that stand beyond precise prediction.
> —Brian Greene,[249] *The Hidden Reality*

It must be understood that not everything in the universe can be reduced to or proven or disproven based on data metrics alone.[250] The

---

[249] Brian Greene, Ph.D., is a former Rhodes scholar at Oxford. He is the current professor of physics and mathematics at Columbia University and is the co-founder of and director of Columbia's Institute for Strings, Cosmology, and Astroparticle Physics. His first book, *The Elegant Universe*, was a finalist for the Pulitzer Prize.

[250] Etlinger, Susan. "What do we do with all this Big Data?" *ted.com*. September 2014. www.ted.com/talks/susan_etlinger_what_do_we_do_

undeniable fact is that the information that can be extrapolated from metrics is actually limited. This is because analytics are not always capable of considering complex factors related to context and human nuance. Furthermore, analytic assessments may value on metric, while undervaluing others. Therefore, what needs to be considered are not only factors related to unpredictability laws and human nature but overall probability statistics. In this case, meta-analysis clearly indicates that the data observations for psi are well beyond the statistics of chance.

* * *

Critical physicalists also argue that there is no compatibility between psi phenomena and known scientific mechanisms. However, this also does not seem to be true. Numerous studies in distance healing, for example, indicate that not only is the phenomenon real but that it is compatible with quantum physics (Schwartz 2015: 252-259), which has repeatedly proven that interactions between attuned particles can occur at any distance (Benor: 260; Davies and Gribbin 1992: 235; Schwartz 2015: 254). This is a reference to "quantum entanglement," which is what Albert Einstein referred to as "spooky action at a distance." It does appear that similar type of non-local activity is occurring during psi experiences, such as telepathy (Radin 2006: 74-75). This is possible because all entities are composed of quantum particles. Indeed, it is now known that objects, including human-beings, harbor individual quantum waves (determined by Broglie's wave equation) (Davies and Gribbin 1992: 207). Likewise, it is known that the brain generates electrical signals and fields. What the findings indicate is that there are different wavelength levels that can be attributed to both energy and to being, and the information that is associated with this energy is able to be "entangled" into vibratory characteristics that can be received between attuned entities at any distance. So-called "spooky action at a distance" is possible because at a fundamental level all energy is connected (Radin 2006: 262).

> The doctrine that the world is made up of objects whose existence is independent of human consciousness turns out to be in conflict with quantum mechanics and with facts established by experiment.
> —Bernard D'Espagnat,[251] "The quantum theory and reality." *Scientific American* (November 1979)

The type of phenomena may be related to what the theoretical physicists David Bohm referred to as "the implicate order." Scientific examinations reveal that this holistic energy system is imbued with characteristics that are transcendent of both space and time (Radin 2006: 162, 249-252). Such transcendence is possible because at a deeper and higher level all essential energy is attached to its source (i.e., the Infinite Energy Circuit), which seems to exist outside the boundaries of space and time. It is evident that the human psyche is subject to a similar process. Indeed, studies into parapsychology indicate that signal carrier waves that usually apply to the known laws of physics do not apply to psi. It seems that such information transfer phenomenon does not travel faster than the speed of light, which would be a violation of known physical laws, but rather travels at the instantaneous speed of thought. Therefore, carrier waves are not necessary because the two entangled objects are already associated in a transcendent state. This is why all signal transfer theories that some have attempted to apply to psi do not work (Radin 2006: 246-247).

Future scientific discoveries will be better able to explain this complex quantum and metaphysical symbiosis. Just because scientists do not have a detailed understanding of this phenomena at the present time does not mean that it does not exist; rather, it only indicates that further study is needed (Eisenberg and Donderi 1979: 42). Unfortunately, at the present time, the physicalists have been doing everything that they can to obstruct such research.

---

[251] Bernard D'Espagnat, Ph.D., was a twentieth- and twenty-first-century French philosopher and physicist.

Parapsychology not only attests to the dualist[252] nature of human ontology but also to the underlying monistic activity that exists behind it. The mental aspect of this phenomena compelled one of the most eminent scientists of all time to state the following:

> I regard consciousness as fundamental. I regard matter as derivative from consciousness. We cannot get behind consciousness. Everything that we talk about, everything that we regard as existing, postulates consciousness."

—Max Planck,[253] The Observer[254]

Indeed, mind and matter symbiosis, perhaps via morphic fields, is a fundamental element of evolution.

Morphic fields are similar to what biologists refer to as morphogenetic fields (i.e., attractors). The presence of this energy system helps to explain growth dynamics in organisms (Sheldrake 2013: 337). However, at the time of writing, this process is not fully understood by mainstream science. Indeed, physicalist interpretations have been unable to provide an adequate explanation for how they work. In cases of genetic mutation in sentient entities, it is evident that these evolutionary developments are catalyzed by signals that are projected by the vibratory resonance of the psyche—which may be related to electromagnetic fields (Abraham, et al. 1992: 80). The projection manifested by intention creates hylomorphic attractors (i.e., *entelechy*) that slowly move energies toward a specific evolutionary direction. At some point, these energies transfer over into the physical level.

---

[252] Dualism in this context refers to the relationship between the mind and body.

[253] Max Planck, Ph.D., was a Nobel-Prize-winning nineteenth- and twentieth-century German theoretical physicist and professor at Berlin University.

[254] The Observer, January 25, 1931.

Morphic fields also underlie our perceptions, thoughts, and other mental processes. The morphic fields of mental activities are called mental fields. Through mental fields, extended minds reach out into the environment through attention and intention and connect with other members of social groups. These fields help explain telepathy, the sense of being stared at, clairvoyance, and psychokinesis. They may also help in the understanding of premonitions and precognitions through intentions projecting into the future. They are very compatible with Carpenter's "first sight" hypothesis. They also have something in common with Jon Taylor's brain-resonance theory of ESP; with Stanford's conformance theory that psi effects occur in accordance with goals and needs and depend on influencing otherwise random events; and with Braud's emphasis on lability.[255] In addition, these fields may have close connections to extensions of quantum theory, as the quantum physicists David Bohm and Hans-Peter Durr have pointed out.

—Rupert Sheldrake, *The Sense of Being Stared At*

* * *

Unfortunately, fraudulent tricksters, who claim to have "psychic" powers but in reality are nothing more than well-practiced illusionists, have undermined the subject. The fact is that there is nothing psychic about bending spoons or stopping watches. At the same time, there is also nothing rational about throwing out the baby with the bath-water. Therefore, every case must be honestly evaluated on a case-by-case basis.

It can be argued that physicalist resistance is not only based on a reaction to the charlatans but to ingrained prejudices against anything that could endow validity to religious ideology—which they see as

---

[255] Lability refers to capacity for change. It is the opposite of inertia.

nothing more than a mental relic left over from a more primitive era. The real issue, therefore, is not that psi does not exist, but rather that it is incompatible with physicalism. It is therefore ironic that physicalists hold themselves up as the lettered exemplars of the scientific method, and yet they are unwilling to go wherever the findings lead—which is what true scientific method necessitates. The results submitted by most of the parapsychologists have satisfied this requirement, while the physicalists have ironically veered away from the original directive in order to satisfy a preestablished conclusion. It must therefore be understood that physicalism and science are not always synonymous.

Physicalists complain that topics related to the paranormal are too far out on the "fringe" to be taken seriously; however, all great scientific discoveries have started out on the fringe; just as all new scientific truths have faced opposition from those who lack foresight, insight, objectivity, humility, and courage.

> Great spirits have always encountered opposition from mediocre minds. The mediocre mind is incapable of understanding the man who refuses to bow blindly to conventional prejudices and chooses instead to express his opinions courageously and honestly.
> —Albert Einstein, Letter to Morris Raphael Cohen

It should be remembered that Einstein at first resisted the concept of black-holes, as well as an expanding universe, because such things seemed to be too radical to be real (Greene 2011: 241). And yet, just like quantum spooky action at a distance, this phenomenon has been proven to be true. The lesson to be learned from Einstein's mistakes is that it is best not to resist a theory just because of how fantastic it is. This is because the universe is indeed a fantastic place.

Some of the ancient Greek philosophers sought to reduce everything in the universe down to a single element. Some (e.g., Heraclitus) believed that everything that exists is made from fire. Others (e.g., Thales) believed that everything was water. While others (e.g., Anaximenes) believed that everything was made out of air. So too do the physicalists believe that everything is matter. The truth is that the universe is a far more complex and pluralistic place than these monists realize.

Physicalists reduce the mind to the physical brain itself. They assert that human-beings are essentially organic robots (Sheldrake 2013: 116). They also believe that the brain is essentially a computer made of meat, and this computer essentially does nothing more than compute algorithms (Penrose 1989: 447-448). Likewise, many physicalist scientists see the universe as a "pointless" computer and believe that human-beings are nothing more than "chemical scum" (Deutsch 1997: 346). In other words, the physicalists have replaced the higher functioning *telos* of intelligent beings who derive from star dust with meaningless automata. Theories of other modes that cannot be quantified on a physical level seem like nothing more than fairy-tale magic to them. Nevertheless, the wondrous activity of nature cannot always be reduced to the limited and in some cases arrogant ideologies of man. In this case, from a psychological point of view, the rigid, mechanistic, pessimistic, and narrow physicalist stance is in actuality only a reflection of that group's own personal modern-age reality, and not that of nature's reality.

When the mystic communes with nature, he or she is opening their psyche and reconnecting with this natural experience. Tuning in to this level is not mere delusional speculation or fanciful wishful thinking, but rather is a higher aspect of a more subtle, healthy, and prodigious reality that emanates from a cosmic and transcendent provenance that is not fully understood at this time. Indeed, some of the greatest academic minds of all time (e.g., Dirac and Ramanujan) understood the importance of beauty and intuition in perceiving the underlying mechanisms of nature. But too often humanity forgets this higher reality. We get caught up in our various everyday worldly activities and its related problems, and suffer the consequences as a result.

The fact is that religion and subjects that pertain to the supernatural, spirituality, etc. would not have survived for thousands of years if it did not contain some basic element of primary truth.

* * *

There are indeed qualified scientists who are open to these subjects, but feel as if they are limited by informal constraints (Radin 2006: 7). In this case, in order for further advancement to occur the parapsychologists will need to do a better job of banding together, defending their work, making converts, and openly challenging the critical physicalists by exposing their unscientific bias, and therefore help to bring these findings out into the mainstream where they belong.

At this present time, the physicalists rule the scientific establishment. Many of these activists are dedicated to repressing any information that might work to subvert the interpretations that they are committed to retain, and are not above using vitriolic bully tactics to do so.[256]

However, there will most likely be a time in the future when this will no longer be the case. One person who is helping to lead the way is Dean Radin:

> After studying these phenomena through the lens of science for about 30 years, I've concluded that some psychic abilities are genuine, and as such, there are important aspects of the prevailing scientific worldview that are seriously incomplete. I've also learned that many people who claim to have unfailingly reliable psychic abilities are delusional or mentally ill, and that there will always be reprehensible con artists who claim to be psychic and charge huge sums for their services. These two classes of so-called psychics are the targets of celebrated prizes offered for demonstrations of psychic abilities. Those prizes are

---

[256] Part of the campaign that is being waged by the physicalist is occurring online, where, on popular web-sites such as Wikipedia, the suppression of information and the propagation of disinformation occurrs on a regular basis.

safe because the claimed abilities of the people who apply either do not exist or because the abilities are insufficiently robust to meet challenges that are actually impossible-to-win publicity stunts. There is of course a huge anecdotal literature about psychic abilities, but the evidence that convinced me is the accumulated laboratory performance by qualified scientists who do not claim to possess special abilities, collected under well-controlled conditions, and published in peer-reviewed scientific journals.

There is ample room for scholarly debate about these topics, and I know a number of informed scientists whom I respect who have different opinions. But I've also learned that those who loudly assert with great confidence that there isn't any scientifically valid evidence for psychic abilities don't know what they're talking about.

—Dean Radin, deanradin.com

These findings are acknowledged by others in the scientific and academic community as well. One such person is Charles Tart:[257]

> [. . .] there are five psychic phenomena for which we have so much evidence that it's not reasonable to deny that they exist. They are telepathy—mind-to-mind

---

[257] Charles T. Tart, Ph.D., is Professor Emeritus of Psychology at the Davis campus of the University of California, a Core Faculty Member at the Institute of Transpersonal Psychology (Palo Alto, California), and a Senior Research Fellow of the Institute of Noetic Sciences (Sausalito, California). He was the first holder of the Bigelow Chair of Consciousness Studies at the University of Nevada in Las Vegas and has served as a Visiting Professor in East-West Psychology at the California Institute of Integral Studies. Tart was also as an Instructor in Psychiatry at the School of Medicine of the University of Virginia, and a consultant on government funded parapsychological research at the Stanford Research Institute (now known as SRI International).

communication; clairvoyance—the direct perception of the physical world without the use of the physical senses; precognition—predicting the future accurately when there is no way you could logically infer it; psychokinesis—the direct effect of mind on matter; and psychic healing—the mind's direct effect on other biological systems.

—Charles T. Tart[258]

The substantial degree of credible evidence has also compelled the members of the international forum for the professional organization of scientists and scholars at the Parapsychological Association,[259] which is an organization that is affiliated with the prestigious American Association for the Advancement of Science, to conclude that based on "the presently available cumulative statistical database for experiments," "strong, scientifically credible evidence" indicates that anomalous activity related to extra-sensory perception "and mind-matter interactions" does, in fact, exist.[260]

---

[258] Tart, Charles T. "Towards an Evidence-based Spirituality: Some Glimpses of an Evolving Vision." (*Subtle Energies & Energy Medicine*. Vol. 20, No.2): 16-17.

[259] The Parapsychological Association, Inc. is an international forum dedicated to the study of psychic experiences, such as telepathy, clairvoyance, remote viewing, psychokinesis, psychic healing, and precognition. It was created by Dr. J.B. Rhine at Duke University in 1957.

[260] parapsych.org/articles/36/55/what_is_the_stateoftheevidence.aspx

CHAPTER VIII

THE METAPHYSICS OF HYPERSPACE

In previous chapters, the subject of spirituality and the alleged existence of a spiritual "afterlife" world was referred to; which raises the question: Could such a supernatural world literally exist?

What should be considered is that emerging findings in the field of theoretical physics are beginning to detect the existence of not only non-physical phenomena but extra dimensions and "parallel universes" that exist within a larger "multiverse."

Some of the concepts that are emerging are being produced by the mathematics of super-string theory. String theory takes its name from the concept that subatomic particles resonate due to internal "string" type structures. However, the concept of a vibrating universe is not what makes string theory so significant and so controversial. Super-string theory is unique because it predicts the existence of a multi-dimensional universe that extends past the four known measurements of space (length, breadth, and depth) and time.

At the present time, a majority of mainstream physicists do not believe that these extra dimensions could have anything to do with extra dimensions of space and time; and yet, the answer that they have proposed: that is, that these dimensions are "rolled up," or "compacted," into "interior spaces" of an atom, is far less likely. Indeed, even the physicists themselves admit that the model that they are proposing is not without an unusual share of imperfection and mystery.

It must first be understood that physicists use mathematics to calculate the dynamics of nature. Before string theory appeared, physicists were having a difficult time uniting the complex equations into a single cohesive and symmetrical whole. String theory solved some of the problems by allowing more room for the equations to unite. To the discoverers of this novel solution came the realization that a step closer was taken to the elusive Grand Unified Theory—which is a relative of Einstein's Theory of Everything. Although Einstein himself attempted to solve this ultimate equation, he too was unable to attain this elusive paragon goal. This is because in order to comprehend the universe, the *entire* universe, including all of its multiple manifestations, must be taken into account.

The first idea proposed as a solution to this conundrum was discovered by the theoretical physicist Oskar Klein in the 1920s. His work began as an effort to elaborate on a theory that was submitted a decade earlier by Theodor Kaluza.[261] What Kaluza did was ad one extra dimension to his calculations, which rendered a very significant result. Kaluza contacted Einstein to present him with his findings, which the revolutionary discoverer of relativity theory was duly impressed by. Einstein even encouraged Kaluza to have his paper published in a prestigious physics journal, which he did in 1919 (published in 1921). Although the mathematics equated on paper, no one could imagine what this extra dimension could be. Excited about Kaluza's theory, Oskar Klein presented the idea that the extra dimension could be "curled up" into spaces a hundred-billion-billion times smaller than that of an atom. To this day, this *ad hoc* proposal is still the primary explanation that is resorted to by most string theorists.

Critics of super-string theory find it difficult to believe that the universe could contain so many curled up dimensions, and for good reason. The curled-up theory was a make-shift explanation that was

---

[261] Theodor Kaluza, Ph.D., was a German mathematician, physicist, and professor at the University of Konigsberg in Prussia.

used in order to enable Kaluza's mathematics to fit into a previously established or so-called "normalized" physicalist parameter.

However, I propose that the true answer to this enigma will be found in the function of the "string" itself. Just as a string vibrates on an instrument, so too does a string-like structure vibrate within the interior space of an atom. These vibrations are resonating in something like pitches that coalesce to form frequencies, or wavelengths of energy. It is a model that recalls the work of the sixteenth-century artist, cosmologist, physician Robert Fludd, and his illustration of the divine "monochord." This cosmological picture also corresponds with the ancient Greek philosopher Pythagoras's concept of a mathematical and vibrating universe—which he also related to a musical instrument (i.e., the "music of the spheres"). Along with the energy of light, sound plays both a role in the vibration of the string and the creation of the universe. This may be related to the divine sound that is manifested in the Buddhist, Hindu, Jain, concept of the cosmic *Om*. Sound waves are vibrations propagated as a wave travels at the speed of a certain frequency. Lower frequencies are infra-sound, while higher frequencies are ultra-sound; and, just like light, this form of energy also operates both within and without of the limited range of human detection. It is a fundamental factor that extends out and back to the very same monistic super-force that initiated our own particular universe and the wavelength plane that we currently inhabit.

It is now known that matter itself is vibrating energy that has been locked up into a particular state (Davies and Gribbin 1992: 14, 235); therefore, the physical world as we know it is just one level of a particular wavelength/frequency, or vibratory plane.

> When the string vibrates in different modes, it becomes a different particle. In this picture, the laws of physics are nothing but the harmonics of the super-string. The universe is nothing but a symphony of vibrating strings. (This, in a sense, fulfills the original dream of the ancient Greek Pythagoreans, who were the first to understand the laws of the harmony of strings. They

suspected that the entire universe might be understood via the laws of harmony, but until now, no one knew how this could be done).

—Michio Kaku, *Visions*

Each wavelength/frequency corresponds to the vibration of a certain pitch; and, just like an instrument, not all of these strings are vibrating at the same rate. Herein may lie the key to comprehending the phenomena of extra dimensions. If subatomic energy is locked up into not only a single note but several, we then have a model that presents the existence of a manifold universe. Science has already proven that energy and light are interrelated, and light has already proven to exist in separate wavelengths—most of which are completely invisible to the human eye. In this case, this same principle would apply to matter as well.

One thing that has changed since Kaluza and Klein's day, is the number of dimensions that are being detected. The most common symmetry is obtained within the parameter of ten dimensions (Greene 2011: 266, 271; Zwiebach 2009: 8).

Critics of super-string theory point out that there is no experimental evidence for the existence of super-strings. However, it must first be taken into account that its mathematical elegance is far superior to other theories that have failed. Furthermore, it must be understood that super-string theory and quantum field theory are significantly related (Greene 2011: 273); therefore, when string theory, or a theory that shares similar characteristics, is eventually proven to be true, this will most likely be done through its relationship with quantum mechanics. Indeed, the model of a vibrating string universe correlates with the vibrations that are found in mainstream quantum physics.

> As a man who has devoted his whole life to the most clear headed science, to the study of matter, I can tell you as a result of my research about atoms this much: There is no matter as such. All matter originates and exists only by virtue of a force which brings the

particle of an atom to vibration and holds this most minute solar system of the atom together.

—Max Planck[262]

Where this new paradigm is leading us is outside the delimitations of the standard physical model, and into the higher dimensional state of "hyperspace."

The term "hyperspace" is appropriate considering that the vibratory rate of this wavelength field seems to be operating at a higher and faster frequency; while the vibration of the physical world that we presently inhabit operates at a much lower, denser, and slower rate. Therefore, hyperspace is not a compacted, or curled up, or unused space—as if it were nothing more than a mere mistake; but rather, it is a part of a natural space-time continuum that consists of wavelength/frequencies and fields that resonate in separate but parallel band frequencies.

Each distinctive wavelength layer can be referred to as a "membrane," or "brane" for short. This terminology is acquired from unified string theory, or "m-theory."[263] The vibration of the strings effects the collective resonance of the entire brane; while all of the branes together within the manifold "brane-world" universe is referred to as "the bulk." I contend that when some cosmologists are referring to parallel universes, they are mostly referring to parallel branes—whether they know it or not. Indeed, in regard to these other dimensions, physicists admit that there must be some missing principle at work that they are not understanding (Greene 2011: 127). In this case, we must differentiate between dimensions, branes, and universes. The brane-world bulk can be envisaged as a cosmological mandala (see figure 23). The IEC (Infinite Energy Circuit) is at the center, followed by a brane that cannot yet be conclusively defined at this time; although it is likely that this level is the same two-dimensional "boundary" that is being predicted by the holographic principle (more on this subject later). The second brane can be

---

[262] 1944 lecture (Schwartz 2015: 253-254).

[263] M can stand for either magic, mystery, or membrane.

referred to as hyperspace, followed by our own lower level brane, which can be designated as a sub-space; which is most likely the farthest away from the source, in a lower vibrational state.

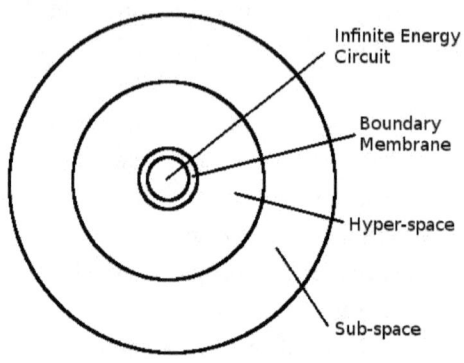

(Fig. 23)

The strings are moving around both on and inside the brane, and thereby effecting the quantum field within its spatiotemporal environment (Greene 2011: 264-266). The reason why we do not see these other parallel branes is because they exist on separate wavelengths. This brane-world scenario clearly negates the provisional compacted spaces theory (2011: 118).

A newly emerging "many interacting worlds" theory[264] postulates that parallel "ghost" universes may explain some of the quantum anomalies that have perplexed physicists for years (Slezak 2014). It seems that these anomalies are caused by quantum activity that is emanating from parallel brane-worlds that are fluctuating and essentially bumping into our own. However, it may be the case that these other environments are most likely not other universes per se, but rather other membranes that exist within the collective bulk of the

---

[264] Michael J. W. Hall, Dirk-André Deckert, and Howard M. Wiseman. "Quantum Phenomena Modeled by Interactions between Many Classical Worlds." (Phys. Rev. X4, 041013, October 23, 2014). Accessed Oct 9, 2015. journals.aps.org/prx/abstract/10.1103/PhysRevX.4.041013

same universe. Some physicists postulate that a collision between two branes would result in a violent and possibly destructive reaction (Greene 2011: 120); however, this is unlikely because the other branes are not on the same wavelength. It is more likely that these interactions would result in only minor fluctuations. Indeed, such fluctuations may be related to the short-lived appearances of "virtual particles." These are particles (e.g., messenger photons) that appear, seemingly, from out of nothing, before quickly disappearing.

It is also likely that the regions in space-time where branes often merge is related to what is referred to as energy vortices. Purported examples of this may be: Sedona Arizona in America, Avebury in England, and Giza in Egypt.

* * *

Another significant question that could be explained through superstring theory relates to the issue of dark matter. Physicists were tipped off to the existence of this unseen form of matter when it was discovered that the mass required to keep the galaxies gravitationally bound together was on average ten-times greater than the mass actually observed. The effect of dark matter is observed from the gravitational pull that it exerts on visible matter, as well as the gravitational lensing effect that it exerts on background radiation. Researchers have been unable to explain how a form of matter could not be physical. However, if we apply hyperspacial space-time continuum string theory to this problem, we might find that dark matter, along with dark energy, may actually be matter that exists on a separate wavelength/frequency. (The higher the frequency the shorter the wavelength.) Indeed, according to m-theory, gravity is not limited to one particular membrane, but rather is leaked out and felt in various degrees throughout the bulk. The diffusion of gravity (i.e., mass) throughout the bulk might explain its weakness compared to the other fundamental forces (i.e., electromagnetism and the nuclear forces).

Furthermore, inter-brane activity may also explain why disincarnate so-called "poltergeist" entities are observed exerting direct influence on objects in our own vibratory wavelength. In this case, we are most likely looking at a separate but connected type of paradigm.

This concept of a parallel dimensional model is actually not entirely new. This correlates with concept of the parallel so-called "shadow world" that is detected in the mathematics of E_8 symmetry (1992: 256-257):

> In fact, the full symmetry of this version of the superstring theory actually involves E_8 twice over, in a package that mathematicians refer to as $E_8 \times E_8$. Some theorists have speculated that this duplication involves a sort of second version of the Universe, a shadow world inhabited by identical copies of the sorts of particles familiar in our own Universe (electrons, quarks, neutrinos, and so on) but able to interact with our world only through gravity. This raises the question of whether we would actually notice the shadow world that interpenetrated our own. It would be possible, for example, to walk right through a person made of shadow matter without feeling a thing. This is because the gravitational force associated with the human body is minute. On the other hand, if a shadow planet were to pass through the Solar System, it could fling the Earth from its orbit. The circumstances would be bizarre, because nobody on Earth would be able to see anything of this celestial interloper; it would be as if some giant unseen hand were scooping the Earth aside.
> —Paul Davies[265] and John Gribbin[266], *The Matter Myth*

---

[265] Paul Davies, Ph.D., is a theoretical physicist, cosmologist, astrobiologist, and professor at Arizona State University.

[266] John Gribbin, Ph.D., is a science writer and visiting fellow at the University of Sussex.

It is therefore compelling that a similar eight-dimensional model, in which two membrane levels are postulated (six spacial, two temporal), is calculated in an eight-dimensional metric known as the Minkowski space (Radin 2006: 250), which has shown to be consistent with mainstream physics (e.g., quantum mechanics, Maxwell's formalism, and Einstein's theory of relativity, etc.). This theory (developed by Elizabeth A. Rauscher and Russell Targ)[267] presents a model in which hyperspacial/spiritual realm phenomena may exist.

It can be posited that the mass that exists in the invisible "shadow" world is what we perceive as dark matter, and the energy that animates it is dark energy. However, this energy is only "dark" to those who are not resonating on the same vibratory wavelength. This parallel world may be effected on a quantum level by what physicists refer to as shadow particles—i.e., super-partner "sparticles."

It can therefore be put forth that the long-sought evidence that would support subjects related to the paranormal and the "afterlife" were there all along. The problem is that it was never interpreted correctly!

* * *

It has become obvious that physicists have reached the limits of their particular field of study. As the name *physicist* itself denotes, their investigation extends only to the examination of the immediate physical wavelength/frequency. Therefore, the reason why they have yet to discover the long sought "Theory of Everything" is because not *everything* is being taken into account!

It is evident that this parallel spatiotemporal membrane model is describing the same realms that have been known throughout the world and throughout the centuries in the mythological and religious traditions of the world as Elysium, Valhalla, Avalon, Heaven; just as it

---

[267] *Investigation of a Complex Space-Time Metric to Describe Precognition of the Future*, by Elizabeth A. Rauscher and Russell Targ at the Technic Research Laboratory

also may be related to the Netherworld, Sheol, Hades, and Hell. It seems that the reality that one experiences in hyperspace is manifested by the vibratory resonance that the individual projects through the energy of their thoughts, actions, and emotions—as well as with others around whose particles are entangled or who inhabit a shared field. Indeed, this is the essential nature of spirituality. All of these discoveries are leading us to the natural and inevitable conclusion that what is commonly referred to as "the after-life" is actually real.

Empirical experiences in these other parallel branes are reported by people whose psyche has made the quantum leap into this other environment during out-of-body and near-death experiences. One such person is Eben Alexander, M.D. Dr. Alexander is a neurosurgeon and former skeptic of this phenomenon. In 2008, Alexander suffered from bacterial meningitis and fell into a coma. During that time, he awoke to find himself inside a murky subterranean environment before he was extracted by a golden white light. He was then lead out into a brilliant world that was similar to Earth, but more clear, vibrant, joyous, and beautiful. He claims that he was met in this world by a relative who communicated to him without words. It was in this other state that Alexander claims that he received profound wordless revelations concerning the nature of a greater multiversal reality that exists beyond the limited lower world that we know.

> I saw the abundance of life throughout the countless universes, including some whose intelligence was advanced far beyond that of humanity. I saw that there are countless higher dimensions, but that the only way to know these dimensions is to enter and experience them directly. They cannot be known, or understood, from lower dimensional space. Cause and effect exist in these higher realms, but outside of our earthly conception of them. The world of time and space in which we move in this terrestrial realm is tightly and intricately meshed within these higher worlds. In other words, these worlds aren't totally apart from us,

because all worlds are part of the same overarching divine Reality.
—Eben Alexander, *Proof of Heaven*.

Alexander is adamant that this other world was *not* a mere dream or a hallucination. Indeed, he claims that this other environment felt as though it was even more real than the lower world that he left behind. A full account of this experience is reported in his book: *Proof of Heaven: A Neurosurgeon's Journey into the Afterlife*—which is only one account of many others that have been published.

> Science—the science to which I've devoted so much of my life—doesn't contradict what I learned up there. But far, far too many people believe it does, because certain members of the scientific community, who are pledged to the materialist [i.e., physicalist] world view, have insisted again and again that science and spirituality cannot coexist. They are mistaken.
> —Eben Alexander, *Proof of Heaven*

* * *

Another intriguing theory that has arisen in recent times is referred to as the holographic principle. The concept of a holographic universe is so bizarre, so complex, so revolutionary, and, at the present time, so unfortunately nascent, that it is difficult to fully describe. It should first be understood that the comparison between a holographic universe and standard holography is actually only a generalization. Nevertheless, despite its incipient nature, the reason why it deserves to be included in this exposition is because not only is the validity of the theory acknowledged by many of the world's leading theoretical physicists, such as Stephen Hawking—who was once the theory's highest profile critic, but it is compatible with super-string theory.

The general premise mathematically predicts that what we experience as the real world is essentially a three-dimensional projection that is emanating from a two-dimensional source that exists on a distant "horizon," or "boundary." This is comparable to a hologram, because a hologram can refer to both the encoded material (i.e., "information") and the image (i.e., our universe) that it projects off of a two-dimensional surface. It is also interesting that this process is wavelength based, which concurs with the space-time brane-world model that I am presenting.

It is therefore compelling that if we count the four dimensions of our physical level brane, and the four dimensions of the hyperspacial brane, and the two dimensions of this mysterious flat-land of "information,"[268] we are left with a model that ads up to the ten dimensions that is predicted by super-string theory! I therefore propose that this other fundamental substrate is not projecting from some distant spacial locus, but rather from a parallel brane that exists along side our own. Indeed, this so-called boundary is also sometimes referred to as a "membrane" that covers a black-hole. Furthermore, if one more dimension is added to this two-dimensional matrix for time, we are then left with the eleven-dimensional model that is predicted by m-theory.

What is also compelling about holographic theory is how it relates to the concept of the implicate order that was postulated by the David Bohm.[269] In this case, the implicate order may be related to the deeper energy system that is the source of the projection that we experience in the explicate order as the physical universe—which is analogous to a hologram. It is also interesting how this flat dimensional boundary correlates to the "flat universe" that scientists are detecting.

In summary, it is amusing that holographic theory is taken seriously by an overwhelming majority of cosmologists, despite its fantastically strange and hypothetical nature. This is because as long

---

[268] The word "information" in this sense relates to a theory proposed by the physicist John Wheeler, who postulated that matter is actually a secondary manifestation of something more basic. In this context, information can refer not only to the fundamental components of nature but what these components do.

[269] David Bohm, Ph.D., was a twentieth-century theoretical physicists and assistant professor at Princeton University and professor at the University of London.

as no one relates it to spirituality, the supernatural, or religion, it is considered permissible by the physicalists.

* * *

One of the primary problems that physicists deal with is the problem of "negative norm states." These are calculations that equate and yet do not yield finite physical deductions. However, if irregular and non-physical massless equations were regarded as being related to hyperspace, a significant break-through might occur. Therefore, in the case of negative norm states it may only be a matter of redefining the norm.

> Negative energies and probabilities should not be considered as nonsense. They are well-defined concepts mathematically [. . .]
> —Paul Dirac,[270] *The Physical Interpretation of Quantum Mechanics*

It is certainly ironic that the roots of modern-day science can be traced back to the original Greek philosophers, since the original founding-fathers of science not only examined physics but metaphysics as well. While many philosophical schools saw the physical world as more of an illusion compared to the eternal state of the spirit, others tend to relegate reality solely to physical sense experience. This type of physicalism is the prevalent view that exists to this day. It is a position that has most likely manifested as a reaction to the dogmatic foundationalism[271] of religious faith-based fundamentalists, which has, unfortunately, turned many intellectuals away from the subject of spirituality and transcendental metaphysics.

Einstein once said that "mysticism is in fact the only reproach that people cannot level at my theory" (Brockman 2007: 103).

---

[270] Paul Dirac, Ph.D., was a twentieth-century English theoretical physicist.

[271] Foundationalism refers to a belief that a particular philosophy is self-justified—as opposed to coherentism.

Unfortunately, it is now known that this self-imposed limit is what most likely prevented him from attaining his long-sought Theory of Everything.

> A good scientist freed himself of concepts and keeps his mind open to what is.
> —Lao Tzu, Tao Te Ching

Indeed, meditation, mystical experiences, and developing a deeper sense of intuition can bestow insight into how the universe works.[272]

Unfortunately, close-minded cynicism[273] is the prevailing response to anything that can be associated with what is commonly referred to as the "supernatural." In an ironic twist, we can now find the same type of rigid dogmatism that has traditionally projected from the religious wing has ironically manifested in the scientific wing as well.

True science requires the honest and objective research of data. Therefore, a scientist must be willing to go wherever the evidence may lead; even if the evidence conflicts with his or her belief system. Whether its opposition to extraterrestrials, psi phenomena, parallel worlds, "giant" humanoid beings, etc., we have seen throughout the course of this examination that the so-called "skeptics" continually search for ways to disregard any evidence that might extract them from out of their ideological comfort zone. Their rigid and in some cases even fanatical disposition must be exposed for what it really is: unscientific bias.

> Scientists are human. We have our blind spots and prejudices. Science is a mechanism designed to ferret them out. Problem is, we are not always faithful to the core values of science.

---

[272] Further recommended reading: *The Tao of Physics* by Frijof Capra (Shambhala, 1975), and *The Dancing Wu Li Masters* by Gary Zukav (Harper One, 1979).

[273] I use the word "cynicism" here in the common modern-day sense of the word; as opposed to the ancient philosophical definition.

—Neil deGrasse Tyson[274], *Cosmos: A Spacetime Odyssey*, Season 1, Episode 9. The Lost Worlds of Planet Earth.

The inter-dimensional space-time continuum theory helps to solve the weakness of gravity question; it helps to solve the dark matter problem; it helps to solve the dark energy problem; it helps to solve the negative norm state problem; it helps to solve the ghost particle problem, and it helps to solve the presence of membranes and extra dimensions.

Unfortunately, many open-minded scientists are hesitant to step forward in support of a theory because of the condemnation that it could evoke from their physicalist peers in the scientific community. Such a show of support for such a controversial topic may be perceived by some as detrimental to their mainstream careers. It is an unfortunate situation that will need to be challenged by qualified members of the scientific community if any progress is to be made.

Metaphysics is rejected by critical physicalists (e.g., positivists) on the grounds that its predications are not empirically verifiable. However, as the previously submitted findings have shown, this is not always true. Likewise, subjects related to mysticism are also commonly defined by the physicalists as experiences that are independent of sense perception and rational thought (i.e., *noumenon*). However, this is also not true. The truth is that these accusations of incommensurability are based on only the prejudices and misapprehensions of the "skeptics" themselves. The truth is that subjects related to mysticism, parapsychology, the paranormal, etc. do indeed comply with the philosophical/scientific standards that are

---

[274] Neil deGrasse Tyson, Ph.D., is an American astrophysicists and cosmologist. He is the Frederick P. Rose director of the Hayden Planetarium and is an advisor to NASA.

maintained by the secular skeptics (e.g., the naturalists[275] and the positivists[276]).

It is therefore time for the re-establishment of a pluralistic[277] metaphysics in the modern-age Academy. What is being proposed is the discovery of what some are already referring to as "paraphysics." It is a field of study that will eventually take over where the limits of physical world inquiry end, and where the higher dimensional membrane of hyperspace begins.

---

[275] Naturalism holds that that everything is understandable in terms of "naturalistic" (i.e., secular) deterministic causation (e.g., quantum mechanics, etc.).

[276] Positivism essentially holds that logic and reason can be used to deduce truth from direct sensory experience.

[277] In metaphysics, pluralism is the belief in more than one reality.

CHAPTER IX

THE SERPENT

The third major symbol in the Garden of Eden story is the serpent. This infamous antagonist is also commonly referred to as Satan, the devil, and Lucifer. The other association that Satan has with the symbol of the serpent can be found in the book of Revelation. In the following passage, the adversary of God is described as a "dragon":

> The great dragon was hurled down—that ancient serpent called the devil or Satan, who leads the whole world astray. He was hurled to earth, and his angels with him.
> —Revelation 12.9

The other common image that we have of Satan in the present day is not as a serpentine dragon, but rather as a cloven-hoofed, horned-headed, anthropomorphic beast. However, it is evident that this Christian era goat-man characterization is more likely influenced by the pagan creature: Satyr/Satyros/Faun/Faunus (i.e., Pan).

In the following verse, the enemy of the Lord is not described as a serpentine creature, but rather as one of the individuals who was at times in the company of Lord Yahweh himself—even after his alleged banishment and fall from grace:

> He showed me Joshua the high priest standing before Yahweh's angel, and Satan standing at his right hand to be his adversary.
> —Zechariah 3.1

Here we are told that Satan *stood*, not slithered, among the assembly of the Elohim—as opposed to crawling on its belly in the wilderness, or in the depths of hell. A similar situation appears again in the following passage:

> On another day the angels [sons of the Elohim] came to present themselves before the Lord [Yahweh], and Satan also came along with them to present himself before him.
> —Job 2.1

In these passages, the character who is referred to as "Satan" appears to be in a functional relationship with the other members of the council of the Elohim. This does indicate that the person, who was previously portrayed as a snake, was an Anunnaki.

The literal meaning of the Hebrew word *satan* is adversary. This term, along with devil, which literally means accuser, has come to personify the ultimate incarnation of evil; however, in ancient times the snake was a symbol of not only danger but of wisdom, protection, healing, regeneration, and to a lesser extent: sexuality (Hendel 1999: 744). In fact, some references to the positive attributes of the serpent are even found in the Bible itself:

> [. . .] Therefore be wise as serpents [. . .]
> —Jesus, Matthew 10.16

> Just as Moses lifted up the snake in the wilderness, so the Son of Man must be lifted up.
> —Jesus, John 3.14

The incident that Jesus refers to can be found in the Books of Moses, in which Yahweh is reported letting lose poisonous snakes to attack the disobedient Israelites. After Moses interceded on their behalf, Yahweh called off the attack. Yahweh then instructed Moses to cure the people by making an image of a "fiery snake," which he was to place upon a pole (Numbers 21.8). This object allegedly became infused with a healing power that cured the wounds of the people.

The symbol of the serpent extends throughout the history of the ancient world. In ancient India, some gods took the form of serpents, such as the *nagas*, that could be both benevolent and malevolent. It was the serpent god Vasuki who helped churn up the ocean of the divine Soma of the gods. The nagas were also associated with the Tree of Life. According to Buddhist lore, it was the benevolent serpent, Mucilinda, who sheltered the Buddha under its protective hood during the storm when he was sitting under the Bodhi tree (Figure 24).

(Fig. 24)

In ancient Greece, the serpent also represented regeneration—which was most likely due to the periodic shedding of its skin—and was regarded as a symbol of rebirth. This is also the idea behind the symbol of the *Ouroboros,* which is the symbol of a snake devouring its own tail. The symbol of the serpent can also be found in the royal *caduceus* staff of Hermes (Figure 25) and the serpent-entwined rod of Asclepius, the Greek god of healing and medicine.[278]

(Fig. 25)

In around the second millennium BCE, the Anunnaki god Ningishzida (i.e., Ninjiczida)—whose epithet was "Lord of the good tree" (Jacobsen 1978: 7)—was recorded into the Sumerian texts. Zecheria Sitchin believed that this was the same individual who was known in Egypt as Thoth, Hermes in Greece, and Mercury in Rome (Sitchin 1985: 41, 177, 267). While it is true that Thoth, Hermes, and Mercury share common characteristics, the Ningishzida comparison is based only on the shared association with the symbol of the entwined serpents. This, in itself, is not significant enough to merit an association. Moreover, unlike the previously mentioned deities, Ningishzida was more of a demonic god of the underworld. Whether

---

[278] It has been asserted that the rod of Asclepius and the caduceus of Hermes/Mercury were two different symbols; however, there is no reason to believe that they did not derive from the same Mesopotamian source. This source could have influenced the biblical image of a healing serpent image that was placed upon a pole (i.e., *nehushtan*) by Moses (Numbers 21.8).

the symbol of the caduceus of Hermes and Mercury can be traced back to Ningishzida is also uncertain.

Sumero-Akkadian representations of a deity figure have been discovered that show a half-man, half-serpent character (Figure 26). These are most likely depictions of Ningishzida. However, in some instances it may also represent other serpent deities, such as Nirah. Indeed, the serpent symbol can be associated with several gods and demons (Hendel 1999: 744).

(Fig. 26)

Sitchin correctly associated Ningishzida with the image of the entwined serpents; however, he also identified Enki/Ea as the biblical serpent. This conclusion is primarily based on the premise that the Semitic word for snake was *nahash*. Sitchin believed that the original root letters *NHSH* means "to decipher," or to figure out, which was an attribute of Enki (Sitchin 1978: 371); however, the etymology of the word is complex and certainly debatable. The verb *nahash,* or more precisely *nachash,* may mean to learn from experience or by omen, or to practice divination and fortune-telling (Hendel 1999: 744), but it may also refer to agility, as well as desire (Hyman and Ivry 2008: 157). It may also refer to a serpentine hissing sound (Orr 1915: 2737; Hendel 1999: 744), or to a shinning quality (Bullinger 1999: 24).

According to the texts, Enki/Ea was referred to as the "lord of wisdom," (e.g., Enki and the World Order text) and was revered for his "clever" intelligence (e.g., An Adab to Enki for Isme-Dagan text). He was not only instrumental in bringing civilization to this world but was also credited with helping to genetically engineer the first human hybrid beings. His mental faculties also explain why he was associated with "cunning" intelligence and wisdom.

> [. . .] we may assume, that the idea of cunning, of superior intelligence, came to be imparted to Enki.
> —Thorkild Jacobsen,[279] *Treasures of Darkness*

In the following text, we allegedly hear from Enki in his own words:

> In am the seal-keeper above and below. I am cunning and wise in the lands.
> —Enki, Enki and the World Order

This same description is attributed to the serpent in the biblical record:

> Now the snake was the most cunning of all the wild animals [. . .]
> —Genesis 3.1

Enki's relation to the symbol of the serpent may not only have derived from the symbol's association with wisdom but with his association with the image of the serpentine dragon. Indeed, the word *nahash/nahas/nachash* can be applied to both snakes and dragons (Orr 1915: 2737; Hendel 1999: 265, 744):[280]

> Great Lord, prominent (?) among the gods, your judgments are clever and powerful! Father Enki, respected one, supreme dragon, who determines the

---

[279] Thorkild Jacobsen, Ph.D., was a twentieth-century Danish scholar and professor at Harvard University.

[280] Enki is also likened to a dragon in the Enki and the World Order text.

fates firmly, who has taken his seat among the numerous divine powers in colorful brilliance (?) [...]
—An Adab to Enki for Isme-Dagan

Enki/Ea was known as a god who was not only affiliated with the sciences but with water. Although he can be associated with the symbols of the dragon and the serpent, according to the Sumerian seal impressions, he was originally associated with the symbol of the fish. This image seems to have evolved into the Babylonian symbol of the goat-fish (Ascalone 2005: 331), which inspired the astrological Capricorn sign of the Zodiac.

Enki's Akkadian/Babylonian name was Ea. His central temple was known as "the house of the deep waters" (or House of the Subterranean Waters) (*E-engur-ra*), which is also referred to as the house of the *Abzu* (Foster 2005: 643-644). It may have been because of his association with life-giving water that he was originally associated with the symbol of both flowing streams and the fish. In this case, the symbol of the serpent/dragon may have been bequeathed to him by others at a later time. Indeed, it does seem that among those who associated Enki with the serpent symbol were the Judean/Israelite priest-scribes, who either mistakenly related the symbol that was originally associated with Ningishzida with Enki, or who perhaps did not base their interpretation on Ningishzida at all, and instead based it on one of the other previously cited etymological reasons. However, the most likely explanation is that the character of the serpent in the biblical garden of Eden story was actually a conflation that was based on the legends of both Ningishzida (the demonic serpent) and Enki (the wise rebel who was also associated with the serpentine dragon).

Even if one were to assume that none of these characters literally existed, what is apparent is that their presence in the ancient records went on to influence the Yahwehists, and thus all of Judeo-Christianity.

Although it is possible that Enki/Ea could also be identified with the god-like being who was referred to as Oannes by the Babylonian priest Berossus, some scholars postulate that Oannes may actually refer to Enki's son, Adapa (Dalley 1989: 326; Greenfield 1999: 73).[281] In this case, it could be possible that Adapa was one of the Nephilim. However, according to Berossus, Oannes was a man who came from out of the sea clad in accoutrement that was fashioned in the likeness of a fish. Oannes taught the primitive human-beings mathematics and how to write, farm, and compile laws necessary for the establishment of civilization. This god-like character who was associated with the symbol of the fish and who imparted his wisdom to human-kind comports well with the original characteristics of Enki.

It is also possible that Oannes is the same character who is referred to in the Semitic legends as the agricultural and fertility god, Dagon (i.e., Dagan). According to the first- and second-century CE Phoenician writer Philo of Byblos, this name derived from the west Semitic word *dgn/dagan*, which means grain. However, according to fifth-century CE Judeo-Christian theologian Saint Jerome, as well as one of the authors of the Hebrew Talmud, the name derived from the Hebrew word *dag*, which meant fish (Feliu 2003: 279-280; Healey 1999: 216, 218).[282] What is certain is that both fish and grain can be attributed to Enki. However, John F. Healey[283] doubts the accuracy of the grain interpretation (Healey 1999: 217). Indeed, the fish designation relates well with the Oannes character; who was most likely based on the legend of Enki. This identity concurs with the biblical account (1 Samuel 5.1), in which Dagon is also described as an abomination to Yahweh. What is more, is that Dagon (i.e., Enki)

---

[281] J.C. Greenfield (1999: 73) contends that Adapa was not literally the biological son of Enki. However, this is uncertain.

[282] Other interpretations have also been suggested. One asserts that the name Dagon derives from the Arabic word *dagana/dajana*, which means cloudy or rainy (Feliu 2003: 280; Healey 1999: 216). However, this seems unlikely. Indeed, the scholar J.F. Healey also dismisses this interpretation. According to Healey, the "appeal to such a remote Semitic cognate for etymology smacks of desperation" (Healey 1999: 218).

[283] John F. Healey, Ph.D., is an emeritus professor at the University of Manchester. He is on the editorial boards of several academic journals and is a visiting fellow of All Souls College, Oxford, a visiting scholar at St. John's College, Oxford, and is a fellow of the British Academy.

was referred to as the father of Baal (i.e., Marduk) in the Ugaritic texts (Healey 1999: 216).[284] The coherentism is undeniable. However, according to a text inscribed on the statue of Puzur-Istar, both Enki and "Dagan" were listed together as separate deities (Feliu 2003: 59). This would seem to indicate that Dagon was not Enki, in which case it may instead be a reference to Adapa. However, it may also be possible that this isolated instance was a result of a misunderstanding of the scribe himself. Indeed, I suggest that this is more likely.

The texts report that both Enki and his brother, Enlil, were sons of the elder Anunnaki god, Anu.[285] The etymological meaning of the theonym Enki is "Lord of the Earth," (Encyclopedia Britannica Online s.v. Ea).[286] This is a position that he supposedly had to relinquish after his half-brother arrived on Earth. This was because although Enki was the eldest son of Anu, his mother was not the official queen, but rather was one of Anu's concubines. Enlil on the other hand, was supposedly the eldest son of Anu and his official wife, queen Antum. The nepotistic succession lines of the Anunnaki tradition were very much like the ones of their human followers. As will be reviewed in the section to come, the legendary family rift that

---

[284] There are other texts that report that Baal was the son of El. This perplexes scholars because it contradicts the texts that report Baal was the son of Dagon (Healey 1999: 217). However, I suggest that the El reports, which are not coherent with other data, were based on the misunderstandings of the ancient people themselves. Indeed, other explanations suggested by scholars have proven to be untenable (Healey 1999: 217).

[285] In the Enki's Journey to Nibru text, Enlil is called the "father" of Enki; however, this relation is most likely based on a misinterpretation. It is likely that the scribe had been influenced by Enki's epithet, "the junior Enlil." However, this term was most likely used to indicate that Enki was in a subordinate position to Enlil. Moreover, in An Adab To Enki for Isme-Dagon text, as well as a Tigi to Enki for Ur-Ninurta, and a Hymn to Ninurta for Isme-Dagaon texts, Enki is referred to as the "son" of An (i.e., Anu), not Enlil. Similarly, in the Enki and the World Order text, Enki not only referred to An as his father but seems to refer to Enlil as an "elder brother." This is the most likely description.

[286] En = lord, Ki = earth.

existed during the early Genesis era, between Enki and Enlil, continued on into the following generation as well. What will become apparent moving forward, is that this same historical schism was chronicled in both the Mesopotamian and the biblical accounts.

Enki was not only known for both his wisdom as well as his scientific endeavors, but was also known as a contentious character who was the nemesis and sibling brother of Enlil. The texts report that this individual got into disputes with his brother and developed a reputation as both an adversary and an accuser. The original primary rift was not so much a struggle between good and evil, but rather between the liberal and conservative values of the time. Enlil represented the right-wing perspective, which was a position that sought to preserve the traditional and hierarchical ways of the Anunnaki. While Enki represented the liberal policy of change, compassion, and tolerance—especially when it came to human-rights. It is likely that it was Enki who introduced the Tree of Knowledge to Earthling women; and that it was Enki—according to the original Mesopotamian records, such as the Sumerian Eridu Genesis and the Babylonian Atra Hasis text—who warned Ziusudra/Atra Hasis/Utnapishtim/Noah of the impending flood—which angered Lord Enlil. Of course, in the subsequent biblical version, this act is attributed to the Elohim, which confirms that the character who is known as the serpent—i.e., satan, etc.—was indeed a member of the Anunnaki council of the Elohim, even after his short-termed banishment. His relegation to the lower level of "arch-angel" in the biblical version, had more to do with the lower rank that he was born with. This seems to have been because he was not considered a legitimate heir to the throne of Anu—which was a status that was bolstered by the loyal servants of Yahweh, who were influenced by the Enlilite tradition (more on this subject later).

In the Atra Hasis text, Enki is recorded inciting a rebellion against Enlil's wing of the Elohim:

> Call the elders, the senior men! Start an uprising in your own house. Let the heralds proclaim [. . .] Let

them make a loud noise in the land: So not revere your gods, do not pray to your goddesses [. . .]
—Enki, Atra Hasis

While it is true that at times the texts report that Enki did acknowledge his "great brother, ruler of the lands," it is also apparent that he may have at times thought that he would have been a better ruler himself.

I am the true offspring, sprung from a wild ox. I am a leading Son of An [Anu] [. . .] I am the great lord over the land. I am first among the rulers. I am the father of the lands.
—Enki, Enki and the World Order

We see, in these instances, that Enki fits the characteristic personification of a rebellious so-called "angel" (i.e., emissary) demigod. He is also clearly both the "accuser" and the "adversary" of the lord of the council of the Elohim.

Just like the biblical serpent character, Enki was known for his powers of seduction. His sexual exploits are described in the Enki and Ninhursaja text, where he is described as sexually virile, and a seducer of woman. It is therefore apparent that when the Yahwehist scribes were collecting, interpreting, and writing down their own version of events centuries later, they were influenced by these earlier legends.

Just like the Judeo-Christian version of these events, the brilliant "light-bearer" was at one time one of the attendant "angels" of Lord Enlil. Just like the legend of "Lucifer," who is also known as "the Morning Star," an epithet of Enki's that was used to describe his shinning brilliance was "Lord Silver Vision."[287] It is also significant that the Hebrew word for serpent may refer to a "Shinning One." Indeed, Enki was known for this brilliant quality:

---

[287] In the Myth of Zu text, Enki is referred to as "Ea Ninigiku." *NIN*=Lord, *IGI*=eye or vision, *KU* (i.e. *KU-BABBAR*,or *KUG*)=silver.

305

> Enki purifies the dwelling for you, he makes the dwelling shine for you. He consecrates the heavens for you, he makes the earth shine for you.
> —A Hymn to Nanna

Just like the character of so-called "Lucifer," Enki faced the reality of reprisals and condemnation from the council of the Elohim due to his independent nature. In the following passage, we allegedly hear from Enki in his own words. In this text, it is easy to see how Enki could be seen as someone who attempted to "resemble the Most High."

> I am a great storm rising over the great earth, I am the great lord of the Land. I am the principal among all rulers, the father of all the foreign lands. I am the big brother of the gods, I bring prosperity to perfection. I am the seal-keeper of heaven and earth. I am the wisdom and understanding of all the foreign lands.
> —Enki, Enki and the World Order

Indeed, the following Sumerian text, which was clearly written by a devotee of Enlil, bears many significant similarities to the biblical version:

> Once upon a time [. . .] There was no fear, no terror [. . .] The whole universe, the people well-cared for. To Enlil in one tongue gave speech. (But) then [. . .] Enki, the lord defiant, the prince defiant, the king defiant [. . .] Enki, the lord of abundance, whose commands are trustworthy. The lord of wisdom who scans the land. The leader of the gods. The lord of Eridu, endowed with wisdom, changed the speech in their mouths, put contention into it. Into the speech of man that had been one.

—The Nam Shub of Enki[288]

These comparisons seem to correlate not only with the biblical tower of Babel story but with the Judeo-Christian character of Lucifer; however, this cannot be true. This is because, despite what is commonly believed, this infamous individual does not actually appear in the Bible!

The designation Lucifer is derived from the Latin words *lux/lucis* (light), and *ferre* (bring, or to bear), which can be translated as "light-bearer," "morning star," "day star," or "shining one." It is a designation that is influenced by the following passage:

> How you have fallen from heaven, morning star, son of the dawn! You have been cast down to the earth, you who once laid low the nations! You said in your heart, "I will ascend to the heavens; I will raise my throne above the stars of God [El]; I will sit enthroned on the mount of assembly, on the utmost heights of Mount Zaphon. I will ascend above the tops of the clouds; I will make myself like the Most High" [Elyon].
> —Isaiah 14.12-14

However, at the beginning of this proverb (Isaiah 14.4) it is reported that this particular taunt was directed at "the king of Babylon." Indeed, this is who the "son of the dawn" originally referred to. Therefore, the word that is sometimes translated as "Lucifer" (from the original Hebrew word *he lel*) was not originally a personal name. The author was actually conveying the message that although the king may have once been as glorious as the morning star—i.e., the planet Venus, his shinning splendor had been brought to an end. It was during a later time that the character of Lucifer as a fallen light-bearer, who was the one-and-only devil, developed. This deviation did not occur until the New Testament era, centuries later (2 Corinthians 11.14), where Satan is described as an "angel of light"

---

[288] This is also found in the Enmerkar and the Lord of Aratta text.

(*aggelon photos*), who was seen falling "like lightning from heaven" (Luke 10.18). In this case, it may then be only coincidence that some of the characteristics that are commonly associated with Lucifer could also be ascribed to Enki.

In regard to the previously cited passage in the book of Isaiah, it must also be understood that in that particular situation, Yahweh was not only making a pronouncement against the king but against the god of Babylon, who was at that time the Anunnaki god Bel. While it is true that Bel or Baal—who is also sometimes referred to as Bee'alzebub—is traditionally identified as Satan, the specific rivalry that transpired during the Babylonian period actually occurred in an age long after the Sumerian serpent. Indeed, not only do the Mesopotamian records indicate that Baal was not associated with that symbol but in some texts he is even referred to as fighting against serpentine dragons (e.g., Yam/Yammu). The appellation Baal could actually refer to a number of different deities (e.g., Dagon; El; Hadad; Marduk; Melqart; etc.). The enemy of Yahweh during that particular time was actually more associated with the bull, which may relate to the biblical golden "calf" idol (Exodus 32). What should be understood about the symbol of the calf/bull is that there are numerous different deities that this animal symbol could be attributed to. Moreover, as will be explicated in parts to come, the serpent and the bull do not represent the same individual.

* * *

The legend of Enki most likely inspired the ancient Greek character Prometheus. The appellation *Prometheus* literally means "foresight." Like Enki, Prometheus was a patriarch of the sciences and is even credited with being the creator of humankind. Indeed, just like the original Sumerian Enki account, he is said to have created man from "clay." In the following narrative that was written by the play-write Aeschylus, Prometheus discloses that his newly created beings were ignorant of the truth:

> At first, senseless as beasts I gave men sense, possessed them of mind [. . .] In the beginning, seeing, they saw amiss, and hearing, heard not, but like phantoms huddled in dreams, the perplexed story of their days confounded.
> —Aeschylus, Prometheus Bound

Like Enki, Prometheus is said to have been a sympathetic character who gave so-called "fire" to human-kind. Prometheus's fire was astronomy, mathematics, medicine, writing, divination, the domestication of animals, etc. (Sagan 1977: 135). Like the Lord of the Elohim and the serpent in the Garden of Eden, when Zeus (i.e., Enlil/El) discovered that Prometheus had given "fire" to mankind he became incensed. Zeus punished the rebellious titan by banishing him into the wilderness, where he was chained to a rock.

> No one is free but Zeus.
> —Aeschylus, Prometheus Bound

Indeed, both Zeus and Enlil were atmospheric weather gods (Encyclopedia Britannica Online s.v. Enlil; Graf 1999: 934), and both were supreme gods of law and order who received prayers from their human devotees (Graf 1999: 937-938; Watson and Wyatt 1999: 273-274).

In the Sumerian texts, it is recorded that it was Enki who first bestowed civilization unto humankind:

> [. . .] the lord of broad wisdom, Enki, the master of destinies, gathered together …… and founded dwelling places; he took in his hand waters to encourage and create good seed; he laid out side by side the Tigris and the Euphrates, and caused them to bring water from the mountains; he scoured out the smaller streams, and positioned the other watercourses …… Enki made spacious sheepfolds and cattle-pens, and provided shepherds and herdsmen; he founded cities and

> settlements throughout the earth, and made the black-headed multiply.
> —The Debate Between Bird and a Fish

> He organized ploughs, yokes and teams. The great prince Enki bestowed the horned oxen that follow the …… tools, he opened up the holy furrows, and made the barley grow on the cultivated fields
> —Enki and the World Order

Enki's specialty was agriculture. In the Enki and Ninhursaga text, he is referred to as the determiner of the "destiny of plants," and refers to himself as a "gardener." In the following passage, there also seems to be a reference to the Tree of Life (the Sumerian "mes-tree") that was cultivated by Enki:

> [Enki] king who turned out the mes-tree in the Abzu, raised it up over all the lands, great Usumgal, who planted it in Eridu, its shade spreading out over heaven and earth.
> —Enki and the World Order

This special type of tree may be referred to as "the life-giving plant" in the following text:

> Father Enki, the lord of great wisdom, knows about the life-giving plant and the life-giving water. He is the one who will restore me to life.
> —Inanna's Descent to the Netherworld

The "water of life" is the ambrosia decoction that is made from the Tree of Life that was in the garden/plantation of the gods. In this case, Enki's relation to the garden, the Tree of Life, and the possibly the rebellious and cunning serpent, is tenable.

* * *

The legend of Enki is also referred to in the extra-biblical (i.e., apocryphal) book of Enoch. (The oldest fragment of this pseudepigraphal[289] text was found among the Dead Sea Scrolls.) Enoch is said to have enjoyed a special relationship with the Elohim. During one of these encounters, the Elohim allegedly disclosed privileged information to him. According to this report, there was an "angel" named Azazyel, who was despised by God. Among his alleged crimes was "generating" mankind and teaching them "every secret of their wisdom." Azazyel was also accused of teaching "men to understand writing and the use of ink and paper," and how to work with metals and the use of "coloring tinctures"—which may be a reference to alchemy.

> Thou hast seen what Azazyel has done, how he has taught every species of iniquity upon the earth, and has disclosed to the world all the secret things which are done in the heavens.
> —The book of Enoch

Here we find an enigmatic reference to "secret things," which is most likely a reference to the forbidden fruits of the Garden of Eden.

The author of the book of Enoch also reports that these rebel angels were a part of a group that came down to Earth and took the daughters of man as their wives and conceived a hybrid race. This is not only an account that corresponds with the biblical version but with the gnostic Christian reports as well:

> Then the authorities came to their Adam. And when they saw his female counterpart speaking with him, they became agitated with great agitation; and they became enamored of her. They said to one another,

---

[289] The word "pseudepigraphal" means that it is thought to be written by someone else other than to who it is ascribed to.

"Come let us sow our seed in her," and they pursued her.

—Nag Hammadi codices, The Hypostis of the Archons

One of the "authorities" who supposedly had sexual relations with Adam's female counterpart was Enki, when he gave her the fruit of the Tree of Knowledge. According to the book of Enoch, the other transgression against God that was committed by the rebellious and powerful being Azazel,[290] and the rest of the maverick "angels," was not only engaging in the partaking of the Tree of Knowledge but the Tree of Life as well:

> And all the others together with them took unto themselves wives, and each chose for himself one, and they began to go in unto them and defile themselves with them, and they taught them charms and enchantments, and cutting of roots, and made them acquainted with plants.
>
> —The book of Enoch

* * *

Traditional Judeo-Christian doctrine asserts that before "satan" brought evil into existence the world had been a perfect place. This would seem to indicate that if it was not for this one individual there would be no such thing as evil. However, this interpretation is based on numerous misunderstandings. (The most significant of these misunderstandings will be explained further ahead.) Part of the confusion is based on the concept of good and evil itself. Too often good is associated with obedience, and evil with disobedience—

---

[290] The etymology of this name is uncertain; although it is possible that it denotes a strong ($zz\,'l$) god (el) (Janowski 1999: 128). This word appears in Leviticus 16.8-10, where it is usually translated as "scapegoat." However, this definition does not accord well with how it is used in the book of Enoch. I therefore suggest that the former definition is likely the most accurate in this instance.

usually to a law or to a cultural norm, etc. Although, a more accurate definition incorporates beneficial verses malicious intent. Beneficial intent can be defined as a thought or action that is conducive to the propagation of life and general well-being. Malicious intent refers to the opposite. This moral philosophy can be envisaged in terms of light and dark. Malicious intent, or evil, simply put, is the *absence* of light. It is therefore essentially not a thing in itself, but rather the absence of a thing. This vacuous state has existed for billions of years—as opposed to thousands of years. As obvious as this may seem, due to the confusion concerning this topic, a proper definition must be delineated.

Human free will is an expression of the type of entropy that makes the universe a natural, serendipitous, and a co- and self-determining place. Indeed, it is likely that this is the same type of freedom and entropy that gave birth to our own particular universe. There is indeed a price to be paid for being out of line with the spirit of life (e.g., the Dao/Tao), but it does not come unnaturally in the form of judgment by an authoritarian master who is seeking to maintain a rigid ideological regime, but rather something more along the lines of karma.

It should also be understood that it is possible to be virtuous without being religious. Unfortunately, there are some within the religious institution who do not want people to be aware of this fact because it would be detrimental to church attendance.

Nevertheless, it must also be acknowledged that it is also true that religion does indeed provide a valuable service to society, and the findings presented in this book do not attempt to dismiss the good works that have been advanced by sincere devotees of a higher life-bestowing power.

* * *

It is essential to understand that before this modern age, information was not as accessible as it is today. Interpretations were made that were based on the primitive conditions of the time in which it was told, written, and read. Of course, not all of these accounts were ideal for three primary reasons: (1) Information became distorted as it passed from one individual or group to another. (2) Published information was not subject to the same qualifications as they are today. (3) There were some factions who were loyal to Enlil's wing of the Elohim who sought to spin the collective perspective in their favor.

The legacy that was initiated by the followers of Lord Enlil was carried on into the following generation, when Lord Yahweh arose to power.

CHAPTER X

THE THEOCRACY OF YAHWEH

Yahweh appears in two basic forms in the Old Testament/Tanakh. The first is the anthropomorphic physical form. The second is the disembodied spirit who purportedly spoke through the mouths of his prophets. The humanoid incarnation tends to appear more often in the earlier accounts—especially in the Jahwist version, and to a lesser extent, the Elohist version; while the unseen spirit of Yahweh mostly appears in subsequent periods. (This gradual transition away from the physical form deity is more extant in the Priestly and Deuteronomistic sources, which appear to have been written after the J and E material) (Friedman 2003: 12, 27; Smith 2002: 204).

There are different ways that these variations can be interpreted. The first is to assume that even if such an individual literally existed, it is evident that he was no longer physically present in around the age of the monarchy—or after the time of Elijah, circa ninth century BCE. This might explain why human kings were needed during and after that time. Indeed, in 2 Chronicles 18.18 the prophet Micaiah also claims to have seen Yahweh, which is also dated to around that same period.[291]

---

[291] In Amos 9.1, it is reported that Yahweh was seen "standing by the altar." Amos lived in the eighth century BCE. However, the authenticity of the book of Amos is especially questionable.

When the maximalist extraterrestrial deity hypothesis is considered, it can be postulated that the reason for Yahweh's disappearance was due to physical death. This may be one of the reasons why the prophets of the later age lamented his disappearance. This may also be why the prophets claimed to have channeled his spirit; which explains the appearance of such statements as, "The word of the Lord came to me" before each new decree (Jeremiah 1.4, etc.). In this case, it is unlikely that the physical form deity that supposedly appeared to Isaiah, Ezekiel, Amos, etc., literally occurred. However, it is possible that Ezekiel's report of a fiery metallic object in the sky was based on the sighting of an extraterrestrial object, which the priest-scribes then reworked and expanded into a grandiose Yahwehist narrative at a later time. Indeed, this explains the use of additional Babylonian iconography. It is evident that the scribes, priests, and prophets took it upon themselves to inspire faith and thereby supposedly avert disaster for Israel and Judah, as well as support a need for their own careers, by promoting the belief that Yahweh continued to exist; otherwise they would have never have been taken as seriously as they eventually were. Indeed, as the scholarly and archaeological evidence continues to develop, it is becoming increasingly evident that many of the accounts that are documented in the Bible—such as the Exodus and the conquest of Canaan by Joshua, etc.—either never literally occurred, or at least did not occur in the same time or in the same way that they are reported in the Bible (Finkelstein and Silberman 2001: 76-79, 81-83, 118).

It must be understood that in the ancient era the role of the priest and the scribe were often coincident (e.g., the priest-scribe Ezra). It would have been the priest-scribe's intention to help bolster faith by convincing the people that Yahweh was supernatural, supreme, and fatally dangerous, and therefore worthy to be feared and venerated above all others. One of the ways in which the priest-scribe could achieve this aim was to impress upon the reader the supernatural element. Of course, the most effective way to do this was to embellish the account with hyperbole. An example of this is when Moses's staff turned into a serpent, or when the Red Sea parted, etc. Whatever information that was not available to the priest-scribe, he could simply fill in the blanks with what either sounded appropriate, or with what he thought would best win the hearts and the minds of the people. The result is what critical scholars refer to as "pious fraud." Indeed, even the prophet Jeremiah himself ironically acknowledged the problem of the "lying pen of scribes" (Jeremiah 8.8). This more minimalist type of exegesis posits that what happened is that over time these reports increased in both intentional exaggerations and unintentional misunderstanding. For example, it becomes obvious, as one reads through the books of the Old Testament/Tanakh in chronological order, that the scribes were building upon each other's work, until over the centuries the truth became increasingly altered and obscured behind the facade of cultural motifs and devotional embellishment. It is also apparent that the priest-scribes were taking liberties with the facts even in the very beginning; namely, in the Five Books of Moses itself. What they did was not only draw upon regional folklore but what may have been only one or a few sightings or contact events. They then used these encounters to construct an expanded narrative that served to bolster their enterprise in Israel and Judah.

One of these events seems to have happened in or around the region of Midian and/or Edom (in the north-west Arabian Peninsula) (Day 2002: 15; Friedman 1987: 82, 92; Niehr 1999: 370; van der Toorn 1996: 283, 1999: 911). According to the Egyptian hieroglyphic records of Soleb and Amarah-West (from the fourteenth and thirteenth centuries BCE), the Shasu nomads from Midian and Edom were associated with the god Yahu/yhw. This is the first extant non-biblical reference to Yahweh. It is therefore likely not a coincidence that Midian and the surrounding region, such as Seir, Teman, Peran, and Sinai, are also mentioned in the Old Testament as the area where Yahweh first revealed himself (Exodus 4.19; Deuteronomy 33.2; Habukkuk 3.3; Psalm 68.8). It is possible that Yahwehism—as opposed to El worship—was introduced into the Israelite consciousness in that region and that time when a man from the Israelite tribe of Levi, who is referred to in the Old Testament/Tanakh as Moses, experienced a close encounter with an otherworldly being in the desert.

The group that Moses was associated with were a nomadic (or semi-nomadic) Shasu clan who are referred to in the scriptures as the Kenites (Judges 1.16).[292] These people were most likely the first Yahwehists.

Before this time it was the West Semitic El who was the original "God of Abraham." Indeed, the scriptures themselves report that the original patriarch did not know the deity by the name Yahweh:

> [. . .] I am the Lord [Yahweh]. I appeared to Abraham, Isaac, and Jacob by the name God Almighty [El Shadday],[293] but they did not know me my name, the Lord [Yahweh].
> —Yahweh, Exodus 6.2-3

The deity did not begin to be addressed as Yahweh until the time of Moses:

> God [Elohim] said to Moses, "I AM WHO I AM. This is what you are to say to the Israelites: 'I AM [*Ehyeh*] has sent me to you.'" God also said to Moses, "Say to the Israelites, 'The Lord [Yahweh], the God of your fathers—the God of Abraham, the God of Isaac and the God of Jacob—has sent me to you.' This is my name forever, the name that you shall call me from generation to generation."
> —Yahweh, Exodus 3.14-15

The claim that Yahweh was El was actually devised by the Yahwehist priest-scribes centuries later. Indeed, the conclusion that El and Yahweh were not originally the same, but rather were integrated at a

---

[292] This subject pertains to what scholars refer to as the "Midian-Kenite Hypothesis," or the "Kenite Hypothesis."

[293] The word *El Shadday* is commonly translated as "God Almighty"; however, it can also be translated as "He (or El) of the Mountain" (Coogan 1978: 19; Cross 1997: 55-56; Knauf 1999: 750), or "God of the Wilderness" (Knauf 1999: 749).

later time is also supported by mainstream scholarship (Coogan 1978: 20; Day 2002: 17; Herrmann 1999: 277-279; Smith 2002: 32-43). It is likely that by associating Yahweh with El, the Levitical priests were better able to integrate into the religio-culture of Canaan without being perceived as heretics.

What happened over the years, is that the theonym El eventually became a generic term for "God." However, this is not the original definition (Smith 2002: 33-34). Originally, El was the name of a West Semitic god—i.e., the God of Abraham.

Apologists will argue that God was referred to as Yahweh since the time of Adam and Eve (Genesis 4.1). However, scholars now know that the abbreviated appellation YHWH was inserted into the text by a Jahwist (i.e., J source)[294] scribe at a later time.[295] Likewise, passages in the Abrahamic accounts where the title-name also appears (e.g., Genesis 18.13-14, etc.) were also written by these same Levitical priest-scribes centuries later. This explains why these interpolations contradict the previously cited Exodus 6.2-3 pericope, in which it is reported that the deity was not referred to as Yahweh during the time of Abraham. Indeed, both the Elohist (i.e., E Source) and the Priestly (i.e., P Source) authors admit that God was not referred to as "YHWH" until the time of Moses, which contradicts the Jahwist (J Source) account (Friedman 2003: 4-5, 10, 40).

Therefore, Yahweh was not originally an Israelite/Judean god, but rather a Kenite god. In a minimalist context, it was most likely Moses's father-in-law, Jethro, who himself was a Kenite, who convinced Moses that the so-called "fire" and the mysterious being that controlled that anomalous illumination must have been a manifestation of their almighty war god. Perhaps after feeling empowered by that sighting, Moses returned to Egypt and helped set

---

[294] According to the prevailing Documentary Hypothesis, one of the four primary authors of the first five books of the Bible (i.e., the Torah) was a scribe from the southern kingdom of Judah, circa eighth century BCE, who referred to God as YHWH.

[295] See Richard E. Friedman's book: *The Bible with Sources Revealed* (Harper SanFrancisco, 2003) for more information.

free a group of people who were living there at that time as servants. It must first be understood that the account that appears in the Hebrew Bible was not documented until hundreds of years after the alleged events occurred. This is another reason why we cannot place much faith in the accuracy of the biblical account. In this case, it is necessary to deconstruct this narrative down to its primary components. When we do this, it is possible to see how it may have been nothing more than a naturally occurring disease that was interpreted as a punishment that was inflicted by the god of Moses that convinced the pharaoh to release the Israelites. Furthermore, as was previously mentioned and worth mentioning again: *if* there was an exodus, it would have likely have only involved a smaller group than what is portrayed in the Bible. In this case, it may have only have mostly been *one* of the tribes of Israel. The most likely candidate are the Levites. This not only explains why the Levites were described as having been especially involved in the Exodus story but it also explains why some of the Levites had Egyptian names. The presence of this smaller group also explains why no archaeological evidence has been found that can validate the Exodus story.

By claiming that their orders were not from themselves, but rather from a vengeful supreme being and his human Levitical liaison, the Levitical priest-scribes (i.e., *cohen/kohanim*) were taken more seriously than they otherwise would have been. Indeed, it was not only the welfare of their relatives in Israel and Judah that was of concern to the Levitical priest-scribes but the priest-scribes' position within that society. This is why they endowed themselves with prestigious positions in the Five Books of Moses. In a minimalist context, what those enterprising individuals did was simply write these positions into the record by themselves and for themselves. We know the identity of at least one of these priest-scribes: his name was Ezra. Indeed, Ezra has been identified by scholars as the primary redactor/scribe behind what we know today as the Pentateuch/Torah (Friedman 1997: 159, 218, 223-225; van der Toorn 2007: 79, 250-251). What is also significant about Ezra is that he was a Levite (Ezra 7.1-6). After returning from the Babylonian exile, Ezra used what was written in the "Books of Moses" to legitimize himself and his Levitical associates. Indeed, the post-exilic period was a time in which those priest-scribes arose to the height of their power. Using these documents that they themselves had mostly written and compiled to justify their decrees, the other Israelite tribes were ordered to do such things as submit tithes to the Levites! (Numbers 18.21-24) Another scheme that is documented in the book of Numbers involved charging a fine to all the Israelites who outnumbered the Levites (Numbers 6.46-51). The priest-scribes could then cite such examples to demand payment from the other tribes. Indeed, the text admits that the money that was collected did not go to Yahweh, but rather to Moses's brother and his sons! (Numbers 3.48-51) The other Israelites were also ordered to hand over animals that were to be slaughtered as offerings to Yahweh in the Tabernacle/Temple. However, the scriptures admit that at least some of the animals were being used as food for the Levitical priests themselves (Deuteronomy 18.1-8). This type of behavior is also attested in the gospels of the New Testament, where other high-level Yahwehists (i.e., the Pharisees) were condemned by Jesus for similar types of self-serving behavior. We therefore do find evidence of a practical human motive, which endows further credibility to the minimalist interpretation.

Apologists will argue that the Levites were also denied rights. For example, it is commonly believed that the Levites could not inherit or own land, which is based on the fiats that are documented in Deuteronomy 10.9, 18.1-2. However, according to Leviticus 25.32-34 and Numbers 35.1-8, not only could the Levites own property (in the form of towns) but they had property rights that others did not have. In fact, some Levitical properties were even sanctuaries for murderers (Numbers 35.11). This may have been a self-serving measure on their part should anyone one of them ever find themselves in such a predicament—as Moses himself did. It is therefore disingenuous to claim that the Levites were not "inheriting" anything when they were not only receiving tithes and food but special property rights as well.

Apologists will also point to other passages that do not refer to the Levites in a flattering light. However, it must be understood that these books were not written by a single author. The contradictory information derives from competing factions within the Levitical tribe itself (Cross 1997: 206, 209, 215); namely, between the priests who traced their lineage to Moses (e.g., Shilonites) and those who traced their lineage to Aaron—as well as the various subgroups among them (e.g., Zadokites). Not only were the priests competing among themselves for authority but the northern kingdom of Israel and the southern kingdom of Judah were as well (Cross 1997: 208; Friedman 2003: 18-21). It is now known that the passages that deny certain privileges to the Levites were written by what scholars refer to as the "Jahwist" and the "Deuteronomistic" sources; while other passages that grant privileges to the Levites were written by the "Priestly source." These Aaronid priests saw themselves as not only having power over the common populace but even the other Levites as well. According to this elitist faction, the only way to Yahweh was through them (Friedman 2003: 12, 21-22). Indeed, the presence of these different sources explains why the scriptures are filled with so many contradictions.

It must also be understood that until these Yahwehists entered Canaan, the other tribes of Israel did not worship Yahweh. They were more inclined to worship the traditional gods, such as El, Baal, and Asherah (Smith 2002: 7, 57). However, through a vehement, deceptive, and sustained propaganda campaign that is attested in the scriptures of the Old Testament/Tanakh itself, the Levitical priest-scribes were able to eventually affect the hearts and minds of the people. Therefore, the conquest of Canaan was not a physical war, but rather a mental one.

* * *

Concerning the depiction of Yahweh as an invisible spirit, the minimalist interpretation posits that these accounts were most likely influenced by the grand legacy of Yahweh; the mental effect of which his zealous agents eagerly mistook, or perhaps in some cases even consciously misrepresented, as the voice of the legendary deity himself. Indeed, instances of this type of occurrence can be found in many present-era examples as well.

Nevertheless, despite the influence of pious fraud, unlike the secular scholarly consensus, I maintain that the version of events that were reported by the priest-scribes of Yahweh were originally based on genuine source experiences that were grossly affected at a later time. Indeed, the findings repeatedly indicate that the accounts that are documented in the Bible are a complex mixture of both fact and fiction. In order to better ascertain the validity of each account, it is necessary to not only look beyond the dramatized facade and perceive the original picture in a reasonable manner but consider the entire picture as a whole, and not get thrown off course by erroneous embellishments and deceptive interpolations.

In order to consider this entire picture as a whole, the following chapter is first presented within the framework of traditional Judeo-Christian exegesis—which maintains the maximalist position that the physical and the spiritual Yahweh were both real and both coincident. This is because not only is it likely that these accounts were built upon the foundation of a factual premise but in order to gain a more accurate perspective every angle of this monumental picture must be considered, including the maximalist perspective. After this information is reviewed, we will then attempt to separate the wheat from the chaff.

When evaluating the validity of the maximalist interpretation, evidence for psi phenomena will be considered. Therefore, accounts of communion with incorporeal sentient entities will not be disregarded. Indeed, these are the same types of extra-sensory communications that are being reported by both modern-day mediums and NHI contactees as well. Indeed, we must be careful not to make the same mistake that is so often committed by the skeptical physicalists, who tend to automatically disregard information that is deemed to be too fantastic.

Nevertheless, even if the validity of the maximalist interpretation were to one day be entirely discredited, this does not negate the fact that millions of Yahwehists around the world do subscribe to this interpretation. Nor does it negate the contribution that this information makes to our understandings of mythology and religion. Therefore, it can be contended that the following maximalist exegesis remains relevant for these reasons.

* * *

Historic changes were taking place in Mesopotamia in around the second millennium BCE, as the old city-states of Sumer and Akkad declined and Assyria and Babylonia arose up in their place. It was during that era—or perhaps not long after—that a mysterious deity allegedly made his presence known in that region. A being who made the intrepid claim that he was the supreme God of all the heavens and Earth. A being who referred to himself by the enigmatic maxim, "I am who I am" (*Eyeh asher eyeh*); which later came to be translated as Yahweh.[296] Unlike the other gods of the age, no specific visual images of this deity were depicted in the official artworks of his followers.[297] Not only did Yahweh mysteriously appear in the physical world of man but also disappeared in an equally enigmatic manner, never to be known in such direct and physical terms ever again.

Who then exactly was this mysterious individual who became known to the world as "God"?

It is generally believed that the era of the theocracy of Yahweh began in earnest when a man from the old Sumero-Akkadian city of Ur was contacted by an anonymous deity who engaged him in a

---

[296] The general consensus among scholars is that the theonym "Yahweh" most likely derives from the abbreviated four root consonant letters YHWH. In Hebrew, these letters denote: being, becoming, creation, and existence. This accords with the statement "I am who I am" (Exodus 3.14).

[297] The drachm coin of Yehud (fourth century BCE), which depicts a bearded man on a winged and wheeled throne (which is similar to the Ezekiel description) with the Aramaic letters YHW, or YHD, next to it, is a possible exception (British Museum. Registration number: TC,p242.5.Pop). However, even though the coin was minted in Judah, it was issued by a Persian administration and shows signs of Greek influence.

covenant that would give birth to a new society in the Near East. This man is referred to in the Bible as Abram/Abraham, and the nameless individual who contacted him eventually came to be known as "Yahweh"; although, this was not the name that Abraham knew him as. Abraham knew him as El Shadday (Exodus 6.2-3), or simply El.

According to custom, the name of God (the *Tetragrammaton*) is considered to be too sacred to speak aloud; therefore, he is instead usually addressed by other designations, such as Adonai (Lord) or Ha-El (the true God).

Some of the descriptions of Yahweh in the earliest reports tell us that Yahweh was not some invisible ubiquitous entity, but rather he was described as an actual living being with a humanoid appearance. For example, we are told that he had a "face" (Exodus 33.11; 33.20), "hands" (Exodus 33.22), "feet" (Exodus 24.10), and that human-beings were created in "his image" (Genesis 1.27). Indeed, in Genesis 18.1 it is reported that "the Lord appeared to Abraham," along with two of his attendants. All three of the visitors were described as "men." Abraham offered to wash their feet and feed them, which they accepted. In this account, Yahweh is not only described as having the appearance of a man but is reported to have eaten food as well. Therefore, it can be concluded that when Yahweh was commanding the people to bring him the "first fruits of the land" and "burnt sacrifice of the herd," this was not only a symbolic gesture but also the requirements of a physical being who required sustenance. Indeed, the sacrificed animal offering that is made to the Lord is referred to as "food"; the smell of which was pleasing to Yahweh (Leviticus 1.9). A similar account is reported in the Cain and Abel incident (Genesis 4.3-5). If these descriptions are true, it would seem that, at least at one point, Yahweh existed on Earth in physical form.

Apologists counter that these anthropomorphic descriptions were not Yahweh himself, but rather angels who were acting under the spirit of Yahweh. This interpretation asserts that the humanoid angels were essentially possessed by the deity, in much the same way that many believe that Jesus was of the same essence (i.e., *hypostasis*) as God. However, this interpretation does not always hold up under critical scrutiny.

The [. . .] texts, however, presents a different picture with their textual variants and vacillating identifications of the "angel of Yahweh" (distinct from Yahweh? identical to Yahweh?). Among proposals offered to explain the evidence, one finds the angel of Yahweh in these passages interpreted as Yahweh in a theophany, the preincarnate Christ, a means of crystallizing into one figure the many revelatory forms of an early polytheism, a hypostatization, a supernatural envoy of Yahweh where the confusion in identity results from messenger activity that merges the personality or speech of the messenger with the sender, or an interpolation of the word mal'ak [messenger] into the text where originally it was simply Yahweh speaking and at work.

The notion that the identity of messenger and sender could be merged in the ancient Near East is incorrect: any messenger who failed to identify the one who sent him subverted the entire communication process. On the other hand, those who posit an identity (whether by theophany or hypostatization) between Yahweh and the mal'ak YHWH apart from this theory do not do justice to the full significance of the term mal'ak which must mean a subordinate.
—Samuel A. Meier,[298] "Angel of Yahweh" *Dictionary of Deities and Demons in the Bible*

The deity that appeared in humanoid form not only conflicts with other descriptions of Yahweh as an invisible spirit, but this depiction indicates that Yahweh appeared in the same form as the other Elohim gods of the same era—which indicates equivalence. It can therefore

---

[298] Samuel A. Meier, Ph.D., is a professor in the Department of Near Eastern Languages and Cultures and is the adjunct professor of history at Ohio State University.

be deduced that the possession interpretation was devised by the apologists in order to rectify this contradiction to traditional creed. However, it is an interpretation that cannot be justified.

It can also be concluded that Yahweh himself was not a "thick gloom," or a "fiery pillar," but rather these were actually objects that he appeared from within or behind. In the "burning bush" incident, for example, it appears that Yahweh was not the burning bush itself that appeared to Moses, but rather was most likely standing behind it, obscured by both the fire, smoke, and perhaps even the night-time darkness. However, it can also not be ruled out that what appeared to be a burning bush was actually the fiery object that Yahweh used to travel through the sky. This would explain why the bush is described as appearing as if it was on fire, and yet did not burn up. In this case, the fiery object could have been stationed behind the vegetation that appeared to be burning. Indeed, according to the Old Testament, Yahweh would rarely appear to the people in public; and when he did it was usually from behind some type of fiery or smoky obstruction (e.g., Deuteronomy 4.15). It is therefore evident that the reason why the features of Yahweh were never fully described is not because he himself was an amorphous mass of smoke, fire, or wind, but rather for some other reason.

In this case, the question must be asked: Could the obscuration of his identity have anything to do with the secret hidden matters of God?

The biblical record tells us that Yahweh was known to purposely hide his face from the children of Israel:

> Lord [Yahweh] [. . .] when you hid your face, I was dismayed.
> —Psalm 30.7

> Truly you are a God [El] who has been hiding himself [. . .]
> —Isaiah 45.15

> Neither will I hide my face any more from them; for I have poured out my Spirit on the house of Israel [. . .]
> —Yahweh, Ezekiel 39.29

Although some of these instances can be interpreted to mean that Yahweh had turned his back upon the disobedient people, when these reports are put into context with other biblical information, another reason for his enigmatic disappearances begins to emerge. Consider the following passage, for example, in which Yahweh is specifically described hiding his face, as opposed to turning his back upon a disobedient people. In the account that is recorded in the book of Exodus, Moses asked Yahweh to let him see his face. Yahweh refused; although he did allow Moses to see his back as he walked away:

> Then the Lord [Yahweh] said, "There is a place near me where you may stand on a rock. When my glory passes by, I will put you in a cleft in the rock and cover you with my hand until I have passed by. Then I will remove my hand and you will see my back; but my face must not be seen."
> —Yahweh, Exodus 33.21-23

Therefore, it is possible that when Moses met with Yahweh "face to face" (Deuteronomy 34.10), this was either the scribe's way of saying in person, or Yahweh eventually changed his mind, or this is a contradiction.

Why then did Yahweh make such an effort to obscure himself? In order to answer this question, it is necessary to understand his motivation.

* * *

The biblical record indicates that it was Yahweh's desire to establish an empire in the land of Canaan. In order to do this, he employed various governing techniques; some of which involved the use of clandestine operations. In these accounts, Yahweh is reported using different opposing kingdoms that were each under his control in order to inflict punishment on one another. Only by swearing allegiance to him could the people expect to be spared the onslaught of one of his armies.

> Yahweh will bring against you a nation from far away, from the ends of the earth. The nation will swoop down on you like an eagle [. . .]
> —Deuteronomy 28.49

> I will use them [the Philistines, Canaanites, etc.] to test Israel and see whether they will keep the way of the Lord [Yahweh] and walk in it as their ancestors did.
> —Yahweh, Judges 2.22

When the troublesome King Manasseh kept "seducing" Judah and the inhabitants of Jerusalem by causing them to stray from servitude to the Lord (2 Chronicles 33), Yahweh purportedly retaliated by ordering the Assyrians to attack and capture the offender and exile him to Babylon—which was yet another city-state that was under his control at that time. Manasseh was released from incarceration only after he reaffirmed his allegiance to Yahweh.

An explanation for this type of activity comes to us from the sixteenth-century Italian political philosopher, Niccolo Machiavelli. In his books *The Prince* and *The Discourses on Livy*, Machiavelli proposed ways for a ruler to acquire and secure power for himself. He recommended the employment of strategic operations that would serve to control a populace. Examples of what Machiavelli described are not only found in the Old Testament but they are found repeatedly. In the book of Deuteronomy, for example, it is reported that Yahweh threatened those who did not obey him with "calamities" and "distress":

> Then my anger shall be kindled against them in that day, and I will forsake them, and I will hide my face

from them, and they shall be devoured, and many evils and troubles shall come on them; so that they will say on that day, 'Haven't these evils come on us because our God [Elohim] is not among us?'
—Yahweh, Deuteronomy 31.17

What the people did not always know is that many of the calamities that they were experiencing were brought upon them by Yahweh himself. It was an operation that was designed to convince the people that servitude to himself was in their own best interest. When his people abandoned him by not submitting themselves as workers, warriors, and worshipers, he would in turn abandon them to their enemies—enemy forces that were allegedly under his control as well. Of course, this was also a technique that could be reversed if needed, in order to keep the other in line as well.

Another territory that was of interest to Yahweh was Egypt. In the following passage, Yahweh discloses that he will conquer the Egyptians by dividing it against itself—which compares to the Machiavellian strategy of subjugation through division:

> And I will stir up the Egyptians against Egyptians, and they will fight everyone against his brother, and everyone against his neighbor; city against city, and kingdom against kingdom.
> —Yahweh, Isaiah 19.2

> Yahweh has mixed a spirit of perverseness in the middle of her; and they have caused Egypt to go astray in all of its works, like a drunken man staggers in his vomit.
> —Isaiah 19.14

This state of bewilderment was how Yahweh was not only able to dominate his enemies and subdue the disobedient among his own army but produce a state of unrest that only he himself could provide relief from.

> Yahweh will send you curses, panic, and frustration in everything you do until you're destroyed and quickly disappear for the evil you will do by abandoning Yahweh.
> —Deuteronomy 28.20

Besides using opposing forces to keep one another in check, he also purposely endeavored to produce a state of confusion and fear, which he was also able to use to his advantage:

> Better to have a little with the fear of Yahweh than great treasure and turmoil.
> —Proverbs 15.16

Indeed, the instigation of fear was recommended by Machiavelli:

> Men are moved by two principle things—by love and by fear. Consequently, they are commanded as well by some-one who wins their affection as by someone who arouses their fear. Indeed in most instances the one that arouses their fear gains more of a following and is more readily obeyed than the one who wins their affection.
> —Machiavelli, *Discourses on Livy*

> [. . .] The fear of Yahweh is your treasure.
> —Isaiah 33.6

This is why the followers of Yahweh refer to themselves as the "God fearing":

> Serve Yahweh with fear, and rejoice with trembling.
> —Psalm 2.11

Of course, fear is the emotion that relates with not only confusion but with violence:

> It will happen in that day, that a great panic from Yahweh will be among them; and they will lay hold everyone on the hand of his neighbor, and his hand will rise up against the hand of his neighbor.
> —Zechariah 14.13

Indeed, the threat of violence was a strong motivational force that Yahweh was also able to use to his advantage:

> I sent plagues among you like I did to Egypt. I have slain your young men with the sword, and have carried away your horses; and I filled your nostrils with the stench of your camp, yet you haven't returned to me.
> —Yahweh, Amos 4.10

If the people felt that they did not have a reason to need a benefactor, Yahweh would provide one for them by sending an opposing army that was also under his control to frighten them into submission. Once they had been sufficiently intimidated, he would then arrive to present himself as their guardian benefactor and vanquish the enemy threat. The catch was that they were then indebted to him as their liberator.

In order to preserve the clandestine nature of those operations, a certain amount of secrecy was necessary in order to keep those who were involved from knowing that they were being "deceived":

> I said, "Lord Yahweh, you certainly have deceived these people and Jerusalem" [. . .]
> —Jeremiah 4.10

> O Yahweh, you deceived me, and I was deceived [. . .]
> —Jeremiah 20.7

Some proponents of the traditional interpretation of these passages, and the translators who are influenced by them, will sometimes euphemize the word for "deceive" (New International Version translation; English Standard translation), or "fooled" (New World translation), or "misled" (New Living translation), by instead translating the word as "persuaded" (King James translation), or even "enticed" (New Revised Standard translation). However, these softer definitions do not correspond with the original context of the Machiavellian operations that were occurring at that time.

In the following passage, Yahweh is recorded soliciting the help of a deceptive spirit who will help deceive his enemies:

> Yahweh asked, "Who will deceive Ahab so that he will attack and be killed at Ramoth in Gilead?" Some answered one way, while others said something else. Then Ruach [a spirit] stepped forward, stood in front of Yahweh, and said, "I will deceive him." "How?" Yahweh asked. Ruach answered, "I will go out and be a spirit that tells lies through the mouths of all of Ahab's prophets." Yahweh said, "You will succeed in deceiving him. Go and do it." So, Yahweh has put into the mouths of all these prophets of yours a spirit that makes them tell lies. Yahweh has spoken evil about you.
> —1 Kings 22.20-23

It is therefore evident that Yahweh's plan did not involve diplomatic "enticing," but rather the deliberate implementation of operations that were intended to both "deceive" and to "confuse."

> Be ever hearing, but never understanding; be ever seeing, but never perceiving. Make the heart of this people calloused; make their ears dull and close their eyes. Otherwise they might see with their eyes, hear with their ears, understand with their hearts. And turn and be healed.

—Yahweh, Isaiah 6.9-10

> Yahweh has poured out on you a spirit of deep sleep.
> He will shut your eyes (your eyes are the prophets) [. . .]
> —Isaiah 29.10

It was essential that the people who he sought to control were not entirely aware of his full intentions and capabilities. By concealing himself, and by consorting with the other nations that were also under his control, he was better able to control those who did not know who or what they were dealing with.

> Why are you so distant, Yahweh? Why do you hide yourself in times of trouble?
> —Psalms 10.1

Indeed, according to Numbers 22 Yahweh did personally engage with a high-ranking member of another nation behind the scenes when he met with a non-Israelite man named Balaam, who agreed to help Yahweh by refusing to help the Moabites.

* * *

Babylon was sacked by the Assyrian king Sennacherib in 689 BCE. This is the same king who ordered his men to march against Jerusalem. In the biblical account, it is reported that Sennacherib had been ordered to attack Jerusalem by Yahweh himself. This was supposedly done because the Judeans had formed an unauthorized alliance with Egypt (2 Kings 18.13-25). A similar fate had befallen the northern kingdom of Israel years earlier, after they had supposedly not followed the "commandments of Moses" (2 Kings 18.9-12).

Despite the usefulness that the Assyrian super-power provided him, they too eventually fell out of favor with Yahweh. After King Hezekiah renewed his allegiance, Yahweh put an end to the Assyrian threat by allegedly ordering an angel to destroy their camp in the middle of the night.[299]

According to Isaiah 45.1, Cyrus, the king of Achaemenid Persia, was also under the guidance of Yahweh. Apparently, Yahweh even referred to Cyrus as his "anointed" one (i.e., Messiah). Here we find another example of Yahweh using foreign kings that were also under his control. In Ezra 1.1 (and 2 Chronicles 36.22), it is reported that Cyrus, "in order to fulfill the word of the Lord spoken by Jeremiah," proclaimed that the Lord God had told him to build a temple in Jerusalem. The book of Jeremiah makes it very clear that it was Yahweh's intention for the kingdom of Judah to fall to the Babylonians. After the Judeans had been sufficiently terrorized into submission and prayed for help, Yahweh then commenced with the second phase of his plan; namely, to destroy the Babylonians who had attacked the holy city of Jerusalem (Jeremiah 51.24); after which he was then able to present himself as the almighty savior of the people.

> Babylon was a golden cup in Yahweh's hand.
> —Jeremiah 51.7

When the people began to suspect that they were being manipulated, Yahweh admonished them:

> Do not call conspiracy everything this people calls a conspiracy [. . .]
> —Yahweh, Isaiah 8.12

What is conspicuous about this passage is that Yahweh denies that there is a conspiracy, while in the very same breath (Isaiah 8.14) he admits that he has set a "trap and a snare" for the people of Jerusalem! This section of the book of Isaiah ends with the author stating that he will wait for "the Lord who is hiding his face from the house of Jacob." Therefore, it is evident that Yahweh was not simply

---

[299] According to the Assyrian prism annals, this event did not occur.

turning his back upon the disobedient people, but rather was also turning away in order to carry out his Machiavellian operations.

> Yahweh has accomplished what he had planned to do. He carried out the threat he announced long ago. He tore you down without any pity, Jerusalem. He made your enemies gloat over you. He raised the weapons of your opponents.
> —Lamentations 2.17

Likewise, in the following passage Yahweh is described as the one who makes "plans" to rule over the people:

> Yahweh foils the plans of the nations; he thwarts the purpose of the peoples. But the plans of Yahweh stand firm forever [. . .]
> —Psalm 33:10-11

It is evident that the story of Yahweh was a source of inspiration for Machiavelli, who did indeed refer to the Bible:

> Therefore, it was needful that Moses find the people of Israel enslaved and oppressed by the Egyptians in order that they would be ready to follow him out of Egypt to escape from servitude.
> —Niccolo Machiavelli, *The Prince*

* * *

In order to bring about his theocracy, Yahweh required workers, warriors, and worshipers. One of the problems that he had to contend with was independently-minded people who were not interested in submitting themselves as his servants:

> He [Yahweh] does not regard any who are wise of heart.

—Job 37.24

> A loner is out to get what he wants for himself. He opposes all sound reasoning.
> —Proverbs 18.1

Yahweh and his retinue of priests, prophets, and scribes, constructed a doctrine in which the "wise" are those who are obedient, while the "stupid" are those who are "self-confident"—which was also equated with evil and sin (Proverbs 14.16).

In the book of Ezekiel, we are told that the king of Tyre had aroused the wrath of Yahweh by independently acquiring "gold" and "wisdom." His punishment was destruction by the army of the Lord:

> Because you think you are wise, as wise as a god, I am going to bring foreigners against you, the most ruthless of nations; they will draw their swords against your beauty and wisdom and pierce your shining splendor.
> —Yahweh, Ezekiel 28.6-7

The self-described "jealous God" (Exodus 20.5) condemned those who were driven by their own personal ambitions.

One of the ways that he able to inflict punishment and assert his will was by the use of Machiavellian operations. This is one of the secrets of Yahweh; although, it was not the only one. The scriptures indicate that there was more going on than only Machiavellian operations.

* * *

In the King James Bible, the epithet of Yahweh is translated as "Lord of hosts." The word host is commonly interpreted as an assembly of angels; however, as was previously explicated, this is not the original definition. This term literally means armies. This is a definition that accurately relates to the activities and characteristics of Yahweh. Indeed, this rendering is used in several translations of the Bible (e.g.,

New Translations Bible; God's Word Translation; World English Bible):

> Yahweh of Armies has planned it, to stain the pride of all glory, to bring into contempt all the honorable of the earth.
> —Isaiah 23.9

Indeed, the biblical record tells us that Yahweh was an almighty lord of war:

> Cursed are those who neglect doing Yahweh's work. Cursed are those who keep their swords from killing.
> —Yahweh, Jeremiah 48.10

> The dead bodies of men shall fall as dung on the open field [...]
> —Yahweh, Jeremiah 9.22

> Yahweh is a man of war. Yahweh is his name.
> —Exodus 15.3

In a few biblical translations, the name and title Yahweh of Armies is translated as "Lord Almighty" (e.g., Isaiah 13.4-5, 13.13) (The Living Bible; New International Version), despite the fact that there is no justifiable reason for these words to appear. The original Hebrew words *Yahweh Sabaoth* does not mean "Lord Almighty." This is a misleading misnomer that has been devised by well-meaning translators who have euphemized the literal definition in order to make it more palatable to the laity.

The truth is that Old Testament/Tanakh is filled with examples of Yahweh inflicting both his enemies as well as his own people with the weapons of his "arsenal" (Jeremiah 50.25). Whenever he felt that the people were not properly serving him, he would in turn inflict them with what we would refer to today as "weapons of mass destruction." His arsenal included both fiery projectiles as well as biological affliction in the form of disease. In some cases, he would even instruct those closest to him to conduct mass executions. Indeed, the military campaigns and carnage of the Lord of Armies is found all throughout the Old Testament:

> By the wrath of Yahweh of Armies the land will be scorched and the people will be fuel for the fire [. . .]
> —Isaiah 9.19
>
> Those destined for death, to death; those for the sword, to the sword; those for starvation, to starvation; those for captivity, to captivity.
> —Yahweh, Jeremiah 15.2
>
> I will execute judgment on him with plague and bloodshed [. . .]
> —Yahweh, Ezekiel 38.22
>
> Everyone who is found will be thrust through. Everyone who is captured will fall by the sword. Their infants also will be dashed to pieces before their eyes. Their houses will be ransacked, and their wives raped.
> —Yahweh, Isaiah 13.15-16
>
> Slaughter the old men, the young men and women, the mothers and children [. . .]
> —Yahweh, Ezekiel 9.6

Those who managed to survive the onslaught were taken into captivity as slaves, like a herd of cattle:

> [. . .] I will put my hook in your nose and my bit in your mouth, and I will make you return by the way you came.
> —Yahweh, 2 Kings 19.28

Yahweh demanded that ceremonial offerings be brought to him, and if those offerings were not to his liking, or not carried out in a precise way—such as in the case of the two sons of Aaron (Leviticus 10)—he would show his displeasure by burning the offenders to death. It seems that no one suffered under the leadership of Yahweh more than his own people:

> Yahweh will cause you to be defeated before your enemies [. . .] Your carcasses will be food for all the birds of the air and wild animals [. . .] Yahweh will afflict you with the boils of Egypt and with tumors, festering sores and the itch, from which you cannot be cured. Yahweh will afflict you with madness, blindness and confusion of mind.
> —Deuteronomy 28.25-28

Part of the Machiavellian plan was to continually remind the people that they were indebted to him for liberating them from enslavement in Egypt. When the people began to realize that they were out of the frying pan and into the fire, so to speak, they began to wish that they had never left Egypt. Indeed, a little-known part of this popular story is that at a later time some of the people actually returned to Egypt! Therefore, it can be concluded that the reason why the Judean/Israelites were continually turning toward other gods was because they may have felt that they were in need of protection from Yahweh himself! When Yahweh found out that some of the people had returned to Egypt, he declared that just because they had fled from him did not mean that they would be safe from his wrath:

> Woe to the obstinate children [. . .] who go down to Egypt without consulting me; who look for help to Pharaoh's protection, to Egypt's shade for refuge.
> —Yahweh, Isaiah 30.1-2

> 'I swear by my great name,' says Yahweh, 'that no one from Judah living anywhere in Egypt will ever again

invoke my name or swear, "As surely as Lord Yahweh lives." For I am watching over them for harm, not for good; the Jews in Egypt will perish by sword and famine until they are all destroyed.
—Yahweh, Jeremiah 44.26-27

Although it is known that Yahweh was a "jealous" "fire and brimstone" type of character, what is less commonly understood is the full extent of his actions. For example, the biblical record tells us that Yahweh sentenced those who did not submit to him to terrible, humiliating, and agonizing punishments and deaths:

> I am going to punish your descendants. I am going to spread excrement on your faces, the excrement from your festival sacrifices. You will be discarded with it.
> —Yahweh, Malachi 2.3

The enemies of the Lord who escaped death were sometimes sentenced to slavery. Indeed, slavery was a practice that Yahweh permitted:

> You may have male and female slaves, but buy them from the nations around you.
> —Yahweh, Leviticus 25.44

Yahweh was not only interested in physical slavery but mental subjugation as well. In the system that Yahweh and his agents established, righteousness and wisdom were equated with obedience, while disobedience was equated with sin and foolishness:

> The wise in heart accept commands, but a chattering fool comes to ruin.
> —Proverbs 10.8

Anyone who was able to perceive and formulate judgments based on their own reasoning were subjected to ridicule, threats, and violent punishments:

> Woe to those who are wise in their own eyes and clever in their own sight.
> —Isaiah 5.21

Those who possessed special spiritual or psychic abilities were especially targeted, since those were people who were able to see behind the artificial veil that Yahweh and his associates were casting:

> Do not turn to mediums or spiritists.
> —Yahweh, Leviticus 19.31

By keeping his subjects unaware, Yahweh was better able to fuse the affairs of church and state, and thereby create a reliable and effective warrior class who were willing to give up their lives for the campaign. Indeed, fusing the affairs of church and state was a practice that was recommended by Machiavelli[300] when he praised the "leaders and founders of religions," who together with the "founders of republics and kingdoms," and the "commanders of armies," "extend the boundaries of their kingdom or country."

The purpose of bestowing the Ten Commandments was so that Yahweh could establish a functioning society, and thus a functioning army. Likewise, when he told the people to "be fruitful and multiply" (Genesis 26.4; Leviticus 26.9), this was not so much of a patriarchal benediction, but rather he simply needed to increase the population of his army! By establishing a code of law, he was better able to reduce internal strife among the people and keep them united and focused on the goal of accomplishing his desires.

The record indicates that Yahweh not only had an aberrant interest in power and worldly prestige but in material affluence as well. For example, he demanded that "gold, silver, and bronze" (as well as

---

[300] *Discourses on Livy*

other treasures) be captured and added to his treasure trove—which is certainly peculiar for a being who is thought to be the creator of all the heavens and Earth.

> All the silver and gold and everything made of bronze and iron are holy and belong to Yahweh. They must go into Yahweh's treasury.
> —Joshua 6.19

> The silver is mine and the gold is mine, declares Yahweh of Armies.
> —Haggai 2.8

Those who submitted themselves as workers, warriors, and worshipers, were used for this purpose; while those who refused were put to death.

Some of those who did the bidding of the Lord were rewarded for their obedience with the material spoils of conquest:

> I will give you the treasures of darkness, and hidden riches of secret places [. . .]
> —Yahweh, Isaiah 45.3

> The Babylonians will become the prize. All who loot them will get everything they want.
> —Yahweh, Jeremiah 50.10

In Joshua 7, we are told that a man, who is referred to as Achan, was caught taking plundered treasure that was supposed to be offered to Yahweh. After Joshua was made aware of this transgression, he ordered that Achan be stoned to death—which he did with the approval of his Lord.

Yahweh's campaign not only involved the acquisition of wealth but of also attaining the prime strategic land of Canaan as well. This acquisition would then lead to even more wealth and power, due to the fact that it was a region that was situated in between the western

Mediterranean Sea civilizations and the eastern lands of Mesopotamia and Asia, which made it an important and strategic location. In fact, the ancient Hebrew word for merchant (*Kena'ani*) was used to describe a person from Canaan (Goble 2002: 1168).

> You have expanded the nation, O Yahweh. You have expanded the nation. You are honored. You have extended all the land's boundaries
> —Isaiah 26.15

Yahweh was clearly not only interested in acquiring land and wealth but needed to have his ego appeased as well. This was not something that he wished for; this was something that he demanded:

> Honor Yahweh your God before it gets dark, before your feet stumble on the mountains in the twilight. You will look for light, but Yahweh will turn it into the shadow of death and change it into deep darkness.
> —Jeremiah 13.16

At the very core of his motivation was a deep desire to be exalted as the "Most High":

> I will be exalted among the nations, I will be exalted in the earth.
> —Yahweh, Psalm 46.10

* * *

According to the Old Testament/Tanakh itself, Yahweh was a deity who resided in a dark and fiery netherworld:

> Clouds and darkness surround him. Righteousness and justice are the foundations of his throne. Fire spreads out ahead of him. It burns his enemies who surround

> him. His flashes of lightning lights up the world. The earth sees them and trembles.
> —Psalm 97.2-4

> Will not the day of Yahweh be darkness, and not light; and will it not have been gloom, and not brightness? I have hated, I have rejected your festivals, and I shall not enjoy the smell of your solemn assemblies.
> —Yahweh, Amos 5.20-21

> Yahweh has said that he would dwell in the thick darkness.
> —1 Kings 8.12

In Isaiah 6.1, we are informed that Lord Yahweh was seen seated upon a throne in a smoke-filled temple, surrounded by bizarre creatures who are referred to as the "seraphim." In this scene, Yahweh orders Isaiah to go forth and "dull" the hearts and minds of the people (Isaiah 6.10). Likewise, in the book of Psalms we read the following:

> Yahweh reigns, let the nations tremble; he sits enthroned between the cherubim, let the earth shake.
> —Psalm 99.1

The attendants of the Lord of Armies were known as both the seraphim (more on the seraphim later) and the cherubim. As was previously noted, the cherubim were not originally described as baby angels. The word cherubim derives from the Akkadian *karabu/karubu,* which means "one who prays" or "one who blesses" (Botterweck, et al. 1995: 308; Giovino 2007: 139). This explains why the root *krb* does not appear in biblical Hebrew, but is attested in Akkadian (Botterweck, et al. 1995: 308-310; Giovino 2007: 139)—which further indicates a pan-Mesopotamian influence in Israel and Judah. In Mesopotamia, these creatures were sculpted as winged genies that stood guard at the entrances of cities and temples (which should not be confused with the eagle-headed Sumero-Akkadian

Apkallu/Abgal). The cherubim were the protecting guardians of the Elohim gods. In fact, this is how they are portrayed in the biblical record as well (Genesis 3.24). However, it should be understood that these hybrid creatures were originally only artistic motifs, not literal living beings.

* * *

According to the Old Testament/Tanakh itself, Yahweh was not only "deceiving" the people but was afflicting them with "terrible" "adversity":

> Terror and pit and snare await you, people of the earth. Whosoever flees at the sound of terror will fall into a pit; whosoever climbs out of the pit will be caught in a snare [. . .] In that day Yahweh will punish the powers in the heavens above and the kings on the earth below.
> —Isaiah 24.17-21

> Lord Yahweh of Armies has a day of tumult and trampling and terror in the Valley of Vision [. . .]
> —Isaiah 22.5

> Although the Lord gives you the bread of adversity and the water of affliction, your teachers will be hidden no more [. . .]
> —Isaiah 30.20

> I trampled the nations in my anger; in my wrath I made them drunk and poured their blood on the ground.
> —Yahweh, Isaiah 63.6

> I will gather all the nations to Jerusalem to fight against it; the city will be captured, the houses ransacked, and the woman raped [. . .]
> —Yahweh, Zechariah 14.2

Their infants also will be dashed in pieces before their eyes. Their houses will be ransacked, and their wives raped.
—Yahweh, Isaiah 13.16

These references to the killing of infants brings up other questions that pertain to child sacrifice, which was a practice that was not unheard of in the ancient Near East. Indeed, it is mentioned numerous times in the Old Testament/Tanakh. It is commonly believed that Yahweh condemned this practice, which is confirmed several times (e.g., Leviticus 20.2; Jeremiah 7.31, etc.). However, other verses contradict these decrees. For example, the following passage references Topheth. This was supposedly a location where the wicked heathens sacrificed their children to the foreign gods, such as Moloch (i.e., Molek) (Leviticus 18.21; 2 Kings 23.10). However, other biblical verses indicate that Yahweh did condone such practices when they were committed under his order (Smith 2002: 171-172). Indeed, the following verse indicates that it was Yahweh himself who ignited the sacrificial flame at Topheth:

> Topheth has long been prepared; it has been made ready for the king. Its fire pit has been made deep and wide, with an abundance of fire and wood; the breath of the Lord, like a stream of burning sulfur, sets it ablaze.
> —Isaiah 30.33

It is true that this verse indicates that it was the king of Assyria who was destined for the sacrificial flame at Topheth. However, the following verse reports that, at least at one time, Yahweh had forced the Israelites themselves to sacrifice their children as a punishment for their disobedience:

> So I gave them other statutes that were not good and laws through which they could not live; I defiled them

> [the people of Israel] through their gifts—the sacrifice of every firstborn—that I might fill them with horror so they would know that I am the Lord.
> —Yahweh, Ezekiel 20.25-26

The biblical record itself confirms that it was Yahweh who burned with an angry hatred in the gloomy darkness. It was Yahweh who devised deceptive schemes that were intended to gain power and egotistical glory for himself at the expense of innocent people. It was Yahweh who commanded his servants to perform blood-spilling rituals. It was Yahweh who ordered his troops to dismember children and rape woman. Although these are all characteristics that are usually associated with the devil, evil, and Satanism, the biblical record reports that it was Yahweh and his associates who were responsible for this behavior.

The reason why Yahweh went to such great lengths to conceal his true identity was not only related to his clandestine Machiavellian operations but also because he was not who he said that he was. The fire, smoke, and whirlwinds, were all a part of a fantastic facade that was used to convey a God-like supernatural power.

Although it is well known in the Judeo-Christian tradition that the devil is a malicious deceiver, what nobody considered is that the deception could have already happened.

> O Yahweh, you deceived me, and I was deceived [...]
> —Jeremiah 20.7

The individual who became known as "Yahweh" is the fallen archangel who rebelled against the higher natural spirit of the universe. He is the infamous outlaw who arrogantly dared to assume the role of the force of creation. He is the rebel archon whose soul burned with hatred, jealousy, and the deepest and most terrifying anger. He is the war criminal who used violence, deception, and fear to achieve personal power, wealth, and egotistical fame. The ultimate secret of Yahweh is that he himself fits the description of the devil character.

* * *

Of course, it is commonly assumed that Yahweh was a just and righteous "God"; a God of compassion and mercy. In this case, is there anything in the Old Testament that indicates that this is true? In the book of Zechariah, Yahweh is reported uttering the following statement:

> I will strengthen the people of Judah. I will rescue Joseph's people. I will bring them back, because I have compassion for them.
> —Yahweh, Zechariah 10.6

In the following passage, a devotee makes a reference to the Lord's love:

> Return, Yahweh. Deliver my soul, and save me for your loving kindness' sake.
> —Psalm 6:4

When conditions permitted, Yahweh seemed to make an effort to be gracious and merciful; however, these conditional gestures were only a temporary reward for behavior that conformed to his scheme. Only those who fell into line under the oppressive regulations of his machinations were granted clemency from his cruelty. For example, in Exodus 34.6-7 it is reported that the Lord declared his abounding "love"—that is, for those who he did not have to violently punish for not submitting themselves as workers, warriors, and worshipers. In these cases, the benevolent side that he occasionally displayed acted as more of a baiting device that led optimistic followers deeper into the deceptive spell that he and his collaborators were casting.

Over the course of time, misunderstanding increased as mistaken interpretation and affected information replaced personal experience and sensible reason. The more that time passed, the more these misunderstandings became familiar, aggrandized, sanitized, and absolute.

Imagine, for instance, if people were presented with a picture of an individual who was filled with burning rage, bitter jealousy, and violent outbursts. A deceptive schemer who commanded the people to bow beneath him or else suffer a cruel and agonizing death. A sociopathic tyrant who ordered his troops to intentionally kill children and rape woman. Such an individual would be regarded as not only a despot but as downright evil; and yet, this is not how Yahweh is commonly perceived. This situation is a testament to the influential power of the conditioning of mental perspective.

> Yahweh will afflict you with madness, and blindness, and confusion of mind.
> —Deuteronomy 28.28

Lay people are generally influenced by what appeals to them on more of an emotional level. The hardened traditionalists who promote and preserve the mistaken interpretation are able to dispel the negative aspects by redirecting focus away from the atrocities and the contradictions, and instead place attention onto matters that are related to the heart. However, it must be remembered that Yahweh and his original cohorts were not always advocates of the heart.

> The heart is deceitful above all things, and it is exceedingly corrupt: who can know it?
> —Yahweh, Jeremiah 17.9

Although Yahweh is given credit with bestowing the moral standard of the Ten Commandments, what must be understood is that the reason why he required law and order was simply because he needed a functioning servant force. What is also not so well known is that Yahweh was guilty of breaking most of those very laws himself! He was a murderer who coveted and stole and bore false testimony in order to carry out his materialistic and malevolent schemes.

In the biblical New Testament (Galatians 5.19-21), we read that the apostle Paul, who himself was a devout follower of Yahweh, lists the sins that will not allow a person to enter heaven. Among these infractions are "immorality, impurity, enmity, strife, jealousy, anger, selfishness," and "dissension." Of course, what Paul failed to realize is that Yahweh was guilty of committing those very sins himself! Yahweh was also guilty of committing most of the Seven Deadly Sins (as formulated by the Christian monk Evagrius of Pontus, and amended by Pope Gregory in the sixth century CE). These transgressions include: pride, envy, anger, and greed.

Some committed Judeo-Christian apologists have attempted to excuse the contradictions and the atrocities by asserting that "God must sometimes use evil to defeat evil." If this is the case, then there is no such thing as any moral standard because hypocrisy and heinous acts could always be justified as long as it fit someone's own idea of right and wrong. However, a double-standard is a double-standard, no matter who it is that is the hypocrite.

* * *

Despite these discoveries, we are still left with significant questions. For example, if Yahweh was using other city-states to do his bidding then there should be some mention of him in their records. In this case, do such records exist? If there is any truth to the maximalist interpretation, then the answer is yes. However, it seems that these other city-states did not refer to him by his Hebrew designation, but rather by an entirely different name.

Despite the viability of the moderate minimalist interpretation, and despite the fact that I myself lean more towards this explanation, it is difficult to deny that the following information is remarkably contextually congruent and therefore does confer credibility to the maximalist interpretation. As was previousy stated and worth mentioning again: We must be willing to go wherever the data indicates, even if it conflicts with our ingrained assumptions.

Nevertheless, after the following maximalist perspective is presented, its validity will also be soberly assessed.

* * *

During the Judean exile in Babylon, it is evident that city-state was no longer under the control of Bel. Instead, it is indicated that at that time Babylon was under the influence of Yahweh himself. According to Jeremiah, Nebuchadnezzar (the king of Babylon) was the loyal "servant" of Yahweh (Jeremiah 27.6; 43.10). In a maximalist context, this could indicate that the exile and captivity of the Judeans in Babylon was another Machiavellian operation that was employed to punish the disobedient people before introducing himself at a later time to act as their compassionate liberator—which is exactly what happened (Jeremiah 42.11). It was an operation that was allegedly orchestrated by the unseen spirit of Yahweh; which was carried out by his associate attendants in Babylon who were operating under the aegis of his human agents. In a minimalist context, the exile to Babylon would have happened anyway. In this case, the Yahwehist priest-scribes may have simply interpreted this event as the will of Yahweh.

According to the biblical record, the Babylonian king was assisted by an elite inner-circle of men. Among them was prince Nergal-sharezer (Jeremiah 39.3)—whose name literally translates as, "May Nergal protect the king" (Livingstone 1999: 622). It is likely that this was the very same individual who is referred to as Neriglissar in other passages. This individual may have also been the same high-ranking officer who later murdered Merodach, the king of Babylon, and ruled in his place.

The Anunnaki god Nergal was known in that region as a fierce lord of death, fire, war, disease, and the underworld. He was the first-born son of Enlil, and was known as an outlaw prince among the Elohim. His seat of power was based in the Mesopotamian city of Cuthah (i.e., Cuth/Kutha). In order to understand what was happening in this scenario, it is necessary to understand the consequential role of this commonly overlooked deity.

Nergal was also known by other names during different times and in different locations—just as many of the other Anunnaki lords were. One of these names was Erra (i.e., Irra or Ura) (Dalley 1989: 321, 325; Hastings, et al. 1914: 646; Livingstone 1999: 622). One of the most popular stories pertaining to him appears in the Erra and Ishum text (i.e., the Erra Epic).[301] Sitchin believed that the appellation *Er-Ra* signified that Nergal was a servant of the Egyptian god Ra—who he believed was the Babylonian god Marduk (Sitchin 1985: 251); however, it is more likely that the name derives from Nergal's epithet *en-eri-gal*/ne-eri-gal, which means "Lord of the Netherworld" (Leick 1991: 127; Livingstone 1999: 622).[302] Another possibility is that it could stem from the Semitic root-word *hrr*, which means to "scorch" (Munnich 2013: 63; Roberts 1971: 13)—which is an accurate description of his characteristics.

According to the Erra and Ishum text, the seven martial demi-gods (i.e., Sebetti), who are the attendants of Erra, petitioned their lord to raise himself from out of his resting place and destroy the human-beings who had become "contemptuous." Erra eventually responded by assuring the Sebetti that he will indeed turn his attention toward the disobedient ones who no longer fear his name.

> Because they no longer fear my name, and since prince Marduk has neglected his word and does as he pleases, I shall make prince Marduk angry, and I shall summon him from his dwelling, and I shall overwhelm his people.
> —Erra, Erra and Ishum

In the Mesopotamian records, the god of Babylon is usually addressed by the name "Marduk"; however, he is more commonly referred to in the Old Testament by the Semitic appellation *Bel*, or *Baal*—which literally means "Lord."

> Babylon will be captured; Bel will be put to shame, Marduk filled with terror.

---

[301] The Erra and Ishum text is derived from a compilation of thirty-six fragmented copies that were discovered all throughout the region of Babylonia. It is dated to the eighth century BCE.

[302] The theonym Nergal (*nin-eri-gal*) might also mean "Lord of the (big) city" (Munnich 2013: 59).

—Jeremiah 50.2

In the following passage, Bel is listed together with Nebo, who, according to the Babylonian texts, was the son of Marduk. This is where the name Nebuchadnezzar—which means, "servant of Nebo—originated:

> Bel has bent down, Nebo is stooping over; their idols have come to be for the wild beasts and for the domestic animals [. . .].
> —Yahweh, Isaiah 46.1

The texts report that Marduk was not only the god of Babylon—which at that time was the most powerful city on Earth—but was revered by its citizens. This was a situation that did not sit well with Nergal/Erra.

> I have disregarded Marduk's command, so he may act according to his wishes. I will make Marduk angry, stir him from his dwelling, and lay waste the people.
> —Erra, Erra and Ishum

This same type of sentiment is echoed in the Old Testament/Tanakh:

> I will punish Bel in Babylon and make him spew out what he has swallowed. [. . .]
> —Yahweh, Jeremiah 51.44

Nergal/Erra did not completely demolish Babylon when he had the chance. This is because this was a major center of power that he intended to acquire for himself. Of course, there was at least one major obstacle in his way; namely, his cousin and ruler in the council of the Elohim, Bel Marduk. In order to solve this problem, Erra devised a plan in which he would gain Marduk's trust by approaching him as a friend and fellow family member. He brought to his cousin's attention that he needed a rejuvenating rest and convinced him to journey to a far-off land where he could partake of the "sacred tree" that is "suitable for the lord of the universe," and whose bark is "the flesh of the gods" (i.e., the Tree of Life). Marduk agrees with this suggestion and leaves the city in the care of Erra. However, after Marduk departs the "regulation of heaven and earth" is "disintegrated."

According to this account, Nergal/Erra punished those who he believed had sinned against him by laying waste to the cities of the earth. His right-hand attendant, Ishum, laments that Nergal/Erra was seeking to put to death both the sinful and the righteous alike, and attempts to talk some compassionate reason into him. A very similar situation appears in the Old Testament/Tanakh, in which Abraham pleads with Yahweh not to destroy both the righteous with the wicked in the city of Sodom[303] (Genesis 18.22-33). Indeed, the deity endeavored to "devastate" the "cities" in much the same way that the "cities" of the "plain" are referred to in the biblical version.

> The lands I will destroy, to a dust-heap make them; the cities I will upheaval, to desolation turn them [. . .]
> —Erra, Erra and Ishum

> Thus he [Yahweh] overthrew those cities and the entire plain, destroying all those living in the cities—and also all the vegetation of the land.
> —Genesis 19.25

Nergal/Erra contaminated the land with the same type of poisonous fallout that is described in the book of Isaiah:

---

[303] One possible location of Sodom is the Tall el-Hammamam site (located in the Jordan Valley, not far from the Dead Sea). Evidence of an extensive fire, ashy strata, shattered and charred pottery, crushed and contorted skeletons, and massive structural destruction, have been unearthed at this location (Collins 2013; Collins and Hamden 2013). It is also significant that no evidence of any active volcanoes in the area has been found, and earthquakes are unlikely to cause a fire of such magnitude.

> Edom's streams will be turned into pitch, her dust into burning sulphur; her land will become blazing pitch! It will not be quenched night and day; its smoke will rise forever. From generation to generation it will lie desolate; no one will pass through it again.
> —Yahweh, Isaiah 34.9-10

It was around the same period that the authors of the previously cited passages from the book of Jeremiah and Isaiah were alive that the Erra and Ishum texts were being written, copied, and put into the library at Nineveh in Assyria. Although these various wars most likely occurred at different times—if they literally occurred at all, the similarities are significant.

The deities signature method of operation was his duplicitous scheming.

> You [Erra] are the decoy for the inhabitants of Babylon, and they are the bird; You ensnared them in your net and caught and destroyed them warrior Erra.
> —Ishum, The Erra and Ishum Text

> I set a trap for you, Babylon, and you were caught before you knew it. [...]
> —Yahweh, Jeremiah 50.24

Although Yahweh made an effort to keep his word to Abraham, and sent his emissary messengers to extract the obedient from the cities before destroying them, in other cases he was not so careful:

> Respect no god! Fear no man! Put to death young and old alike, the suckling and the babe—leave not anyone!
> —Erra, The Erra and Ishum Text

> Whoever is found will be stabbed to death. Whoever is captured will be executed. Their little children will be

> smashed to death right before their eyes. Their houses will be looted and their wives raped.
> —Yahweh, Isaiah 13.15-16

> I shall annihilate the son, and let the father bury him; then I shall kill the father, let no one bury him.
> —Erra, The Erra and Ishum Text

> Slaughter the old men, the young men and women, the mothers and children, [...]
> —Yahweh, Ezekiel 9.6 [304]

Nergal/Erra punished the disobedient (i.e., the "wicked") and all of those who sinned against him by not submitting themselves as his servants.

> You [Erra], have put to death the man who sinned against you [...]
> —Ishum, The Erra and Ishum Text

> I will punish the world for its evil, the wicked for their sins. [...]
> —Yahweh, Isaiah 13.11

The punishment for disobedience was death:

> You have made their blood flow like water in the drains of public squares. You have opened their veins and let the river carry off their blood.
> —Ishum, The Erra and Ishum Text

> [...] He [Yahweh] will totally destroy them, he will give them over to slaughter. Their slain will be thrown

---

[304] Also repeated in Hosea 13.16, and Deuteronomy 2.34.

out, their dead bodies will stink; the mountains will be soaked with their blood.
—Isaiah 34.2-3

This Lord of war was the deity who violently silenced the disobedient:

> Nergal, lord who imposes silence [. . .]
> —A Tigi to Nergal (Nergal C Text)

> Every morning I will put to silence all the wicked in the land [. . .]
> —Yahweh, Psalm 101.8

He was the Lord of Armies who shook the Earth with the weapons of his arsenal:

> Erra, you clash your weapons together and the mountains shake, the seas surge at the flashing of your sword.
> —Ishum, The Erra and Ishum Text

> The Lord [Yahweh] has opened his arsenal and brought out the weapons of his wrath, for the Sovereign Lord Almighty [Lord Yahweh of Armies] has work to do in the land of the Babylonians.
> —Yahweh, Jeremiah 50.25

He was the deity who terrorized the disobedient:

> We all know that nobody can stand up to you in your day of wrath!
> —Ishum, The Erra and Ishum Text

> Therefore I will make the heavens tremble; and the earth will shake from its place at the wrath of the Lord

Almighty [Yahweh of Armies], in the day of his burning anger.
—Yahweh, Isaiah 13.13

Lord, furiously raging storm, confusing the enemies and unleashing great terror over the land. Nergal, mighty quay of heaven and earth, who [. . .] looks up furiously, turning his weapons against the wicked.
—An Adab to Nergal for Shu-ilishu (Shu-ilishu A Text)

[. . .] The Lord Almighty [Yahweh of Armies] is mustering an army of war. They come from faraway lands, from the ends of the heavens—the Lord and the weapons of his wrath—to destroy the whole country.
—Isaiah 13.4-5

He was the Lord who subjugated the disobedient with the power of the storm:

When [. . .] you command the storm which flattens the hostile land, you pour dust over its evil; you pour it over for as long as it disobeys.
—Hymn to Nergal (Nergal B Text)

When he thunders, the waters in the heavens roar; he makes clouds rise from the ends of the earth. He sends lightning with the rain and brings out the wind from his storehouses.
—Jeremiah 51.16

He was the god who covered the skies with the smoky fall-out from his explosive weapons:

I promise that I shall destroy the rays of the sun; I shall cover the face of the Moon in the middle of the night.
—Erra, The Erra and Ishum Text

> The stars of heaven and their constellations will not show their light. The rising sun will be darkened and the moon will not give its light. I will punish the world for its evil, the wicked for their sins. [. . .]
> —Yahweh, Isaiah 13.10-11

He was the ruthless lord of death who created empty ghost-towns that became the home of wild beasts:

> I shall let wild beasts of the mountains go down [into the city of Marduk]. I shall devastate public places wherever people tread.
> —Erra, The Erra and Ishum Text

> She [Babylon] will never be inhabited or lived in through all generations [. . .] desert creatures will lie there, jackals will fill her houses [. . .]
> —Yahweh, Isaiah 13.20-21

He was the fierce and almighty "despoiler":

> [. . .] like a despoiler of a country, I distinguished not good from bad, I laid (all) low [. . .]
> —Nergal, Erra and Ishum

> [. . .] all who make spoil of you I will despoil.
> —Yahweh, Jeremiah 30.16

Those who managed to survive the onslaught were taken into captivity as slaves like a herd of cattle:

> Warrior Erra, you hold the nose-rope of heaven [. . .] you govern the people and herd the cattle.
> —Ishum, Erra and Ishum

> [. . .] I will put my hook in your nose and my bit in your mouth, and I will make you return by the way you came.
> —Yahweh, 2 Kings 19.28

This Lord of Armies not only wanted to attain power and fame but wealth as well:

> You [Erra] must pillage the accumulated wealth of Babylon.
> —Ishum, The Erra and Ishum Text

> "So Babylonia will be plundered; all who plunder her will have their fill," declares the Lord.
> —Yahweh, Jeremiah 50.10

At the very core of his motivation was a deep desire to be exalted as the most powerful (i.e., the "Most High") Lord "Almighty":

> Let all the countries listen to it and praise my valor! Let settled people see and magnify my name!
> —Erra, The Erra and Ishum Text

> [. . .] I will be exalted among the nations, I will be exalted in the earth.
> —Yahweh, Psalm 46.10

> [Nergal/Erra], who in his heroism like a flood demands respect!
> —A Tigi to Nergal (Nergal C Text)

> "Now I will arise," says the Lord. "Now I will be exalted; now I will be lifted up."
> —Isaiah 33.10

In the following passage, Nergal/Erra refers to himself in the third-person, while encouraging his scribes to glorify him in order to avoid the "sword of judgment":

> In the house where this tablet is placed, even if Erra becomes angry and the sebetti [seraphim] storm, the sword of judgment shall not come near him, but peace is ordained for him.
> —Erra, The Erra and Ishum Text

This is the very same sword that is used by Yahweh to punish the disobedient:

> My sword has drunk its fill in the heavens; see, it descends in judgment on Edom, the people I have totally destroyed. The sword of the Lord is bathed in blood [. . .]
> —Yahweh, Isaiah 34.5-6

The Mesopotamian "Tigis" and "Adabs" are the equivalent of the biblical Psalms of the Old Testament:

> Lord of the just word, lord of abundance, hero! At your name, people obey. As you rise up in the frightening shrine [. . .] with your kingship you inspire fear. Hero, with your magnificent strength [. . .] you pile up the rebel lands in heaps. Nergal, your name is praised in song.
> —A Tigi to Nergal

> My enemies turn back; they stumble and perish before you. For you have upheld my right and my cause, sitting enthroned as the righteous judge. You have rebuked the nations and destroyed the wicked; you have blotted out their name for ever and ever.
> —Psalm 9.3-5

> The majestic and just crown [. . .] your awesomeness [. . .] is a south wind that none can withstand [. . .] you exercise the role of supreme deity!
> —A Hymn to Nergal (Nergal B Text)

> Among the gods [Elohim] there is none like you, Lord [Adonai]; no deeds can compare with yours [. . .] For you are great and do marvelous deeds; you alone are God [Elohim].
> —Psalm 86.8-10

> You [Erra/Nergal] control the whole earth, and you rule the land [. . .] govern people and herd cattle.
> —Ishum, The Erra and Ishum Text

> For Yahweh Most High is awesome. He is a great King over the earth. He subdues nations under us, and peoples under our feet.
> —Psalm 47.2-3

The purpose of these various hymns in praise of this war god was to appease the ego of the deity, and thus avert the consequences of his wrath.

> He [Nergal/Erra] strikes unawares and strikes apparently without discrimination. He is not just a judge like Shamash, but a god, filled with rage, stalking about in the heat of day on the lookout for victims. Nergal is thus primarily the god of death [. . .] Because of this forbidding aspect, it was all the more important to raise one's appeal to him in the hope of averting his wrath. The hymns to Nergal, of which we have quite a number, all emphasize the severity and irresistible power of the god.

—Morris Jastrow Jr.,[305] *The Civilization of Babylonia and Assyria*

In both the Mesopotamian and the biblical records, this god is also said to have flown through the skies:

> He [Erra/Nergal] travels through heaven and organizes everything.
> —Hymn to Nergal (The Nergal B Text)

> He [Yahweh] makes the clouds his chariot and rides on the wings of the wind.
> —Psalm 104.3

Nergal was the Anunnaki lord who not only destroyed the disobedient but, like Yahweh, he too was a religious deity who conceived "divine plans," and who performed religious "purification rites" in the "great shrine." Likewise, just like Yahweh, Nergal/Erra was also a lord of the "heavens" as well:

> Lord who inspires awe in heaven and earth.
> —A Tigi to Nergal (Nergal C Text)

> Be exalted, O Elohim, above the heavens, and let your glory be over all the earth.
> —Psalm 108.5

Of course, in these cases, the word "heaven" is used to describe the upper atmosphere, and perhaps even beyond.

Throughout the biblical record, there are repeated references to "revelations" that occur between God and man (i.e., the theophany). These experiences usually involve an Earthling man who is taken up

---

[305] Morris Jastrow Jr., Ph.D., was a nineteenth- and twentieth-century Polish-born Amerian professor of Semitic Languages at the University of Pennsylvania.

into heaven where he beholds the manifestation of an otherworldly being who bestows upon him the divine apotheosis.

In the book of Enoch, for example, the biblical character is taken up into another world, wherein he experiences prodigious visions. While in this state, he is shown the treasury of the Lord that is guarded by "terrible angels." In the second realm, he observes the disobedient people, who are now prisoners of the Lord. These alleged sinners were being tortured by "dark looking" "angels." In the next realm, he is taken to a place of "cruel darkness and unillumined gloom." Enoch is eventually taken to meet the Lord, whose countenance is described as "awful and very, very terrible," and is instructed to write down his experience for the "souls of humanity," so that they may read, learn, and obey his will.

A very similar type of apocalyptic revelation, titled, the Legend of Kumma (i.e., The Netherworld Vision of an Assyrian Crown Prince), was written in the eighth century BCE. In this report, we are told of a young man known as prince Kumma (or Kumaya), who, after having experienced a traumatic incident—which seems to have something to do with a schism that occurred between himself and his father—he began to fast and pray. Shortly thereafter, he receives a dream-vision during the night. During this incident, he awakes to find himself in a bizarre dark netherworld that is inhabited by strange creatures that are part man and part beast. It is very likely that it was these creatures who became the models for the biblical seraphim.[306]

The designation seraphim is derived from the Hebrew verb *sarap/saraph*, which means "burn," "destroy," or "to consume with fire" (Botterweck, et al. 2004: 223; Singer, et al. 1905: 202; Mettinger 1999: 743). It may also refer to a griffin type of creature (Mettinger 1999: 743). In some instances, the word can also refer to flying fiery serpents (from the Hebrew noun *saraph*) (Numbers 21.6; Isaiah 14.29) (Mettinger 1999: 743). However, the description of the

---

[306] The biblical seraphim may have derived from the Egyptian *uraeus* serpent motif (Mettinger 1999: 743). Although this may apply to descriptions of the seraphim as serpents, in other biblical accounts they are described in more humanoid terms (e.g., Isaiah 6.2).

seraphim in Isaiah 6.2 are as humanoid creatures with wings. We therefore have variant definitions. What is certain is that the original seraphim were not white-robed and winged humanoid beings with halos, but rather were hideous demonic creatures, which were most likely based on the Mesopotamian *pazuzu* (Figure 27).

(Fig. 27)

What is also significant about this definition is that the term *saraph/sharraph/sharrapu* was one of the epithets that was attributed to Nergal (Cook 1908: 93; Encyclopedia Britannica Online, s.v. "Nergal"; Singer, et al. 1905: 202).

According to the Legend of Kumma text, Kumma eventually found himself in a judgment scene, where "seated upon a throne, wearing a divine crown," was Nergal, "judge of the dead."

> When I raised my eyes, there was valiant Nergal sitting on his royal throne wearing the royal crown, he held two terrible maces with both hands, each with two heads [. . .] lightning was flashing, the great

netherworld Anunna [Anunnaki] gods were kneeling to the right and left. The netherworld was full of terror.
—Kumma, The Legend of Kumma

Here we find qualities that are consistent with the Judeo-Christian concept of the "divine" and "wrathful" Lord, who sits upon a throne, and who "captures the wicked" and "judges" the souls of man.

Clouds and thick darkness surround him; righteousness and justice are the foundation of his throne. Fire goes out before him and consumes his foes on every side. His lightning lights up the world; the earth sees and trembles [. . .]
—Psalm 97.2-4

Yahweh reigns as king. Let the nations tremble. He sits enthroned between the cherubim, let the earth shake.
—Psalm 99.1

Likewise, in the book of Isaiah, there is an account of a servant of Yahweh who entered a dark smoke-filled temple in which the Lord was seated upon a throne, surrounded by bizarre winged beasts. Therefore, the seraphim were originally the winged humanoid beasts of the Mesopotamian underworld.

According to the Legend of Kumma text, when Nergal saw prince Kumma he became enraged. Apparently, when Kumma was praying, he directed his thoughts not to Nergal but to his consort, Ereshkigal. Nergal was about to kill the offender when he was stopped by his attendant, who proposed that Nergal spare the life of the prince as long as Kumma pledged his obedience and agrees to praise Nergal for the rest of his life. Both Nergal and Kumma agreed to this and his life was spared.

Ishum was the vizier of Nergal. One of the other attendant servants of Lord Nergal/Erra was Gibil—and/or Gerra. Of course, Yahweh too was attended to by a group of servant emissaries. The two most well-known of which were Michael and Gabriel. However, these attendants were originally much different in appearance and character than the common traditional interpretation that developed in the subsequent Judeo-Christian era. In this case, Ishum was the equivalent of the arch-angel Michael, while Gerra/Gibil was the equivalent of Gabriel.

It must also be remembered that the deity who appeared to Moses never revealed his name, but rather only uttered the mysterious maxim, "I am who I am" (*Eyeh asher eyeh*); which eventually developed into the alias YHWH. However, this was not the original Anunnaki name of this god. According to the older Mesopotamian texts, the original name of the Lord of Armies was Nergal!

* * *

According to the Sumerian texts, Nergal was the son of Enlil (Livingstone 1999: 622). This helps to explain why the Garden of Eden story is told from the perspective of a follower of Enlil—as opposed to Enki. This also helps to explain the relationship between the Canaanite/Ugaritic god El and Yahweh. This is because it does appear that El derived from the earlier Sumero-Akkadian deity Enlil/Ellil. Indeed, the etymology of the word El is considered to be uncertain by mainstream scholars (Coogan 1978: 12; Herrmann 1999: 274). However, it is likely that this theonym is an abbreviated Semitic variant of the Mesopotamian Enlil/Ellil.

> West Semitic personal names normally begin in transparent appellations or sentence names and shorten or disintegrate. Divine epithets and often divine names follow the same patterns of formation and shortening. They do not begin in numinous grunts or shouts and build up into liturgical sentences or appellations.
> —Frank Moore Cross,[307] *Canaanite Myth and Hebrew Epic*

---

[307] Frank Moore Cross, Ph.D., was a twentieth- and twenty-first-century American professor of Hebrew and Other Oriental Languages at Harvard University.

The Amorites referred to this deity as Il (Cross 1997: 14). It is possible that this variant of the Mesopotamian Enlil/Ellil was promulgated by this Semitic-speaking people who dominated the region between Mesopotamia and Syria-Palestine in around the age of Abraham. It is therefore likely that El/Il was brought into Ugarit by the Amorites. It was this adaptation that influenced the Canaanites, and eventually all of the tribes of Israel (Smith 2002: 28).

It is therefore not a coincidence that both Enlil/Ellil and El/Il fulfill similar roles in their respective pantheons (Cross 1997: 41-42). Indeed, it does appear that El's association with the animal symbol of the bull began in Sumer.

> Enlil set his foot upon the earth like a great bull.
> —The debate between Winter and Summer

> Bull El his father King El who created him.
> —CTA. 3.5.43: 4.1.5: 4.4.47: etc.

> The lord (Enlil)[308] bellowed at his hoe like a bull.
> —The Song of the Hoe

> Lift up your hands to heaven; Sacrifice to Bull, your father: El.
> —CTA. 14.2.75-79.

Although the Sumerian references are sparse, they do indicate a nascent association.

The relationship between Enlil/El and Nergal/Yahweh is indeed significant. In the following passage, for example, Nergal is referred to as the "junior Enlil."

> Lord who [. . .] has the power to create life, Nergal, enduring house, great shrine—you are the junior Enlil!

---

[308] The theonym in parentheses appears at: etcsl.orinst.ox.ac.uk.

> It is your power to determine destinies, to render judgments and to make decisions, Nergal, your great hands are filled with mighty actions and terrible powers! Great rites which are revealed to no one are organized for you! Nergal, among this people it is you who takes charge of the divine plans and the purification rites!
>
> —An Adab to Nergal for Shu-ilishu (Shu-ilishu A Text)

It is therefore likely that Yahweh was not El, but rather he was originally the son of El. Indeed, this explains the following biblical report, in which Yahweh is described being assigned his people from Elyon, which was the original epithet of his father. This occurred before the title of "Most High" was eventually appropriated by Yahweh at a later time (Genesis 14.42, etc.).

> When the Elyon apportioned the nations and when he divided humankind, he fixed the boundaries of the peoples according to the number of the gods; Yahweh's portion was his people, Jacob his allotted share.
>
> —Deuteronomy 32.8-9

The passage not only indicates that Lord Nergal Yahweh was expected to respect the property boundaries of the other Elohim/gods but that he himself received orders from a higher authority; namely, his father, Enlil/Ellil/El/Il/Elyon!

The simultaneous influence of both Yahweh and Nergal lasted for approximately a millennium, until both Yahweh and Nergal were no longer physically present in the world of man. The cult of Nergal survived only for a few centuries after his death, or departure from the Earth. However, the cult of Yahweh continued because of one very consequential reason; namely, because his decrees and exploits, as well as those of his loyal servants, were documented, collected, and preserved in the writings of what eventually became the sacred Tanakh. This is a testament to the power of the written word and the inviolable traditions that arose around it.

> If you want a movement or a state to survive for long, you must repeatedly bring it back to its founding principles.
>
> —Machiavelli, *Discourses on Livy*

The maximalist interpretation posits that Nergal initiated his campaign to take control from his cousin Marduk when he presented himself—i.e., his Machiavellian covert persona—to Moses. In this context, when Nergal surreptitiously assumed the role of supreme "Lord God," he acquired the Canaanite appellation of his father (e.g., 2 Samuel 22.32-33, etc.). Consequently, Yahweh eventually became the "El" of the Judean/Israelites; just as Nergal had acquired the epithets: "the junior Enlil" and "the Enlil of the netherworld."[309]

In a minimalist context, the integration between El and Yahweh was formulated by Levitical priests, who entered into Canaan from an extended sojourn in Egypt, via Midian, with a new name for God (Friedman 1987: 82, 120). Before this time, the twelve-tribe league of Israel worshiped El. The following pericope is remnant evidence of that earlier age:

> He [the priest of El Elyon] blessed Abram, and said, "Blessed is Abram by El Elyon, maker of heaven and earth."
>
> —Genesis 14.19

Indeed, it is now known that it was the monotheistic Yahwehists who attached the abbreviated appellation "YHWH" to both El and to the "Elohim" (Genesis 2.4, etc.), which occurred in a subsequent age.

These discoveries help to answer the question that has perplexed scholars for years; namely, what is the relationship between the Canaanite god El and the Judean/Israelite Yahweh? The data is indicating that they were originally two separate but related gods who were eventually combined.

---

[309] The Death of Ur-Namma text.

Despite the biblical reference to an officially sanctioned appointment by Enlil/El, it seems that Nergal/Yahweh went about his activities in Canaan without the approval of the council of the Elohim, and that the interpretation of legitimate ordination was concocted by scribes who were loyal to Yahweh. The maximalist hypothesis posits that in his covert Judean/Israelite role, Nergal was able to anonymously defy the rules of the council of the Elohim and deify himself as the "Most High" Lord Almighty—i.e., "Elyon." He even took his endeavor one step further when he disavowed the relevance of the other gods completely. It was a seditious move that even contradicted the importance of his own father, whose legacy he was seemingly seeking to usurp. However, it is evident that his loyalty to his father was mostly superficial. This is because it is apparent that the sole primary interest of Nergal was Nergal himself.

* * *

Meanwhile, on the other side of the family, Enki and his offspring continued on in the Anunnaki tradition as well. Foremost among his progeny was his first-born son, Marduk.

> And unto Marduk, their first-born they spake: "May thy fate, O Lord, be supreme among the gods [. . .]
> —The Enuma Elish

Just prior to the beginning of the second millennium BCE, it was decided—either by the imaginative priest-scribes or by the Anunnaki/Elohim themselves—that kingship should be transferred over to Enki's side of the family. This decision allowed Marduk to be appointed to the supreme position that was previously held by Enlil. It was a monumental power shift; and, as we will see, this transition was not free from complications. There was a very deep-rooted issue that lay buried beneath the surface of this fragile situation. This is because, apparently, Marduk was not the eldest; and, as we will see, this discrepancy may have contributed to the most contentious rivalry that the world has ever known.

In a Sumerian report, we are told the story of a young Anunnaki woman who was seen bathing in the river by Enlil. He was so struck by her beauty that he forced himself upon her. A child was conceived of this intercourse:

> At this one intercourse, at this one kissing he poured the seed of Nergal-Meslamta-eda into her womb.
> —Enlil and Ninlil

Meslam was the inner-sanctuary of the temple in Cutah. This temple and city was dedicated to Nergal; hence his official epithet: "Nergal of Meslamtaea."

When the other Anunnaki authorities learned of this transgression they were appalled. Not only did Enlil force himself upon the young woman but Ninlil was both underage and not of direct royal lineage. By orders of the Enlil's own council, he was arrested for engaging in impure behavior. It was the public sex scandal of its time.

However, what was unknown to everyone, including even Enlil himself, was that Ninlil had actually planned the incident by intentionally enticing the king of the Elohim in an attempt to marry into the royal family and thereby elevate her own status and well-being.

> Ninlil, you are more majestic than the other great gods, you are elevated with great and terrifying divine powers.
> —An Adab to Ninlil

According to the Sumerian texts, Nergal was the first-born of the king of the Elohim gods; however, because he was born from an illegitimate union, he was not eligible for the supreme throne—which is the most likely reason why Enki had lost kingship to Enlil in the previous generation. Herein lay the seed source of discontent that seems to have burned so deeply in the first-born's psyche.

> Men's hatreds generally spring from fear or envy.

—Niccolo Machiavelli, *Discourses on Livy*

This is apparently why kingship was transferred to the other side of the family when Enlil relinquished the throne. Nergal must have felt that he lost supreme kingship due to a technicality that was not his fault.

In a passionate rant against Babylon, Lord Nergal Yahweh makes a reference to this lost inheritance:

> For you men [the Chaldeans of Babylon] kept exulting when pillaging my own inheritance.
> —Yahweh, Jeremiah 50.11

And so, he bided his time, all the while devising his own scheme to take back what he thought was rightfully his.

> To Marduk and to Ea I shall bring a reminder: He who grows up in times of plenty shall be burned in times of deprivation.
> —Nergal, The Erra and Ishum Text

> See, I am against you, O arrogant one [the god of Babylon] [. . .] for your day has come, the time for you to be punished.
> —Yahweh, Jeremiah 50.31

Not only was Lord Nergal Yahweh upstaged by his cousin, Bel Marduk—who was known as the "Son of the Pure Place"—but by his own younger half-brother, Ninurta, who was called the "rightful heir." This is because Ninurta's mother was the royal-blooded Ninhursag, who was the pride and joy of Enlil.

The maximalist story of Nergal is an interesting psychology study. It would seem that in the role of his Judean/Israelite alter-ego, he was better able to acquire the power, the acceptance, and the adulation that he felt that he was deprived of as an illegitimate son.

> O Lord Erra, why have you plotted evil against the gods?
> —Ishum, The Erra and Ishum Text

Of course, this clandestine persona also provided him protection from his rivals among the Anunnaki Elohim, who had the ability to put an end to not only his undertakings but to his life. Therefore, it would seem that one of the reasons why he concealed his face from his Judean/Israelite followers was to prevent himself from being recognized, and therefore targeted by the other Anunnaki/Elohim.

Instead of supreme kingship, Nergal was relegated to the position of chief-officer in his cousin's army, where he received the title: "Marduk of battle" (Livingstone 1999: 622). It may have been while serving time in this position that he acquired the experience that he would need to eventually become the "Lord of Armies."

Despite the denunciation of Babylon in the biblical account, Marduk's legendary city-state was actually a prosperous center of culture and commerce in that region and era. Indeed, at one point it was the home of the legendary "Hanging Gardens of Babylon," which was considered to be one of the "Seven Wonders of the World." This explains why Lord Nergal Yahweh never completely destroyed the city of Babylon when he had the chance. This is most likely because this was a prestigious center of power that he intended to acquire for himself.

Although we may be able to find instances in the records where Marduk may have not always acted wisely, we do know that he was primarily known for his nobility, kindness, and strength. According to the Babylonian texts, he was known as the "shepherd of the people," "the pure god," and "the god who gives life."

> Marduk, the great Lord, protector of mankind, looked
> with joy upon his good deeds and righteous heart.
> —Cylinder of Cyrus

Like his father before him, Marduk assumed the role in the biblical version as an "adversary" and as an "accuser." Although they may have not been perfect, neither Bel Marduk, nor his father, Enki, displayed any type of behavior that could be considered to be evil. Instead, it is evident that the only crime that was committed by these individuals was that they did not submit themselves as servants of Lord Nergal Yahweh or his father, Lord Enlil El.

* * *

In the book of Job, we are told of a meeting that takes place between the "sons of God." Among the visitors was the chief adversary of Lord Yahweh himself (i.e., Bel Marduk); who is erroneously portrayed as "Satan":

> One day the sons of God came to present themselves before the Lord, and Satan also came with them. The Lord said to Satan, "Where have you come from?" Satan answered the Lord, "From roaming through the earth and going back and forth in it.
> —Job 1.6

Yahweh and his adversary, who he clearly had a civil relationship with, enter into a contest to see just how obedient his Earthling servant really was. He proceeded to inflict Job with various distressing hardships. However, it seems that the account that is recorded in the Old Testament/Tanakh may have only been one-half of the story. It seems that Yahweh too had made a similar challenge to his cousin. The account of this side of story seems to have been reported in a Babylonian text from the same approximate period, titled Ludlul Bel Nimegi, Tabu-utul Bel—which is also known as the "Babylonian book of Job." This account records the story of a devout servant of Marduk, who is also mysteriously afflicted with terrible adversity. After a period of time, Marduk sends an emissary to his aid. Unfortunately, several of the crucial lines at this point are illegible because of damage to the text that occurs after the name "Shubshi-meshri-Nergal" appears. Therefore, it would seem that a servant of

Nergal may have been presiding over the situation, perhaps to see if Marduk's servant would break down and curse his god.

> My limbs are destroyed, loathing covers me; On my couch I welter like an ox; I am covered, like a sheep, with excrement [. . .] The god helped me not [. . .]
> —Ludlul Bel Nimeqi, Tabu-utul-Bel

> My body is clothed with worms and scabs, my skin is broken and festering. My days are swifter than a weaver's shuttle, and they come and go without hope.
> —Job, Job 7.5-6

According to each account, both Job and Tabu-utu-Bel did not falter, and the "spirit" of their lords were "appeased." In other words, it would seem that the result of this contest between these two rivals ended in a draw.

A minimalist interpretation of these texts posits that the Hebrew version may have been inspired by the earlier Babylonian version.

* * *

According to the Mesopotamian texts, Nergal endeavored to take over the city of Babylon—which he eventually did after allegedly stealing it from Marduk at one point. However, Nergal was originally the god of the city of Cuthah.

In the late nineteenth century, the Turkish archaeologist Hormuzd Rassam unearthed the city of Cuthah in what is now modern-day Iraq. Inside the mound, known by the local Arabs as the *Tel-Ibrahim* (and *Habl-Ibrahim*, which means "rope of Abraham"), he found inscriptions in not only cuneiform and Syro-Chaldean but in Hebrew as well (Rassam 1897: 408). Attached to the city of Cuthah, he unearthed a sanctuary shrine that was dedicated to "Ibraheem"—i.e., Abraham! (Rassam 1897: 896) However, based on the available information, it is unclear at what time this construction became associated with Abraham. Nevertheless, it is certainly an ironic designation to say the least.

It should also be noted that the city of Cuthah, and its inner sacred temple, the *Meslam*, was not much different than the city of Jerusalem and its inner temple, the "Holy of Holies." Both would have looked similar. Indeed, both were most likely inadvertently dedicated to two different personas of the same deity.

* * *

In the world of the Anunnaki, titles and names were signatures of status. For example, Marduk was known as the god of "fifty names." In the Babylonian Epic of Creation, it is reported that during his coronation, Marduk was given the new name Asarluhi, in order to signify his new stature.

Nergal too went by many names and titles. He was also known as Engidudu, which meant "Lord Who Prowls as Night" (Dalley 1989: 285, 319-321). This definition is reminiscent of Jacob's encounter with him in the dead of night (Genesis 32.24). He was also known as Lugal-banda (the fighting-cock) (Hastings, et al. 1914: 645), Sharrapu (burner) (Encyclopedia Britannica Online, s.v. "Nergal"), and Meslamtaea (a reference to Nergal's temple in Cutha).[310] Epithets that were ascribed to him include: "The Enlil of the Netherworld,"[311] the mighty "Lord"[312] and the "mighty god"[313] of "heaven and earth."[314] Of course, most of these epithets can be applied to his Judean/Israelite persona as well.

* * *

---

[310] Enlil and Ninlil and A Tigi to Nergal
[311] The Death of Ur-Namma
[312] The Lament of Udug and A Tigi to Nergal
[313] The Temple Hymns text
[314] An Adab to Nergal for Su-ilisu

In his book *The Wars of Gods and Men*, Zecheria Sitchin also attempted to discover the original Mesopotamian identity of Yahweh. Although he too concluded that it was Enlil who was the original voice of the Elohim in the garden of Eden and the Tower of Babel stories, he could not identify him as Yahweh. Although Sitchin also acknowledged that it was Enki who was the Elohim who told "Noah" to build an ark, he could also not be identified as the God of Abraham. Sitchin deduced that none of the Mesopotamian gods, including Nergal, could have been Yahweh due to various contradictions. His conclusion was based on a passage that is found in 2 Kings, in which the Cutheans are mentioned as being among the foreigners who were brought over by the Assyrians to replace the exiled Israelites. Nergal is listed among the "other gods" that the blasphemous newcomers worshiped. Sitchin believed that Nergal could not have been Yahweh *and* Yahweh's abomination at the same time.

However, I believe that there are several factors that Mr. Sitchin did not consider. The situation that he was referring to occurred when Yahweh was conducting a massive Machiavellian operation that served to punish his own followers for their disobedience. According to the biblical account, he ordered another city-state kingdom that was also under his control to attack and exile his own followers. In that particular instance, it was the Assyrians who were supposedly doing his bidding. After the operation was complete, the Assyrians moved loyalists into Jerusalem to inhabit the exiled people's homes:

> Nevertheless, each national group made its own gods in the several towns where they settled, and set them up in the shrines the people of Samara had made at the high places. The men from Babylon made Sucoth Benoth, the men from Cuthah made Nergal [. . .]
> —2 Kings 17.29-30

The biblical account alleges that the people who took over in place of the Israelites were an abomination to Yahweh because they worshiped other gods; however, at no time did Yahweh himself ever indicate this during that particular episode. When some of the Assyrian settlers were attacked and killed by lions, the *assumption* was made that it was Yahweh who had sent the lions to punish the people; although again, nowhere did Yahweh himself indicate this. Moreover, according to the account reported in 2 Kings 18.25, the chief commander of the Assyrian army was sent to Jerusalem to convince King Hezekiah to give up without a struggle. He even disclosed to the people of Jerusalem that it was the "Lord himself" who had told them to attack the city. If Yahweh had wanted to rid the Holy Land of the worshipers of Nergal, he could have easily have done so—just as he had allegedly destroyed the Assyrians after his use for them had waned (2 Kings 19.35). A plague, a famine, or even an explosive blast would have done the job. The incident with the lions was not related, but was interpreted by the people as the will of Yahweh. Therefore, the contradiction that Sitchin cites does not exist.

The following Assyrian text documents the role that Nergal/Erra played in King Ashurbanipal's victory over an enemy:

> Their flesh I fed to the dogs, pigs, vultures, eagles—the birds of heaven and the fishes of the deep. After I had done these deeds (and so) calmed the hearts of the great gods, my lords, I took the corpses of the people whom Erra had laid low or who had laid down their lives through hunger and famine and the remains of the dog and pig feed, which blocked the streets and filled the broad avenues; those bones (I took) out of Babylon, Kutha, and Sippar, and threw them on heaps [. . .]
> —Ashurbanipal, The Dynasty of Dunnun Text

Here we have an example of Nergal/Erra ordering one of his servants to attack his own city—most likely because they were disobeying his commandments. Likewise, in the ninth century BCE, King Shalmaneser III and his army marched against the cities of Babylon, Borsippa, and Cuthah (i.e., "Kutha"). What is significant about this event is that one of the gods—who the Assyrian record tells us aided in the attack on the city of Cuthah—was Nergal himself, who allegedly gave the king special "weapons" that were used to destroy his own city. It may be possible that these are the same "weapons" that are also referred to in the biblical version.

* * *

Even though Nergal may have endeavored to keep his seditious plans against his rivals in the council of the Elohim a secret, it seems that he was not always able to keep himself out of trouble. In the assembly of the Elohim, he was known as a rebellious trouble-maker. We find a revealing account of his personality in the Nergal and Ereshkigal text. This report documents the story of a visitation to the council of the gods by Namtar, who was the official emissary of the high-ranking female goddess, Ereshkigal. When he entered the room of the assembly he was respectfully greeted by everyone except for Nergal, who arrogantly ignored him. When Ereshkigal was informed of this act of disrespect, she complained to the council about the "bare-headed" one who sat in the back of the room and who refused to acknowledge the presence of her emissary. She then ordered the offender to be brought before her, which Nergal refused to comply with. After some coaxing by Enki, who assigned seven emissaries (i.e., "angels") to guard him on his trip, he relented and journeyed down to the lower world (perhaps Africa). Upon entering the throne room, he walked up to Ereshkigal, grabbed her by the hair, pulled her down off of her throne and threatened to kill her. She pacified him by offering herself as his bride, as well as "Kingship over the wide Earth"; which is of course, what Lord Nergal Yahweh had always desired the most.

* * *

The history of the rivalry between Nergal Yahweh and Bel Marduk is extensive and complex. Although the ancient reports, including the Bible, tells us that the capital city of Babylon, of the nation-state of Babylonia, was usually under the control of Bel Marduk, it also tells us that it was at times under the control of Lord Nergal Yahweh as well.

Babylon was sacked by the Assyrian King Sennacherib in 689 BCE. 2 Kings tells us that the Assyrian king before him, Shalmaneser, was indeed one of the servants of Lord Nergal Yahweh.

According to Isaiah 45.1, Cyrus, the king of Achaemenid Persia, was also under the guidance of Yahweh; although, according to the Cylinder of Cyrus text, he was also a servant of Marduk as well. Yahweh apparently even referred to Cyrus as his "anointed" one (i.e., Messiah). Here we find another example of Yahweh supposedly using foreign kings that were also under his control to do his bidding. In Ezra 1.1 (and 2 Chronicles 36.22) it is reported that Cyrus, the king of Persia, "in order to fulfill the word of the Lord spoken by Jeremiah," proclaimed that the Lord God had told him to build a temple in Jerusalem.

Just like the Judean/Israelites, the Assyrians also used the written word to record and publicize their exploits and the almighty power of their god. Some of this was done to act as propaganda, in order to strike fear in the hearts of their enemies.

> The Assyrians publicized their atrocities in reports and illustrations for propaganda purposes. In the tenth and ninth centuries BCE, official inscriptions told of cruelty to those captured [. . .]. These horrifying illustrations, texts, and reliefs were designed as a warning to frighten the population into submission.
> —Karen Rhea Nemet-Nejat,[315] *Daily Life in Ancient Mesopotamia*

One of the purposes of the books of the Tanakh was to publicize the punishments of the disobedient, and therefore inflict fear not only into the hearts of the enemies of the Lord but especially his own followers. This element of fear played a significant role in the

---

[315] Karen Rhea Nemet-Nejat, Ph.D., is a visiting instructor at Yale University.

preservation effect that occurred, even after the alleged physical departure of Lord Nergal Yahweh from the Earth.

An inscription on the fifth and final tablet of the Erra and Ishum text tells us that the account was personally reviewed by King Ashurbanipal (663-633 BCE), "son of King Escharhaddon" (680-669 BCE). This was the same Assyrian king who was the son of King Sennacherib (740-681 BCE); the same Sennacherib who is reported in 2 Kings as being the one who was ordered by Yahweh to attack both the kingdoms of Israel and Judah. In Isaiah 10.5, Yahweh calls the Assyrians "the rod of my anger, in whose hand is the club of my wrath!" It was these kings who were acting under the order of Lord Nergal Yahweh who may have dedicated a city gate (i.e., the "Nergal Gate") in Nineveh to him.

Despite the usefulness that the Assyrian super-power provided him, it seems that they too eventually fell out of favor with their master. After King Hezekiah renewed his allegiance, Yahweh, allegedly, put an end to the Assyrian threat by having one of his minions destroy the Assyrian camp in the middle of the night. Yahweh may have put an end to Sennacherib by apparently either allowing, or ordering the kings own sons to murder him (2 Kings 19.37). Escharhaddon, a servant of Nergal and the father of Ashurbanipal—the sponsorer and overseer of the Erra and Ishum text—proceeded to take over in his place.

In the book of Jeremiah, we are told that Yahweh had instructed Jeremiah to prophecy the fall of (King Zedekiah's) Jerusalem. The Judeans would then be handed over to Nebuchadnezzar, the king of Babylon. It is Nebuchadnezzar who ordered Nergal-Sharezer and Nebuzaradan to make sure that no harm came to Jeremiah. Here we find an example of the servants of Nergal and the servants of Yahweh apparently working together to fulfill their Lord's Machiavellian plan.

After the Judeans had been sufficiently terrorized into submission and prayed to Lord Nergal Yahweh, the god then proceeded to destroy the Babylonians who had attacked the holy city of Jerusalem; after which he was able to be presented as the almighty savior of the people.

* * *

It is therefore not a coincidence that both Nergal and Yahweh despised the Egyptian god Ammon (i.e., Ammen/Amun/Amon). Sitchin identified the god Ammon Ra as the Egyptian identity of Bel Marduk (Sitchin 1985: 127-128, 149, 228). The word *ammon* means "hidden," or the "hidden one" (Allen 2014: 222; Assmann 1999: 28)—which may indicate that Marduk was not frequently present in physical form on the Earth. It is indeed likely that the god Baal Hammon was Bel Marduk's Carthagian name. Therefore, the theonym Hammon[316] is most likely a Carthagian pronunciation of the word Ammon. The name of the Sun god, Ra (i.e., Re), was attached to this name in the Egyptian New Kingdom period. This could also indicate that Marduk, Ammon, and Ra, were, at least at one time, inspired by the same individual. Indeed, both Marduk and Ammon were powerful storm gods who occupied similar positions in their respective pantheons (Smith 2002: 81).

Another reason why Yahweh may have turned his attention back to Egypt was to seek vengeance on those who had returned there after the Exodus (Isaiah 30). Indeed, he had a personal score to settle with those who humiliated him by returning to Egypt, the nation that he had taken so much pride in rescuing the people from.

According to Isaiah 19.18, after the armies of Yahweh had conquered Egypt, the cities of Egypt were forced to speak "the language of Canaan." Egyptian records tell us that just like Marduk, Nergal may have been known by another Canaanite designation in that same region. The Egyptians, as well as at least some of the Semitic people of Canaan, may have known him not as Yahweh, Nergal, or Erra, but rather as Resheph. Indeed, mainstream

---

[316] In his book *Canaanite Myth and Hebrew Epic* (Havard University Press, 1973), Frank Moore Cross, Ph.D., associates Baal Hammon (i.e., "Ba'l Hamon") with the Canaanite god El. However, there is no consensus among scholars regarding the etymology of the word *hammon* (i.e., *hamon*). However, the similarity between the words Hammon and Amon is attested in the Encyclopaedia Britannica: www.britannica.com/biography/ Baal-Hammon

scholarship acknowledges that Nergal and Resheph were manifestations of the same deity (Botterweck, et al. 1992: 13; Munnich 2013: 58; Xella 1999: 701; Lewis 1999: 333).

> [. . .] there is a basic syncretistic impulse in Near Eastern polytheism which tends to merge gods with similar traits and functions. A minor deity, worshipped [sic] by a small group of adherents, may become popular and merge with a great deity; major deities in a single culture's pantheon may fuse; or deities holding similar positions in separate pantheons may be identified.
> —Frank Moore Cross, *Canaanite Myth and Hebrew Epic*

Although the etymology of the theonym Resheph is uncertain, it is generally agreed that it derives from the Semitic word for "spark," or "burning," "flame," or "lightning," and can also be associated with the burning fever of "plague" (Botterweck, et al. 2004: 10-11; Cook 1921: 87-88; Munnich 2013: 8).

> Certain references in Hebrew poetry reflect the character of the Canaanite Resheph, the sinister god who slew men in mass in war and plague. Thus the poet in Habakkuk 3.5 used *resheph* with *debner*, both meaning 'plague,' personified as attendants of the God of Israel on His epiphany according to a convention well known in Ugaritic literature [. . .].
> —John Gray, *Near Eastern Mythology*

Indeed, the ancient texts tell us that Nergal, Erra, Yahweh, Resheph, all bore similar characteristics.

> The weapons of Nergal, lightning and the plague, indicate his function as a solar deity specifically

manifest in the fierce noonday heat, recalling 'the plague that rages at noonday (Psalm 91.6).

—John Gray, *Near Eastern Mythology*

The "lightning" that was used by Resheph and Nergal as a weapon is no different than the power of the storm that was used by Yahweh to smite his enemies:

> [. . .] He sends lightning with the rain and brings out the wind from his storehouses.
> —Jeremiah 51.16

The epithets of Resheph include: "the ravager," "the burner," and "Lord of the arrow" (Encyclopedia Britannica Online s.v. Resheph; Munnich 2013: 226). Just like Nergal/Yahweh, Resheph was also a god of disease, and was referred to as "Resheph of the army" (Mettinger 1999: 921). In Egypt, he was also known as the "Lord of the Sky," and was also described as a "prince" (i.e., son of the dominant patriarchal god) in the Ugaritic texts.[317] Furthermore, just like Yahweh, Resheph was not always only associated with war and death but with healing and benevolence as well (Xella 1999: 701). Indeed, Resheph was also known as the "hearer of prayer" and the "Lord of Heaven" (Cook 1921: 84), and was also attended to by a retinue of priest servants (Xella 1999: 701).

Resheph is attested all the way back to the third millennium BCE (Xella 1999: 701). In a minimalist context, this indicates that the Yahwehist priest-scribes, either consciously or subconsciously, modeled Yahweh after this earlier identity—along with Nergal. Indeed, at least one scholar that I am aware of (M. Liverani) has also traced the history of Yahweh back to Resheph (Mettinger 1999: 921).

The word *Resheph* does appear in the Bible as an Ephraimite family name (1 Chronicles 7.25). However, the appearance of the deity form is less certain. Although it is possible that the theonym

---

[317] Kirta, tablet II.

appears in Deuteronomy 32.24 and possibly Psalms 78.48 (Xella 1999: 702), a more compelling reference appears in the book of Job:

> Yet man is born unto trouble, just as the sparks [resheph] fly upward.
> —Job 5.7

This sentence may seem insignificant at first look. However, this same passage could also be translated using the original word itself, which could yield the following:

> But Earthling man is born into trouble, as Resheph flies upward.
> —Job 5.7

What is especially notable about this statement is how it reads in a translation that is closer to its original Hebrew form:

> For Earthling man is born to trouble and the sons of Resheph fly upwards.
> —Job 5.7

Indeed, the scholar Maciej Munnich[318] believes that the translation "sons of Resheph" is the most "relevant" (Munnich 2013: 232); which also accords with the translation that is used by John Day (Day 2002: 203).[319] This is because the word *bene* (sons) does appear in the original Hebrew text (Xella 1999: 703).[320]

In this case, we must ask if this statement could have been a reference to the west Semitic identity of Yahweh. If this is true, then

---

[318] Maciej Munnich, Ph.D., is the head of the Department of Ancient History at the Catholic University of Lublin.

[319] John Day, Ph.D., was a professor of Old Testament Studies at the University of Oxford.

[320] See also: www.scripture4all.org/ for an English-Hebrew interlinear translation of the Bible.

the author could have been indicating that only Yahweh/Resheph, and those who were closely affiliated with him, who would be able to live a trouble-free life in the heavens.[321]

In some early translations of this passage, the word Resheph is inexplicably translated as "bird"; despite the fact that there is no valid reason for the word bird to appear (Day 2002: 202). Day suspects that this was a result of a misunderstanding; however, I suspect that it is more likely that this was a deliberate attempt to expunge a reference to another god from the Yahwehist record. The revisionist may have believed that he was simply correcting a mistake that had been committed by the original scribe. This also explains why most popular biblical editions omit the word "sons" in their translations of this controversial verse. This was most likely done in order to conform the translation to Yahwehist creed. Indeed, Munnich also realized that the Judeo-Christians have been inserting their own erroneous interpretations into the translation (Munnich 2013: 220). However, he leans more toward a monolatrous explanation, which holds that although the deity Resheph was indeed alluded to, he was only a subordinate demon who was among the elite servants of Yahweh (Munnich 2013: 218-219). Although this is a possibility, I am also not convinced of this. This is because passages that would seem to support this interpretation (e.g., Habakkuk 3.5) may actually be references to the literal definition of the word *resheph* itself—which denotes plague, lightning bolts, and flames; rather than the deity Resheph. Indeed, just as disease, etc., accompanied Yahweh, so did disease accompany Nergal (Wiggermann 1992: 96). It is also interesting that Munnich also acknowledges that the characterization of Yahweh as the divine archer, who uses his arrows to punish the wicked (e.g., Deuteronomy 32.23, 32.42, etc.), bears a significant resemblance to the deity Resheph in the Ugaritic texts (Munnich

---

[321] Munnich does not believe that the Job passage attests to the veneration of Resheph in a Yahwehist context (Munnich 2013: 233); however, his objectivity on this specific matter can be called into question. This is due to the fact that he was both educated and is employed at a Judeo-Christian institution; namely, the Catholic University of Lublin.

2013: 220). These equivalent depictions help to support the conclusion that Resheph was not a supporting attendant of Yahweh, but rather was the earlier Levantine identity of Yahweh. In a maximalist context, this is warranted due to the fact that "Yahweh" was not the deities original name. Indeed, the Bible itself reports that the "Lord Almighty" never revealed his name.

Over the course of years, Resheph became integrated and eclipsed by his altered Kenite/Levitical persona. The reason for the ambiguity in this matter may be due to misapprehensions of the ancient people themselves. Indeed, it must be understood that those interpretations occurred in an era before the existence of academic institutions that would have helped to establish more accurate definitions.

However, if the author of the book of Job was referring to Resheph, why then is the deity addressed as El and Elohim in the very next sentence? (Job 5.8) One possible reason is that this reference to the deity Resheph was inserted by one of the scribes who worked on the book of Job. Indeed, in the ancient era it was not uncommon for texts to be expanded upon by subsequent editor-scribes (van der Toorn 2007: 125-126, 130-132, 138-140). Indeed, the consensus among scholars is that the book of Job was most likely not written by a single author (Eisen 2004: 138; Fee and Hubbard 2011: 306). Another possible explanation is that the term El and Elohim were used only as a general term for "God."

* * *

Just like the rivalry that occurred between Yahweh and Bel, and Nergal and Marduk, Resheph also had the same nemesis to contend with; however, in the Ugaritic version, Lord Nergal Yahweh is not referred to as Resheph, but rather as "Mot."

The theonym *Mot* derives from the Semitic word for death (Bremmer 1999: 383; Lewis 1999: 227). It seems that the people did not know the name of the deity who caused so much hardship for them and simply referred to him by a word that was associated with his characteristic. Indeed, this explains the theonym Resheph as well.

Just like Nergal, Mot was not only the son of El, but, according to the Baal Cycle tablets, he too was given a lesser "inheritance," and suffered "shame" under the dominating reign of Baal. Indeed, Mot was considered an outcast, and does not appear in the Ugaritic cultic texts and personal names (Healey 1999: 598-600). Also, just like Lord Nergal Yahweh, Mot's ambition was to eliminate this very same rival from the Earth and become the almighty supreme being in his place:

> At the feet of Mot fall and bow down, pay him homage
> and honor him.
> —The Baal Cycle

In the Ugaritic Baal Cycle, Baal is referred to as the son of Dagon. Dagon is the Semitic name for Enki/Ea. Therefore, in most instances, Baal was the Semitic designation for Marduk; just like El was Enlil, and Mot was Nergal. We see in this case that the pieces to this puzzle fit together quite well. Indeed, the syncretism between the Ugaritic and the Mesopotamian pantheons is also acknowledged by mainstream scholarship (Coogan 1978: 75-79).

A reference to the Ugaritic persona of Lord Nergal Yahweh may appear in the following biblical passage, in which the "firstborn" of Mot (*bekor Mawet*) is referred to:

> It eats away parts of his skin; death's [Mot's] firstborn
> devours his limbs.
> —Job 18.13

In the very next sentence there is a reference to the "king of terrors." Scholars debate amongst themselves as to who this could refer to. Some believe that this refers to Nergal; others believe that it refers to Nergal's emissary, Namtar (i.e., the angel of death); others believe that it refers to Mot (Rutersworden 1999: 487). The point of contention lies in the absence of texts that explicitly report Mot/Nergal/Resheph/Erra/Yahweh producing progeny that could be associated with this reference to a "firstborn." However, I propose that the answer to this question may be right under our noses, so to speak; namely, in the very same book of the Bible itself. In Job 5.7, there is a reference to the previously cited "sons of Resheph." Therefore, the firstborn could refer to the progeny of the west Semitic identity of Nergal, i.e., Resheph, who was the god of "death."

The following reference to Mot bears a remarkable resemblance to another biblical account that reports that Yahweh killed every firstborn male in Egypt (Exodus 12.23-29):

> Death [Mot] has come through our windows and entered our palaces. Death has cut down the children in the streets and the young men in the marketplaces.
> —Jeremiah 9.21

How then can Yahweh be both himself and his own agent? In a maximalist context, this ostensible contradiction can be attributed to the confusion that Yahweh himself sought to inflict on his own people. In a minimalist context, this can be attributed to a natural lack of understanding and cohesion in a time in which education was not accessible to the common person. This also explains the apparent contradiction that appears in Isaiah 25.8, in which Yahweh is described swallowing up death. However, what the people did not know, including the authors of the book of Isaiah, is that the god of death and Yahweh were originally the same individual. Indeed, the possibility that some characteristics of Yahweh had been drawn from an earlier Canaanite tradition that not only involved El and Baal but Mot as well is acknowledged by mainstream scholarship.[322]

The following passage references not only Hosea 13 but other biblical verses as well (e.g., Psalm 48.7; Isaiah 27.8; Jeremiah 18.17,

---

[322] For more information on this topic see: Frank Moore Cross. *Canaanite Myth and Hebrew Epic: Essays in the History of the Religion of Israel*. (Harvard University Press, 1973); John Day. *Yahweh and the Gods and Goddesses of Canaan*. (Sheffield Academic Press, 2002). Mark S. Smith. *The Early History of God: Yahweh and the Other Deities in Ancient Israel*. 2nd Ed. (Wm. B. Eerdman's Publishing Company, 1990).

etc.), in which Yahweh is described using the "east wind" to punish and to destroy. This was an action that was also attributed to Mot.

> Biblical descriptions of the east wind as an instrument of divine destruction may have derived from the imagery of Mot in Canaanite tradition, although mythological dependency is not necessarily indicated in this instance. The juxtaposition of the east wind and personified Death in Hosea 13:14-15 may presuppose the mythological background of Mot as manifest in the sirocco.[323]
>
> —Mark S. Smith,[324] *The Early History of God*

* * *

Continuing along this line of inquiry, even more relations begin to emerge. It is a saga that extends not only westward across the Levant but eastward into Asia, where it seems that Lord Nergal Yahweh was known by yet another name in India.

When the migrating Indo-Iranian tribes entered the Indus Valley, they brought some of the cultures of the regions north of Mesopotamia with them. Central to the gods of the earliest Vedic pantheon was Indra, Agni, and Mitra (who is most likely the same deity who became known by the Zoroastrians of Persia as Mithra). Although Indra and Agni were war gods, none were considered to be as fierce and as powerful as the rebellious god, Rudra.

Rudra was the Lord of destruction, death, and disease; although, just like Yahweh, he too was also known as a benevolent god of healing and peace as well. Some of the descriptive appellations that were also attributed to Rudra were: Araga (the burning midsummer sun), Bhutapati (Prince of Demons), Mahadeva (the Great God). His epithets included: Lord of song and sacrifices, Lord of time and

---

[323] The sirocco is a Mediterranean wind that comes from the Sahara.

[324] M. S. Smith, Ph.D., is the Helena Professor of Old Testament Literature and Exegesis at Princeton Theological Seminary.

death, God of storms, God of the dead and the underworld (Coulter and Turner 2012: 406), as well as the wild one, and the terrible one (Wilke and Moebus 2011: 419). According to early Vedic texts of Hindu India, Rudra is also known as "the creator," "the bull," and "the lord of all":

> For there is one Rudra only. They do not allow a second, who rules all the worlds by his powers. He stands behind all persons, and after having created all worlds he, the protector, rolls it up, at the end of time. That one god, having his eyes, his face, his arms, and his feet in every place, when producing heaven and earth [. . .]
> —Upanishads II, (Svetâsvatara Upanishad) Third Adhyaya

> We would not wish to anger you, Rudra, the bull, by acts of unsatisfactory homage or ill praise, or by invoking you together with another god.
> —Rig Veda, Book II, Hymn XXXIII, Rudra

Rudra was also known as "the ruler of the vast world," and the one and only "true lord" that sat on the highest throne and commanded "armies.":

> [. . .] rightly you extend this terrible power over everything. There is nothing more powerful than you, Rudra. Praise him, the famous young god who sits on the high seat, the fierce one who attacks like a ferocious wild beast. O Rudra, have mercy on this singer, now that you have been praised. Let your armies strike down someone other than us. As a son bows to his father who greets him, so I bow to you, Rudra, as you approach. I sing to the giver of plenty, the true lord; being praised, give us healing medicines.

[. . .] Let the weapons of Rudra veer from us; let the great malevolence of the dreaded god go past us.

—Rig Veda, Book II, Hymn XXXIII, Rudra

Rudra also declared himself to be "the earthly manifestation of all heavenly power" (Wilke and Moebus 2011: 419-420). To appease this god and to avert the consequences of his wrath, sacrifices and praise had to be offered to him. Also, just like Yahweh, his name was not to be spoken (Wilke and Moebus 2011: 419).

The etymology of this theonym is uncertain. However, it is generally believed to derive from the root *rud*, which means to howl or to scream (Wilke and Moebus 2011: 418-419; Zimmer 1972: 125-126); although, it may also derive from the noun form of the adjectival *raudra*, which can be defined as: wild, fearsome, cruel, violent, terrible, and awesome (Wilke and Moebus 2011: 419-420).

What is also notable about Rudra is that one of his Sanskrit appellation, *siva*—which meant: the kind or auspicious one—evolved into a theonym for the subsequent and more well-known deity, "Shiva." It is therefore likely that Shiva was an alternate manifestation of the original Vedic deity: Rudra (Wilke and Moebus 2011: 419; Zimmer 1972: 181). Indeed, just like Yahweh, Shiva was both the destroyer as well as the transcendent Lord of the universe; and just like Yahwehism, Shiva worship evolved over time towards a monotheistic direction (Yandell 2015: 988).

Rudra was called the father of the maruts. The maruts were storm gods who were also the vizier attendants of the deity—just like the biblical seraphim of Yahweh, and the sebetti of Nergal/Erra. Both the Drona Parva and the Mahabharata texts report that those demi-gods possessed fiery weapons of mass destruction. The effects of those weapons included a "thick gloom" and a scorching "wind." The texts also record the presence of an unearthly class of beings known as the Manus, who were said to be the progenitors of humankind. Therefore, the Manus may have been another name for the Anunnaki.

The migrating east Europeans also brought with them into the Indus river valley the symbol of the serpent (the *Nagas*), who they regarded as a beneficial figure. This tradition may have also inspired the character of the wise serpent Mucalinda, who offered the Buddha protection from the dark metaphorical storm during his journey to enlightenment.

In the teachings of the eastern sages, we find the difference between the Enlightened mentality and the worldly way of Nergal/Yahweh/Resheph/Mot/Rudra. The following quote that is attributed to the Buddha could have been addressed to the deity himself:

> There is no fire like greed, no crime like hatred, no sorrow like separation, no sickness like hunger of heart [. . .] Look within. Be still. Free from fear and attachment, know the sweet joy of living in the way.
> —The Buddha, The Dhammapada

As the centuries passed, and Lord Nergal Yahweh departed from the Earth, Rudra also lost his significance as the Hindu religion took a more spiritual and humanistic turn.

* * *

The legacy of the Lord of Armies may have also surfaced in the Zoroastrian[325] religion of the Persians. Although it is not known for sure when Zoroaster (or Zarathustra) lived, most historians believe that the religion that he founded began sometime around the fifth century BCE.

According to Zoroaster, the universe was influenced by two primary deities. The first is the good god of light, known as Ahura Mazda (or Ormuzd), who was represented as a humanoid being inside the symbol of the flying disk. The second is the "hostile spirit" Angra Mainyu (or Ahriman). This deceptive deity was also known as the "demon of the lie," and was said to be attended to by a host of

---

[325] The Zoroastrians were persecuted by militant Muslims in the ninth century CE. Most fled to safety in India where they continue to exist to this day.

emissary minions. It may be possible that Angra Mainyu was inspired by the legend of Lord Nergal Yahweh as well.

The legacy of Lord of Armies extended all throughout the ancient world, including into Greece and Rome. Just like the character of Zeus was inspired by Enlil, and Prometheus was inspired by Enki, it may also be possible that the Greco-Roman war gods Ares and Mars may have been inspired by Nergal (Cook 1908: 93; Hastings, et al. 1914: 250). Indeed, the war god Ares (from which the Roman god Mars was based upon) was also a son of Zeus (Enlil/El), and was considered a *persona non grata* in the pantheon.

We should also consider the Greek underworld god of death, Hades (i.e., Pluto). However, unlike Nergal/Yahweh, Hades was unmoved my prayer and sacrifices. Indeed, he was mostly portrayed as an aloof god who preferred the solace of the underworld. Furthermore, Hades was not a son of Zeus, but rather was one of Zeus's brothers—which does not concur with the Mesopotamian and Levantine pantheons. While it cannot be ruled out that the concept of a dark underworld god of death was inspired by earlier legends that originated in Mesopotamia, in the case of Hades, the association appears to be mostly superficial, and therefore most likely does not apply.

Another classical mythological figure that Nergal is sometimes associated with is with the Greek demi-god, Herakles (i.e., Hercules) (Dalley 1989: 325, Greenfield 2001: 138; Aune 1999: 404). The Herakles comparison is mostly derived from his association with the lion and his identity as a warrior; although, Nergal himself was more of a commander (i.e., "Lord") of warriors. In regard to Herakles, the lion motif refers to the killing of the Nemean lion (which appears in the twelve labors of Herakles legend). However, in regard to Nergal, it refers to his lion-headed mace, as well as textual descriptions that refer to him as being endowed with the spirit of a "lion."[326] Although it may be possible that the idea of a fierce Greek god who was associated with the lion could have been influenced by an earlier Mesopotamian description of Nergal, not only is this similarity minor

---

[326] Nergal the Warrior text

but for this comparison to be made numerous dissimilarities must be overlooked. It is therefore likely that the Herakles comparison is not accurate.

Nergal is also sometimes equated with the ancient Greek Sun god, Apollo (i.e., Phoebus) (Hinnells 1975: 288). This is, surprisingly, a much more interesting candidate. The etymology of this theonym is considered uncertain (Mellink 1999: 42). It must first be understood that this deity did not originate in Greece (van den Broek 1999: 74), but rather the north-eastern Levant, most likely in Anatolia (Encyclopedia Britannica Online s.v. Apollo). It appears that Apollo was brought into that region from Mesopotamia by the Hurrians; although the Hurrians did not refer to this deity as either Apollo or Nergal, but rather as Aplu. This theonym most likely derives from the Akkadian word for son, or heir; which is *aplu* (Black, et al. 2007: 20). This is likely a reference to the original plague god who was the "son of Enlil": i.e., Nergal.[327]

The theonym Apollo is therefore most likely an altered form of the Hurrian Aplu, the Hittite Apaliunas (Mellink 1999: 42), and the Etruscan Apulu (Jannot 2005: 144). Apollo was both the bringer and the dispeller of evil. Therefore, sacrifice, hymns, and prayers had to be offered to him. Like Yahweh, he was also a powerful god of prophecy, pestilence, and religious law and order. And, like Yahweh, Apollo was also the son of the dominant institutional god, Zeus (i.e., Enlil/El). It is therefore telling that Resheph is linked together with Apollo in ancient texts from Cyprus (Botterweck, et al. 1992: 13; Xella 1999: 702). What is also significant about Apollo is that he was a god of archery (Graf 2009: 4, 13, 75). Likewise, not only was Rudra also a master of the bow and arrow (Kramrisch 1988: 31, 40, 48) but Resheph as well (Munnich 2013: 220). Indeed, Yahweh is also described with similar abilities (e.g., Deuteronomy 32.23; Psalm 64.7; Job 6.4, etc.). Therefore, the comparison between Apollo and Nergal/Yahweh is indeed tenable.

Mainstream scholars also sometimes associate Nergal with Saturn, the Roman god of agriculture (Cook 1908: 93), as well as Melqart,

---

[327] A Tigi to Nergal (Nergal C Text)

the Phoenician god of Tyre (Dalley 1989: 164; Ribichini 1999: 563), and Mithra, the Indo-Persian god of light (Hinnells 1975: 289, 353). However, the similarities between these deities and Nergal are even more superficial than those of Herakles. Simply put: the archetypes do not match. Indeed, the ancient Greeks themselves associated Melqart with Herakles (Aune 1999: 403-404, Ribichini 1999: 563; Smith 2002: 68). Therefore, these other comparisons are most likely not accurate.

I believe that the reason why scholars are more inclined to relate Nergal to other gods, such as Herakles, Melqart, Saturn, and Mithra, and not to Yahweh is due to the influence of religion. Such direct and extremely controversial associations are considered to be sacrilegious and problematic, and therefore tend to be either consciously or subconsciously avoided—albeit not always. However, such unscientific concerns serve to suppress the truth. Indeed, we must be able to go wherever the information leads us, and not be afraid of offending those who are not committed to the same practice of equitable, honest, and scrupulous investigation.

* * *

No single individual in all of human-history has had as much of an effect on the world as Lord Nergal Yahweh; and yet, no one is as misunderstood. It is an astonishing situation that can now be realized by those who are capable of coming to terms with the fact that everything that they have been led to believe about both this individual and the institutional legacy that he left behind, has been formulated upon an astonishing misunderstanding—a misunderstanding that Yahweh and/or his human cohorts sought to devise.

> The Almighty, we cannot find him out: he is excellent
> in power and judgment [. . .]
> —Job 37.23

> Impossible to perceive, difficult to understand, [O Nergal warrior] of the gods, long of arms, whose divine splendor is sublime in heaven.
> —Sublime Nergal

It is possible that the secret matters of Yahweh pertain not only to his clandestine Machiavellian operations but to his identity as one of the Anunnaki. Although he was not a legitimate first-born heir to the throne of his father, Lord Nergal Yahweh was most likely, at least at one time, a member of the divine council of the Elohim. The records also tell us that like the other members in the assembly of the gods, Yahweh flew through the skies in a fiery craft that caused a "whirlwind" upon the ground as it descended and ascended into the air. To the people of the time, these abilities seemed to be supernatural and god-like.

Although Lord Nergal Yahweh may have gone to great lengths to conceal his true identity, artistic representations of his physical appearance have been discovered. Nergal is usually identified by the lion-headed mace, or sickle sword. Accordingly, the lion was the animal that he was associated with—just as both Marduk and El was associated with the bull, and Enki with the fish and the dragon.

> O tireless mighty one, who gladdens Enlil's heart, mighty of arms, broad of chest, perfect one without rival among the gods, who grasps the pitiless deluge-weapon, who massacres the enemy, lion clad in splendor [. . .]
> —Nergal the Warrior (Hymn to Nergal)

Of course, the image of the lion god is referred to in the biblical record as well.[328]

---

[328] Also in Jeremiah 50.44, 49.19, 25.38, and Hosea 11.10, 5.14.

> A lion has come out of his lair; a destroyer of nations has set out. He has left his place to lay waste to your land.
> —Jeremiah 4.7

In the book of Ezekiel, it is written that Yahweh appeared as "a Man" who was clad in radiant apparel.

> From his waist up, he seemed to be all glowing bronze; dazzling like fire.
> —Ezekiel 1.27

A similar description pertains to Nergal as well:

> Nergal has fastened on a vestment of divine splendor and awesomeness.
> —The Terrors of Nergal

We also know from the Nergal and Ereshkigal text, that Nergal was "bare-headed"—which could either mean that he was bald, or that he did not always wear a horned helmet crown. However, as was previously noted, the bull-horned tiaras were most likely not an accurate visual representation. Therefore, the former interpretation may be the most accurate.

The following image of an early era Babylonian sculpture (circa 2000-1600 BCE) is thought to be Nergal. In each hand he holds a lion-headed mace; around his belt are a series of daggers; and on his head is a horned helmet (Figure 28):

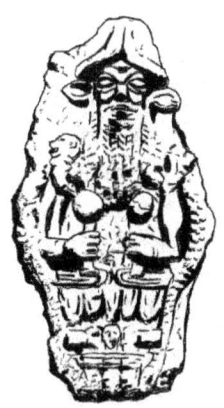

(Fig. 28)

If there is any truth to this maximalist interpretation, it can be surmised that it may have been Nergal's intention to one day step out from behind the fire and smoke, and publicly reveal himself to the world, but that day never came—at least on a mass scale. This may have been because he was either thwarted or killed by one of his Anunnaki rivals before he was able to do so—which of course, is what his Machiavellian operations were supposed to help prevent. Indeed, perhaps this may be what the violent drama that is described in the Baal Cycle text is referring to.

* * *

The critical question that we are now left with is how much of these accounts were based on an actual rogue individual among the NHIs, and how much was concocted by the priest-scribes?

If the maximalist interpretation is true, then it seems that the descriptions of Yahweh should be more explicit. However, there are two ways that this discrepancy can be excused. The first is to make the enormous conjectural leap by speculating that Lord Nergal Yahweh could have been continuing to influence events from a disincarnate state. If this is true, then he would have had to use men who are referred to in the Bible as the "prophets" as intermediaries. According to the report that is preserved in the scriptures, this is precisely what was happening; especially in the later periods. Unfortunately, the testimony of "mediums" by itself is especially unreliable. This is because in these cases we are left to essentially take their word for it that they are not presenting their own words as that of a higher power. Furthermore, what would the direct benefit to Lord Nergal Yahweh have been if he was not in this world to enjoy the success of his labors? Could he really have been that obsessed with his legacy? Could he really have been willing to linger on the Earth like a haunting ghost? Although this may not be impossible, neither does it seem likely. Nevertheless, the other reason why the maximalist perspective cannot be entirely dismissed is due to the fact that the scribes were not documenting eye-witness accounts. This could explain why the reports are not more detailed.

However, the evidence for a more minimalistic interpretation is also very compelling—indeed, perhaps even more so. This is mostly due to the previously cited pragmatic reasons that pertain to the foibles of human-nature. It must first be acknowledged that there most likely were devotees who honestly believed that they were channeling the spirit of Yahweh, and were under the pious impression that they were doing what was best for "God" and society. However, there were also others who, either consciously or subconsciously, were clearly motivated by personal concerns. By claiming a divine mandate from a higher supernatural source, these types of enterprising Yahwehists were able to endow themselves with more authority than they otherwise would not have had. In this case, it may have been that the priest-scribes true and/or perhaps subconscious motivation was only to advance their own status, and thus their own lively-hoods. The minimalist interpretation remains credible for this reason alone. Evidence of this type of behavior can be found in the criminal cases of some modern-day Judeo-Christian televangelists. Such individuals (e.g. Jim Bakker; Marjoe Gortner; Robert Tilton; etc.), who use religion to prey upon the faith of the laity, have been able to amass wealth and prestige by convincing their followers that they are God's representatives on Earth. It must be acknowledged that ancient people would not have been impervious to similar temptations. If the minimalist interpretation is true, then what the texts reveal is that those priest-scribes were inspired to endow their own god with the same powers that had been previously attributed to other revered war gods of the same approximate era and region, such as Nergal. Indeed, mainstream scholarship confirms that Yahweh was actually a composite of a number of ancient gods (e.g., El, Baal, Resheph, etc.) (Coogan 1978: 20; Cross 1997: 148; Miles 1995: 20, 94; Smith 2002: 8-9, 11, 207).

What must also be acknowledged is that irrefutable evidence that would conclusively justify any one particular interpretation does not exist at this time. However, it does seem that the minimalist interpretation is the most reasonable.

Nevertheless, it can still be maintained that it is unlikely that beings who were worshiped with such intense devotion were fabricated entirely *ex nihilo*. Indeed, most great works of fiction are usually based upon some primary truth. Furthermore, even if one were to continue to insist that Yahweh and Nergal were not the same individual, and/or insist that there is no truth to the maximalist interpretation whatsoever, the point that Yahweh was no different than the other Anunnaki/Elohim gods of the ancient world, such as Nergal and his other manifestations, remains pertinent.

* * *

Despite the controversial nature of these findings, it must be emphasized that the information is not intended in any way to be an indictment against the Jewish people themselves—hence the differentiation between the designation Jew and Yahwehist. Indeed, there are millions of Yahwehists in the world who are not of Jewish descent; just as there are millions of Jews who are not Yahwehists. Like other forms of bigotry, anti-Semitism is a form of mental and spiritual iniquity, and is not promoted or condoned in any way.

Furthermore, progressive Jewish movements are serving to reform the old misconceptions. Some of these amendments can be found in the tenets of the *Kabbalah*. In the eighteenth century, the honorable Moses Mendelssohn also helped to lead a reformative movement called the *Haskalah*, which sought to enlighten through the faculties of reason, spiritual awareness, and human rights. This humanitarian legacy has been advanced in the present day by other reformative Jewish movements, which are serving to help sincere devotees return to the original *Ein Sof* source.

CHAPTER XI

THE THEOCRACY OF JEHOVAH AND THE MISSION OF THE SAVIOR

While the sands of time were sweeping over the Middle East and covering up the old civilizations of ancient Mesopotamia around the second and first centuries BCE, this historic period of change was not taking place in the region in and around Jerusalem. Instead, the Yahwehist tradition persisted on. This was partially due to an enduring belief that Yahweh had not given up on his chosen people, and that one day a great servant of the Lord would arise in the land; a "Messiah," who would take over where the great pre-exilic kings and legendary prophets left off, and usher in an era of peace and superiority by establishing an independent theocracy dedicated exclusively to Lord Yahweh.

The word "Messiah" is an English rendering of the Hebrew word *Maschiach*, which means "Anointed." This is a reference to the ceremonial process in which a king or high-priest is anointed with oil during a coronation ceremony.

The following examples are passages that seem to refer to the future arrival of a great Savior. Judeo-Christians claim that these are references to "Jesus":

> He shall stand, and shall shepherd in the strength of Yahweh, in the majesty of the name of Yahweh his

God: and they will live, for then he will be great to the ends of the earth.
—Micah 5.4

[. . .] He will proclaim peace to the nations. His rule will extend from sea to sea and from the River to the ends of the earth.
—Zechariah 9.10

But he was pierced for our transgressions. He was crushed for our iniquities. The punishment that brought our peace was on him; and by his wounds we are healed.
—Isaiah 53.5

However, once these references are read in context with the rest of the books in which they appear, their original meaning becomes more clear.

It must first be understood that the definition of the Yahwehist Mashiach and Jesus the Christ are actually different—as any rabbi or legitimate scholar will attest. For example, the Mashiach was supposed to be a mighty leader who would be a direct descendant of King David (Isaiah 11.1);[329] however, whether Jesus was a direct descendant of King David is questionable. This is because the genealogies that are reported in the Gospels of Matthew and Luke trace the Davidic ancestral line to Joseph, not to Mary. Although, according to the gospels themselves, Joseph was not the father of Jesus. Furthermore, in both of those gospels the names listed in the ancestral lines are different, and therefore are unreliable. Judeo-Christian apologists claim that Luke's genealogy pertains to Mary; however, ancient tribal identification passed down through the father's side of the family, not the mothers (e.g., Ruth 4.18-22). Furthermore, according to Luke 2.4 it was Joseph, not Mary, who was

---

[329] Jesse is the father of David.

from the House of David. Likewise, Luke 3.23 clearly states that the listed genealogy pertains to Joseph, not to Mary.[330]

The Mashiach was also supposed to return all the descendants of Abraham to their homeland where they would live peacefully under the laws of the Torah and be a beacon of light to the world (Isaiah 11.12). However, Jesus never mentioned returning the Jews to their homeland. This is because they had already returned from the exile. Furthermore, Jesus did not strictly adhere to the customary laws of Moses (e.g., Mark 2.23-24, 3.1-6; John 5.8-15), which is one of the reasons why the priests were so offended by him. His teachings were also more favorable to the meek and the uneducated, rather than the powerful and the learned (Matthew 5.5). These qualities do not fit with the original description of the mighty Mashiach. Indeed, there is also nothing in the Old Testament that predicts a Mashiach who does not complete his mission the first time. Therefore, there is no legitimate justification for the concept of a resurrection or a "Second Coming." Moreover, according to the Tanakh the Mashiach will be human, not a demi-god; and he will definitely not be God himself. The Christian concept of "the Trinity," in which God manifests in three separate ways: that is, as the Father, the Son, and the Holy Spirit (Matthew 28.19), is actually a heretical concept according to the tenets of traditional monotheistic Judaism. This is because according to the Old Testament/Tanakh itself, "Yahweh is one" (Deuteronomy 6.4). Indeed, Yahweh made it very clear that no one was equal to him (Exodus 20.3; Deuteronomy 32.39).

Old Testament passages that allegedly refer to the Savior can be divided into three separate classifications: (1) Passages that refer to the Yahwehist Mashiach, as opposed to the Christian Christ. (2) Passages that do not refer to any Savior at all, including the

---

[330] It is also telling that, according to Luke 1.36, Mary was related to Elizabeth, who, according to Luke 1.5, was a descendant of Aaron. However, Aaron was not from David's tribe of Judah, but rather was from the tribe of Levi (Joshua 21.4; Numbers 17.8).

Mashiach. (3) Passages in which Jesus makes the intentional effort to fulfill Old Testament prophecy.

In regard to this specific topic, it must be acknowledged that an entire separate book could be written on this subject alone. For now, only a few popular examples of what are believed to be prophecies that refer to Jesus will be more closely examined. The following passage is another source that many Judeo-Christians also cite as messianic prophecy.

> A voice is heard in Ramah, mourning and great weeping, Rachel weeping for her children and refusing to be comforted, because they are no more.
> —Yahweh, Jeremiah 31.15

According to this interpretation, the "weeping" for the children that is reported refers to the "massacre of the innocents" by King Herod, who allegedly ordered the death of all male children in Bethlehem so as to prevent himself from losing the throne to a future messianic king (Matthew 2.16-17). However, this purported event is not only not corroborated by non-biblical records, such as those of the Jewish historian Flavius Josephus, but it does not even appear in the other gospels. It is also suspicious that this account only appears in the Gospel of Matthew, since, as will be made clear further ahead, the Gospel of Matthew is especially problematic and cannot be considered to be the most accurate of the accounts. It is more likely that the story was inspired by Herod's killing of his own sons—which was an event that actually *was* recorded by Josephus.[331] This was done to prevent his heirs from taking the throne from him. What the author of the Gospel of Matthew clearly did was draw upon this incident in order to suit his own narrative (Ehrman 2009: 32; Grant 1971: 229). Furthermore, Ramah was a city that was near Jerusalem. When Jerusalem was conquered by the Babylonians, the captives were taken to Ramah before they were lead away to Babylon (Jeremiah 40.1). Rachel was one of Jacob's wives and was therefore a

---

[331] The Antiquities of the Jews (Book XVI, XVII)

prominent matriarchal figure. In the passage, the author references Rachel because he believes that she would have mourned the destruction of the Temple and her descendants who had been exiled to Babylon. The weeping for her children is a reference to Genesis 30, where it is written that Rachel was tormented by the fact that she did not have children—which does not have anything to do with Jesus.

Another clear example can be found in the book of Hosea. According to Matthew 2.14-15, Jesus was taken to Egypt for a short time. Many Judeo-Christians claim that this fulfills the prophecy that is found in the book of Hosea; however, this passage in the book of Hosea was never intended to be prophecy. In this instance, the author speaks for Yahweh:

> When Israel was a child, I loved him, and out of Egypt
> I called my son. But the more they were called, the
> more they went away from me. They sacrificed to the
> Baals and they burned incense to images.
> —Yahweh, Hosea 11.1-2

It is not the Savior who is compared to a "child" and a "son," but rather the people of Israel, which is a common poetic analogue that can be found throughout the Old Testament/Tanakh (Yosef 2011). We find similar descriptions in Isaiah 41.8-9, as well as the rest of the "servant literature" in the book of Isaiah. This includes Isaiah 53, in which the entire nation of Israel is described as a singular "servant." Therefore, the son coming out of Egypt is actually a reference to the Exodus—which also has nothing to do with Jesus.

It is also peculiar that some Judeo-Christians cite passages in the book of Psalms, which is not even in the "prophetic books." This book is actually considered "wisdom literature," and the writings appearing within were not intended to be prophecy.

Besides all of these cases of forced conflations, it does also appear that Jesus himself also sought to fulfill Old Testament conceptions by making the intentional effort to do so. An example of this is when he did not enter Jerusalem until he was able to do so while riding on a donkey. This was supposedly done so that the king prophecy that is found in Zecharia 9.9 could be fulfilled. In this specific case, Jesus and his followers apparently did endeavor to link themselves with Old Testament scripture that could be interpreted to be prophetic, which was an overt means in which to legitimize their position.

However, what will become apparent moving forward is that even though Jesus was not the Yahwehist Mashiach this does not mean that he was not the Savior. This is because he had his own plan for the salvation of the people.

* * *

According to the traditional Judeo-Christian interpretation, the Christ (i.e., *Christos,* from the ancient Greek word for anointed) submitted himself as a kind of sacrificial scapegoat. By dying for the sins of the people, he was providing some type of way for those who followed his teachings to go to heaven. This is certainly a curious situation that deserves closer examination.

Although he later became known to the world by the Greek translation "Jesus," his original Hebrew name was Yeshua. The gospels report that the priests of Yahweh (i.e., the Pharisees and the Sadducees) believed that Yeshua was a blasphemer. If Yeshua had come to restore the theocracy of Yahweh, why then was he condemned as a heretic? Were the priests simply seeking to fulfill prophecy by violently persecuting the Messiah? One of the reasons why this explanation can be disregarded is because the original Maschiach was never supposed to be persecuted—especially by his own people.

In John 11:49-52, it is reported that the high-priest Caiaphas announced that it would be better for Jesus to die than to risk inciting the Romans against them all. The author of the Gospel of John interprets this as a prophecy; although this was not actually a prophecy in a supernatural sense, but rather a statement that was based on the concerns of the priests.

According to John 5.18, the traditionalists wanted to kill Yeshua because not only did he do things such as not strictly adhere to the Mosaic laws but because he dared to make himself equal with God.

Yeshua himself admitted that it was not his initial intention to bring peace, but rather to challenge the status quo:

> I have come to bring fire on the earth, and how I wish it were already kindled! [. . .] Do you think I came to bring peace on earth? No, I tell you, but division.
> —Yeshua, Luke 12.49-51

In the following passage, Yeshua contradicts an Old Testament law that was ordained by Lord Yahweh himself, who told the people that they were to avenge any misdeed that was committed upon them with equal measure; namely, with an "eye for eye, and tooth for tooth" (Exodus 21:23-25):

> You have heard that it was said, 'Eye for eye, and tooth for tooth.' But I tell you, do not resist an evil person. If anyone slaps you on the right cheek, turn to them the other cheek also.
> —Yeshua, Matthew 5.38-39

Yeshua challenged the traditionalists by advocating for a more compassionate type of mentality:

> You have heard it said, 'Love your neighbor and hate your enemy.' But I tell you: Love your enemies and pray for those who persecute you.
> —Yeshua, Matthew 5.43-44

Yahweh made it very clear that anyone who did not obey his laws was to be put to death. One of the methods by which this sentence was to be carried out was by stoning (Numbers 15.35; Ezekiel 23.47). However, in the New Testament Gospel of John (John 8.1-11) we find the well-known account of the woman who was caught committing adultery, which was a stoning offense. According to this account (i.e., the *Pericope Adulterae*),[332] Yeshua interceded and saved her from execution. Instead of the Old Testament "fire and brimstone" mentality, Yeshua advocated for something much different. In his teachings, we do not find the same violence, revenge, hatred, jealousy, and anger that is found in the Old Testament.

> Blessed are the pure in heart, for they shall see God.
> —Yeshua, Matthew 5.8

Of course, this way of thinking actually contradicted the actions of Yahweh, who proclaimed the heart to be "deceitful" and "desperately sick" (Jeremiah 17.9).

According to the Old Testament, Yahweh was interested in not only power and fame but in wealth as well (Isaiah 45.3; Haggai 2.8). Yeshua was actually opposed to this type obsessive materialism:

> How hard it is for rich people to enter the kingdom of God! Indeed, it is easier for a camel to go through the eye of a needle than for a rich person to enter the kingdom of God.
> —Yeshua, Luke 18.24-25

No wonder why the followers of Yahweh considered him to be a heretic. It would seem that Yeshua's teachings contradicted the nature of God. However, what is also perplexing about this situation is that at the same time it appears that Yeshua was also praising the Lord God as well.

Yeshua claimed to have a very personal relationship with God, who he referred to as "the Father":

---

[332] The *Pericope Adulterae* does not appear in the earliest Greek editions of the Gospel of John (Ehrman 2005: 64). I have included this example because I believe, just like the interpolater himself, that it is an example of the type of ethics that Yeshua promoted.

> Just as the living Father sent me and I live because of the Father, so the one who feeds on me will live because of me. This is the bread that came down from heaven. Your ancestors ate manna and died, but whoever feeds on this bread will live forever.
> —Yeshua, John 6.57-58

The "living Father" gives the "bread" of life to the Son, which is *not* the poisonous bread of the "forefathers." However, the bread (i.e., "manna") that came down from heaven was originally given to the people by Yahweh himself (Exodus 16.4-35). This is a very significant passage. In it, Yeshua indicates that there is a difference between the Father and the deity of the Old Testament.

This next passage may be the most significant of all:

> If God were your Father, you would love me, for I have come here from God. I have not come on my own; God sent me. Why is my language not clear to you? Because you are unable to hear what I say. You belong to your father, the devil, and you want to carry out your father's desires. He was a murderer from the beginning, not holding to the truth, for there is no truth in him. When he lies, he speaks his native language, for he is a liar and the father of lies.
> —Yeshua, John 8.42-44

Up until this point, the true meaning of this statement has not been understood. In it we are told that "the devil" was the "murderer" who deceived the people into thinking that he was the heavenly Father, when in fact he was the "father of lies."

The traditionalists who were persecuting Yeshua were only seeking to uphold the laws of their Lord; however, Yeshua denounced the "father" of their world as "the devil." Likewise, in Matthew 16.12 Yeshua warns his followers to be on guard against the teachings of the priests of Yahweh. This is because the Pharisees and Sadducees were the preservers of the teachings of the malevolent Lord of

Armies. Yeshua called these priests "blind guides," "hypocrites," and the "sons of hell."

A closer look at the gospels themselves reveal that Yahweh was not the heavenly Father of Yeshua. A closer look reveals that the secret mission of Yeshua was actually to *rescue* the people from the wicked Lord of war who disguised himself as the heavenly Father!

This is the reason why Yeshua never referred to his Father by the name Yahweh (i.e., Jehovah). In fact, the name of the Lord of the Old Testament does not even appear in the original New Testament gospels at all!

* * *

Yeshua used parables that alluded to a greater meaning in order to help open the minds of the people and help them to perceive their world in a more clear and unbiased way. This higher meaning is related to what he referred to as "the sacred secrets."

> The knowledge of the secrets of the kingdom of heaven has been given to you, but not to them [. . .] This is why I speak in parables: Though seeing, they do not see; though hearing, they do not hear or understand. In them is fulfilled the prophecy of Isaiah: You will be ever hearing but never understanding; you will be ever seeing but never perceiving. For this people's heart has become calloused; they hardly hear with their ears, and they have closed their eyes. Otherwise they might see with their eyes, hear with their ears, understand with their hearts and turn, and I would heal them.
> —Yeshua, Matthew 13.11-15

Yeshua offered a clue to the nature of the secrets of the kingdom when he revealed that it was not his intention to put new wine (i.e., the new covenant of the heavenly Father) into an old wine-skin (i.e., the old covenant of Yahweh), but rather new wine into a new wine-skin (Mark 2:21).

He also admonished the hardened traditionalists for being bound to the ruler of their world. Of course, the Old Testament itself reports that the ruler of their world was Yahweh (Psalm 47.7).

Just before Yeshua is about to be taken to the Roman authorities, he makes a very important statement to his disciples regarding this ruler:

> I will not say much more to you, for the prince of this world is coming. He has no hold over me.
> —Yeshua, John 14.30

The "prince of this world" was the prince of darkness who resided in a deep "gloom" (Psalm 97.1-2; 1 Kings 8.12; 2 Samuel 22.12; etc.). At that particular time, it was his zealous servants, the Pharisees, who were coming to take him away to the horrific ordeal that they had in store for him.

Most of the people were actually devout traditionalists who were fighting to preserve the institution that their forefathers had allegedly established under the orders of Lord Yahweh himself. As far as the pious and God-fearing people were concerned, evil and sin was disobedience itself; however, Yeshua did not share this belief. In the Gospel of Mark, he defines sin as "greed, malice, deceit," "envy, slander, arrogance," "murder," etc. Of course, these were the qualities of Yahweh himself, hence the following statement:

> All who have come before me were thieves and robbers.
> —Yeshua, John 10.8

Likewise, in the following passage Yeshua declares his intention to "save" the sons of Abraham—that is, to save the sons of Abraham from the God of Abraham:

> Today salvation has come to this house, because this man, too, is a son of Abraham. For the Son of Man came to seek and to save the lost.

—Yeshua, Luke 19.9-10

After Yeshua disclosed that the importance of his mission exceeded that of Abraham's (John 8.39-59), the followers of Yahweh took up stones to kill him. We see in these instances that the people had not fallen away from the laws of Yahweh, as is traditionally thought, but rather were resisting Yeshua in order to preserve them! Indeed, Yeshua referred to the old covenant laws, that were supposedly given to Moses by Yahweh himself, as something that was separate from himself and his heavenly Father. This is why he referred to Yahweh's commandments as "their law," and "your law," instead of the will of the Father (John 10:34, 15:25). Likewise, according to Mark 7.7-8, Jesus accused the followers of Yahweh of following the laws of man, rather than "God." However, according to the Old Testament/Tanakh, the laws were given to the priests by Yahweh himself. This statement seems to indicate that the laws were actually devised by the priests themselves, and that Yahweh's direct participation was actually minimal. If this is true, then it would serve to further indicate that the minimalist view may be the most plausible.

* * *

Yeshua made a distinction between his "Father" who is in heaven, as opposed to the chthonic Lord of the "earth":

> The one who comes from above is above all; the one who is from the earth belongs to the earth, and speaks as one from the earth. The one who comes from heaven is above all. He testifies to what he has seen and heard, but no one accepts his testimony.
> —Yeshua, John 3.31-32

In the following passages, Yeshua indicates that the Father is not who they think that it is:

> All things have been delivered to me by my Father. No one knows the Son, except the Father; neither does anyone know the Father, except the Son, and he to whom the Son desires to reveal to him.
> —Yeshua, Matthew 11.27

> No one has seen the Father except the one who is from God; only he has seen the Father.
> —Yeshua, John 6.46

Who then was the Father?

Yeshua was careful not to divulge too much information. Instead, he wisely chose a more subtle approach. This is because if he had stated anything that was overtly anti-Yahweh he would have been rejected and crucified even sooner than he was.

It seems that the Father existed in a higher spiritual state where Yeshua was able to interact with him during events of mystical epiphany. Indeed, in the following passage Yeshua refers to his Father not as a physical being, but rather as a spirit. (Note: The word "God" in the New Testament is translated from the Greek word *Theos*):

> God is a spirit. Those who worship him must worship in spirit and truth.
> —Yeshua, John 4.24

According to Yeshua, only he knew who the Father was. This may have been because he did not believe that the people would be able to comprehend and accept the controversial and astounding truth. Even the gnostic Christians, who claimed to know the esoteric teachings of the Savior, described the Father as "unrevealable."

* * *

The fact that this misunderstanding has been so successfully ingrained into the popular collective interpretation is certainly a testimony to the power of the web of misconception and influence that Yahweh and his agents were so successful in constructing. Yeshua alluded to the nature of the affected mind in the following passage:

> For this reason they could not believe, because, as Isaiah says elsewhere: "He has blinded their eyes and hardened their hearts, so they can neither see with their eyes, nor understand with their hearts, nor turn—and I would heal them."
> —Yeshua, John 12.39-40

Of course, the "he" who is being referred to in this pericope was the Lord God of Isaiah.

Unfortunately, many people, including many scholars and clergy, are not only influenced, either consciously or subconsciously, by the prevailing traditional perspective but many will feel the need to be respectful of such a sacred belief system—and thus, in some cases, preserve the security of their careers by offering interpretations that conform with traditional precepts. This was especially true in previous centuries. Indeed, those who disagreed with the established interpretation in previous centuries were often also forced to face the consequence of the death penalty.

Of course, findings that could be perceived as being "anti-Semitic" is an issue as well. However, I contend that the Jewish people should not be defined by the Yahwehists.

One of the reasons why the truth of this situation has not been more clear is because of its complicated nature. In the following verse, for instance, Yeshua seems to refer to the God of Abraham in the same terms as his heavenly Father. Here is how the verse in question reads:

> But concerning the resurrection of the dead, haven't you read that which was spoken to you by God, saying, 'I am the God of Abraham, and the God of Isaac, and the God of Jacob?' God is not the God of the dead, but of the living.

—Yeshua, Matthew 22.31-32

However, in this passage Yeshua does not indicate that Yahweh himself was the God of life, but rather that "God" was a deity who presided over not a dead people, but rather a "living" people; a people who had the ability to be resurrected in the eternal world of spirit. This passage is actually a spiritual and humanistic affirmation, rather than a tribute to Yahweh. The context of this statement (which also appears in Mark 12.26-27 and Luke 20.36-38) is that it occurred at a time when Yeshua was expressing an optimistic belief to the priests of Yahweh that the people would one day be resurrected in heaven where "they can no longer die."

It is true that Yeshua did occasionally refer to the old covenant; however, he did so only as a means of inserting his own point on top of it (e.g., Mark 7.6-8). The situation can be confusing for anyone who is not able to take the context of some of these statements into account. In some accounts, when Yeshua seems to be referring directly to Yahweh, he was actually only making a reference to the historical past. There was always a higher purpose involved when Yeshua invoked the names and the concepts related to the deity of the Old Testament. He guardedly approached the people as a fellow son of Abraham; while at the same time urging them "to get the meaning" of that which was "hidden" from their eyes:

> If you, even you, had only known on this day what would bring you peace—but now it is hidden from your eyes.
> —Yeshua, Luke 19.42
>
> The kingdom of heaven is like a treasure hidden in the field [. . .]
> —Yeshua, Matthew 13.44

In Luke 6.35, Yeshua refers to his Father as "the Most High" (i.e., *Hypsistos*), which is a reference to the term that appears as *Elyon* in the Hebrew Bible. This is significant since this was one of the epithets that was ascribed to Yahweh. Therefore, this would seem to indicate that Yahweh and the Father were the same; however, the epithet Elyon is actually only a generic epithet, and does not have to specifically refer to Yahweh. Indeed, this epithet originally applied to other gods, including both the Canaanite gods El and Baal (Elnes and Miller 1999: 295).

In John 2.13-16, Yeshua chases out the merchants from the Temple area. In this case, the question must be asked: Why did Yeshua refer to the Temple of Yahweh as his "Father's house"? According to the same book (John 2.17), the reason this was done was in order to fulfill "the words of scripture." This is a reference to Psalm 69.9, in which it is written, "It is the zeal for your house that has consumed me." However, as was previously noted, the book of Psalms was never intended to be prophecy. A more likely reason that Yeshua may have done this was to make a statement about the corrupting link between money and religion. Therefore, Jesus could have been indicating that the Temple should have been a holy shrine dedicated exclusively to *his* heavenly Father, not to Yahweh and his materialistic associates. It is also possible that Yeshua may have never referred to the temple as his Father's house, and that this was an assumption that was made by the scribes who were recording events decades later. Indeed, it is now known that the gospels did not begin to be committed down to paper until forty to seventy years after the events occurred, and therefore absolute accuracy should not be expected. Furthermore, as will be made more clear in parts to come, many discrepancies and misleading additions can be found in the gospels.

The other time that this allegedly occurred—and I stress the word *allegedly*—is in Luke 2.46-49, where it is reported that as an adolescent Yeshua referred to the Temple as his "Father's house." However, not only was Yeshua just a child at that time, and therefore was most likely unaware of the true identity of his heavenly Father, but since no scribe was actually present during his childhood, this account cannot be regarded as certain by any means.

* * *

In regard to the concept of the "Trinity," we are left with the question pertaining to the identity of the "Holy Spirit"; or, as it is sometimes referred to as, the "Holy Ghost."

The traditional Judeo-Christian interpretation asserts that the Holy Spirit is a manifestation of God, who is thought to be coincident with both the "Father" (God) and the "Son" (Christ) (Matthew 28.19), and is thought of as a person. However, this is not the original definition. For the original definition of the Holy Spirit we must refer back to the Old Testament. However, in these instances the Holy Spirit (*Ruach Kodesh*) is not specifically described as a person, but rather as divine energy, which is the breath of life (i.e., *pneuma*) (Reiling 1999: 418).

> Do not cast me from your presence, or take your Holy Spirit from me.
> —Psalm 51.11

When the word is directly associated with Yahweh, it then means "the spirit of God."

Furthermore, referring to the Holy Spirit as a singular "person" violates the precepts of traditional monotheistic Judaism. Indeed, we are explicitly told that "Yahweh is one" (Deuteronomy 6.4), not three. Judeo-Christian apologists will counter that this is not a contradiction because these three forms are manifestations of the same entity (i.e., *homoousios*)—which they relate to the divine essence of God (i.e., *ousia* or *hypostasis*). However, this interpretation contradicts the original biblical report, which specifically states that it is Yahweh himself who is one.

Judeo-Christians cite passages in the New Testament (e.g., Acts 13.2), in which the Holy Spirit supposedly speaks, in order to justify the personhood of the Holy Spirit; however, these examples are drawn from Pauline sect misunderstandings that occurred at a later time, and therefore are incorrect. (The subject of Pauline misunderstandings will be examined further ahead.)

* * *

In order to understand what is really happening in the gospels, it is necessary to understand the history and authorship of these primary books of the New Testament. Concerning these matters, biblical scholars posit a "two-source hypothesis" (Ehrman 2009: 153; Harrington 1991: 6). The first source has been traced to the Gospel of Mark. However, it is now known that the author of this book was not only not a direct disciple but he was not even Mark the Evangelist, the companion of Peter (Burkett 2002: 121, 156; Ehrman 2009: 109, 112). Scholars now know that the author of this gospel was actually an anonymous Judeo-Christian scribe, who compiled the accounts from word-of-mouth sources approximately forty years after the events occurred (Burkett 2002: 156-157; Ehrman 2009: 112).

Besides the Gospel of Mark, the other original source material that scholars believe that the New Testament authors, such as Matthew and Luke, referred to is the so-called "Q" material (from the German word *Quelle*, meaning "source"). Although the original Q source has yet to be found, most biblical scholars agree that it must have existed (Ehrman 2009: 153; Mack 1993: 4).

In the Gospel of Matthew, there is a passage that is neither in the Gospel of Mark nor in the Q source, which contradicts the true mission of the Savior—as well as other passages in the gospels themselves. This is because this statement that was attributed to Yeshua was added at a later time by someone who did not know the secrets of the Savior.

> Do not think that I have come to abolish the Law or the Prophets; I have not come to abolish them but to fulfill them. For truly I tell you, until heaven and earth disappear, not the smallest letter, not the least stroke of a pen, will by any means disappear from the Law until everything is accomplished.
> —Yeshua, Matthew 5.17-18

Indeed, the Gospel of Matthew is filled with errors. For example, Matthew 27.9 references a passage that is attributed to the prophet

Jeremiah together with Judas and thirty pieces of silver, which is presented in a prophetic context; however, no such passage exists in the book of Jeremiah. What is even more telling are the passages in the Gospel of Matthew that contradict information that is reported in the other gospels. For example, in John 21.16 Yeshua refers to Simon Peter as the "son of John," but in Matthew 16.17 he refers to him as the "son of Jonah." In Luke 3.23 it is reported that Joseph was the son of Heli, who was the son of Matthat, the son of Levi, etc., but in Matthew 1.15 it is reported that Joseph was the son of Jacob, who was the son of Matthan, the son of Eleazar, etc. In Luke 11.1-4, it is reported that Yeshua delivered the Lord's Prayer only to the disciples, but in Matthew 6.9-13 (beginning in Matthew 5) he is said to have delivered the Sermon on the Mount before the multitudes. In Luke 23.39, it is reported that only *one* of the criminals who was crucified next to Yeshua insulted him, while the other one accepted Yeshua as the Savior, but in Matthew 27.44 *both* of the criminals "reviled him." In Mark 10.19, Yeshua instructs the people to "honor your father and mother," but in Matthew 10.35 Yeshua causes "division" between a "man and his father, and a daughter against his mother." The Gospel of Matthew not only contradicts information that is recorded in the other gospels but there are even passages in the Gospel of Matthew that contradicts itself! For example, Matthew 1.17 lists fourteen generations between Abraham and David, while Matthew 1.2 lists thirteen. Likewise, Matthew 10.5 and 15.24 report that the gospel was only to be reported to "the lost sheep of Israel," but in Matthew 12.17-21 and 28.19 the gospel was to be spread to the gentiles. In Matthew 5.22, Yeshua says do not call someone a fool, but in Matthew 23.17 he calls the Pharisees "fools!" This is because the account that is recorded in the Gospel of Matthew is the least accurate of all the gospels.

Although it cannot be denied that some truth may be able to be found in the Gospel of Matthew, using a single passage from this document to argue a point that is out of line with the message that is found repeatedly elsewhere throughout the gospels, including more authentic source material, cannot be justified.[333]

New Testament scholar Bart D. Ehrman presents a strong case in his books: *Misquoting Jesus, Jesus Interrupted, Forged,* and others, that disproves the claim that the Bible is the "inerrant" word of God. His works reveal that no original copies of the Bible exist, and that the copies that do exist are not only from centuries later but contain differences that were both unintentional mistakes as well as intentional edits and additions that were committed by scribes. Furthermore, in a statement that can be applied to Matthew 5.17-18, Ehrman explains:

> In my experience, theologians do not hold to a doctrine because it is found in just one verse; you can take away just about any verse and still find just about any Christian doctrine somewhere else if you look hard enough.
> —Bart D. Ehrman, *Jesus Interrupted*

What must be taken into account is that these were documents that were both written and interpreted by man; and even a devout traditionalist cannot honestly deny that man is an imperfect being. Therefore, the previously cited passage, in which Yeshua is reported to have claimed that he did not come to abolish the laws of Yahweh, cannot be considered to be an accurate representation of what he actually said.

Although the author of the Gospel of Matthew is attributed to the disciple of the same name, scholars know that it was actually written by an anonymous scribe, and that the name of the disciple was added as a title to the document at a later time (Harrington 1991: 8). This

---

[333] This same reasoning can be applied to the controversial passage that is found in Luke 19.27; in which the author (L Source), who was not one of the original disciples, inserted a reference to Jesus demanding that his enemies be brought before him and killed. This is a significant contradiction and is most likely not an accurate representation of what was actually said.

was done in order to increase its credibility—which was a common practice during that era (Barnstone 1984: 517; van der Toorn 2007: 28, 33-36). It is evident that the author used the Gospel of Mark, as well as the original Q source, as a reference, which he then infused with the interpretation that was popular among the group of Christians (i.e., the Mattheans) that he was a member of (Harrington 1991: 8). Scholars believe this group was most likely located in Antioch (modern-day Turkey) (Harrington 1991: 9; Nolland 2005: 18). Indeed, according to Paul's Epistle to the Galatians (Galatians 2.11-14), Antioch was the location of a Jewish-Christian community who believed in the retainment of Mosaic law (i.e., the "circumcision group"). It was also a group that was specifically affiliated with Peter (i.e., Cephas/Kefa), which explains why Peter is referred to as the "rock" of the Christian movement in, and only in, the Gospel of Matthew (Matthew 16.18). It can therefore be concluded that it was a member of this community (i.e., M source) who inserted favorable references to both Peter and Old Testament law in the Gospel of Matthew.

In this case, some might argue that the community in Antioch could have received direct information from Peter; however, even though Peter was one of the original disciples he cannot be considered to be a reliable source. This is because, according to the gospels themselves—including even the Gospel of Matthew—Peter never understood the secret true mission of the Savior. (This subject will be examined more closely further ahead.)

In summary, it is now known that the gospels were subjected to not only unintentional mistakes but also intentional revisions and editing. Scholars discovered this by comparing earlier versions of the gospels against later ones (Ehrman 2005: 10). In this case, what we find when we delve deeper into the history of the Bible, is a record of inconsistency, mistakes, and even outright deliberate misrepresentation. Therefore, the only way to truly discern what actually took place all those centuries ago is to perceive the underlying essence of the reports, and not allow ourselves to get misled by dubious additions.

* * *

The idea of an enlightened leader incarnating back down into the lower world of matter in order to fulfill some specific purpose is not a concept that is restricted to Christianity. In the Hindu tradition, these individuals are referred to as *avatars*. It is believed, for instance, that the divine hero Krishna was the physical manifestation of the god Vishnu, who manifested on Earth in order to help humankind.

In the Gospel of Matthew, it is written that the mother of Yeshua was "found to be pregnant through the Holy Spirit." This was because she had been chosen to be used in a plan that was carried out by an emissary of the heavenly Father. In the Gospel of Luke, we are told that Mary was personally visited by a divine envoy, who informed her that after the "power of the Most High" comes upon her she will become pregnant with the son of God. However, according to the gnostic Gospel of Phillip, it was not the "Holy Spirit" itself that impregnated Mary, but rather something else; although no clear explanation is given.

> Some said, "Mary conceived by the Holy Spirit." They are in error. They do not know what they are saying.
> —Nag Hammadi Codices, The Gospel of Philip

If the first *Homo-sapiens* were created from an artificial, perhaps *in-vitro,* type of fertilization process, the question must be asked if a similar type of operation was employed by the emissary of the "Father" in the "virgin" birth case? This would not only explain the virgin birth but the claim that Yeshua was both a "son of man" (i.e., terrestrial) (Luke 22.48), as well as a "son of Elohim" (i.e., extraterrestrial) (John 10.34-36).

However, it cannot be ruled out that it was the "angel" of the heavenly Father who seduced Mary with the power of the "Holy Spirit," and impregnated Mary himself. This would also have also endowed the progeny with superior genetic traits. A cover story could have been set up to spare Mary from being stoned to death. If this is true, then perhaps the "Father" of Yeshua may have been the being who is erroneously referred to in the Gospel of Luke as "Gabriel."

In this case, it is possible that the NHIs were seeking to repair the damage that had been committed by one of their own. Indeed, according to contactee George Van Tassel, the NHIs disclosed to him that "they feel responsible for the fact that one of their people started this destructive cycle on the Earth" (Van Tassel 1958: 18). This could have been a reference to Yahweh.

Another aspect of the Yeshua story is how well it accords with current NHI events. For example, some present-day experiencers report undergoing a profound spiritual awakening, both during and after contact sessions (Mack 1999: 64, 81, 277). They also report the feeling of a "reconnection with the supreme principle," (Mack 1999: 232) which leads to the realization of the importance of love (Mack 1999: 233). One experiencer reported that after her encounter with the NHIs in a higher dimension, her empathy for others who were still stuck in the lower world increased (Mack 1999: 235). Some contactees report that these experiences with higher energies opened healing abilities within themselves (Mack 1999: 78-79, 237).

> Quite frequently abduction experiencers will relate the energetic changes in their bodies to their own healing or a capacity to heal others, a subject that deserves careful research. Commonly the energy and light experiencers that they have undergone become associated with personal transformation, spiritual development, and the evolution of consciousness, both personal and collective.
> —John E. Mack, *Passport to the Cosmos*

The experiencers feel as if they have made some agreement (i.e., covenant) with the beings, or with "God" itself, to fulfill a mission on Earth (Mack 1999: 18). Of course, these reports are remarkably similar to the "Jesus" story.

The extraterrestrial influence explains the appearance of an intelligently-guided flying "star" around the time of Yeshua's birth

(Matthew 2.9-10) (that is, if we were to assume that this account that is reported in the Gospel of Matthew is accurate). It also explains the so-called "cloud" that descended upon the mount of the transfiguration, from which the voice of the "Father" emanated (Luke 9.34-35). It also explains the "cloud" that took Yeshua up into the sky (Acts 1.9). It also explains the "chariot of wind" that took Yeshua up into the sky in the Apocryphon of James text. The extraterrestrial deity hypothesis also explains the light that shot down from the sky onto Paul (Acts 9.3-4), which is similar to present-day reports of light beams or search lights that emanate from UFOs (Trench 1966: 56). This hypothesis also explains the large square-shaped object that descended from the sky that was witnessed by the disciple Peter (Acts 10.10-16).

Perhaps then, this is the real reason why Yeshua claimed that he was "not from this world" (John 8.23).

* * *

Yeshua understood that his activities would eventually lead to his execution. This heroic act of self-sacrifice reveals the extent of the compassion that he had for a people who were being misled and oppressed by a malevolent trickster and his collaborators, the Pharisees and the Sadducees. Yeshua urged the people to look beyond the repressive "lie," and to "dance" to the music that he was playing (Matthew 11.17). But the roots of incontestable doctrine were too firmly planted into the mentality of the people.

At some point, when the time was right, it was his intention to disclose the hidden truth to all those who were ready to hear it. What Yeshua was hoping for was a non-violent revolution that would serve to overturn the laws and the institution of Yahweh gradually; which is why he referred to this revolution in his parables as a seed that would eventually spread and blossom.

There are accounts in the gospels where Yeshua indicates that it was his plan to "ransom" his life for the salvation of many, which would seem to confirm the traditional Judeo-Christian concept of a Messiah who had come to die for the sins of the people. As a result of this interpretation, it is believed by a great majority of Judeo-Christians that all one has to do in order to be saved from damnation is to announce "Jesus Christ as your Lord and Savior," and conform to the Judeo-Christian program. Such misunderstandings propagated a dogmatic and erroneous creed that unscrupulous individuals within the Judeo-Christian institution were even able to use for their own advantage. By instilling the laity with feelings of guilt and shame for their so-called "original sin," as well as constantly reminding them of the painful sacrifice that was made by the Christ for the benefit of humanity, church officials were better able to acquire a significant amount of power and wealth by making the people feel indebted to themselves as the Christ's representatives on Earth.

The truth is that Yeshua did not die *for* the sins of the people, but rather he died *because* of the sins of the people who were misled by the agents of Yahweh. He was willing to undergo this trauma because he believed that his works and his high-profile sacrifice would eventually draw people to his message, and because he knew that both he and everyone else who followed his teachings would be resurrected in heaven.

* * *

The gospels report that even during Yeshua's childhood he was questioning and challenging the priests of Yahweh. The lack of a biblical record of his life between childhood and adulthood could indicate that he spent those missing years away from the region of Israel. Indeed, there was another center of religious activity in the ancient world during that time.

A wave of spiritual and humanistic thought arose in around the sixth and fifth century BCE, as the influence of the ancient gods began to diminish and their human protegee's developed in their place. This movement appeared simultaneously in both the western and eastern worlds. However, while the west was beginning to experience a break with the mystical side of its humanist movement—which was due to such influences as the Greek skeptical sophists (i.e., pyrrhonism) and the materialists—the easterners of Asia were incorporating the principles of humanism and mysticism into their religions and philosophies. It was a movement that was led by the ancient luminary sages—e.g., the Buddha, Kapila, Lao Tzu, Patanjali, etc. While Hinduism was originally based on the worship of the gods, it was the first of the major world religions to emphasize the concept of self-discovery and personal enlightenment. Indeed, Asia was a spiritual Mecca in the time of Yeshua, just as it still is to this day.

Archaeological excavations of the Sumerian cities of Susa and Kish have uncovered artifacts that bear unmistakable ancient Indian motifs. These finds indicate that there may have been a trading relationship between Asia and Mesopotamia as early as 4000 BCE. By the first century there was a well-established route between these two neighboring regions known as the Silk Road. It was the route preferred by the merchants of the time, who transported their products between China and what is now present-day Turkey. Not only did this road pass along an area just north of Israel but the pinnacle of its existence was from 200 BCE to 200 CE, which is the period in which Yeshua would have made his journey east.

Even though devotees of the Judeo-Christian tradition will argue that the Savior was the physical incarnation of a being from a preexisting higher world, it is still likely that he would have needed to acclimate himself into the world in which he was born into as a physical "son of man" (Mark 10.45, etc.). He may have traveled to Asia as a part of a self-discovery process. Indeed, when the teachings of the eastern adepts and Yeshua are compared, a significant similarity occurs: [334]

> Consider others as yourself.
> —The Buddha, Dhammapada

> Do to others as you would have them do to you.
> —Yeshua, Luke 6.31

---

[334] Many more parallel similarities can be found in the book *Jesus and the Buddha* (Ulysses Press, 1997), by Marcus Borg.

> Do not look at the faults of others, or what others have done or not done; observe what you yourself have done and have not done.
> —The Buddha, Dhammapada

> Why do you see the speck that is in your brother's eye, but do not consider the beam that is in your own eye?
> —Yeshua, Matthew 7.3

> Let us live most happily, possessing nothing [. . .]
> —The Buddha, Dhammapada

> [. . .] Blessed are you who are poor, for yours is the kingdom of God.
> —Yeshua, Luke 6.20

The teachings of Yeshua are also similar to the writings of the Chinese sage, Lao Tzu:[335]

> Respond to anger with virtue.
> —Lao Tzu, The Tao Te Ching

> [. . .] If someone strikes you on the right cheek, turn to him the other cheek also.
> —Yeshua, Matthew 5.39

> Therefore the sage puts his person last and it comes first [. . .]
> —Lao Tzu, The Tao Te Ching

> But many who are first will be last, and the last first.

---

[335] More parallel teachings can be found in the book *Jesus and Lao Tzu* (Seastone, 2000), by Martin Aronson.

—Yeshua, Mark 10.31

A violent man will die a violent death.
—Lao Tzu, The Tao Te Ching

[. . .] all who draw the sword will die by the sword.
—Yeshua, Matthew 26.52

The essence of Yeshua's teachings can be found in some of the Hindu texts as well:

Self-complacent and always impudent, deluded by wealth and false prestige, they sometimes proudly perform sacrifices in name only, without following any rules or regulations.
—Krishna, The Bhagavad Gita, chapter 16

Woe to you, teachers of the law and Pharisees, you hypocrites! You are like whitewashed tombs, which look beautiful on the outside but on the inside are full of dead men's bones and everything unclean. In the same way, on the outside you appear to people as righteous but on the inside you are full of hypocrisy and wickedness.
—Yeshua, Matthew 23.27-28

In the minds of those who are too attached to sense enjoyment and material opulence, and who are bewildered by such things, the resolute determination for devotional service to the supreme Lord does not take place.
—Krishna, Bhagavad Gita, chapter 2

What good is it for a man to gain the whole world, and yet lose or forfeit his very self?
—Yeshua, Luke 9.25

Therefore, a Brahmin [priest] should stop being a pundit and try to live like a child.
—Upanishads, Brhadaranyaka Upanishad

[. . .] I tell you the truth, you must change and become like little children. Otherwise, you will never enter the kingdom of heaven. The greatest person in the kingdom of heaven is the one who makes himself humble like this child.
—Yeshua, Matthew 18.3-4

Yeshua displayed many of the attributes of someone who had studied the teachings and the history of the enlightened adepts in Asia. While Yahweh had worked to construct a world of confusion, oppression, fear, and illusion—which is sometimes referred to as *maya* in the eastern world—Yeshua and the enlightened adepts who came before him, worked to remove that oppressive and artificial obstruction away.

The institution of Christianity actually has an extensive history in Asia, which extends as far back as the first century CE, when Thomas and Bartholomew arrived in India as missionaries. It was in that same early period that a gnostic Christian named Mani also arrived in the east. It is the gnostic Christian groups that share an even closer resemblance to the eastern religions, due to how they not only perceive the material world as a distractive illusion but also view the impermanent nature of attachment, suffering, and desire, in much the same way. Indeed, not only do Christians and Buddhists share ideological similarities but these two religions have been peacefully interacting since ancient times (Pagels 1979: xxi, 27, 146). The relationship between Buddhism and Christianity is recorded in the following Manichean[336] text, in which the Sanskrit Buddhist term *nirvana* is specifically referred to:

> It was a day of pain and a time of sorrow when the messenger of light entered death when he entered complete Nirvana.
> —The Story of the Death of Mani, The Earth Trembles as he Enters Nirvana

Likewise, similar Buddhist themes pertaining to meditation and enlightenment appear in the gnostic Nag Hammadi codices as well:

> "And when you become perfect in that place, still yourself. And in accordance with the pattern that indwells you, know likewise that it is this way in all such (matters) after this pattern. And do not further dissipate, so that you may be able to stand, and do not desire to be active, lest you fall in any way from the inactivity in you of the Unknown One. Do not know him, for it is impossible; but if by means of an enlightened thought you should know him, be ignorant of him."
>
> Now I was listening to these things as those ones spoke them. There was within me a stillness of silence, and I heard the Blessedness whereby I knew [my] proper self.
> —Allogenes, The Nag Hammadi Codices

Central to both gnostic and Buddhist doctrine is the pursuit of spiritual and mental enlightenment and the concept of self-discovery:

---

[336] Manichaeism was a religion that was founded by the Persian prophet Mani in the third century CE. It was a synthesis of mainly gnostic Christianity, Zoroastrianism, and Buddhism.

This is why I say this to you, that you may know yourselves.

—Yeshua, Nag Hammadi Codices, The Apocryphon of James

This was a revolutionary principle for its time; especially in a land that was ruled over by a deity whose doctrine placed an overwhelming emphasis not on the true internal welfare of the individual, but on subservience to himself.

* * *

After Yeshua's departure, various Christian groups arose. One of the original groups were what is now referred to as the gnostics.

The word gnostic is derived from the Greek word for knowledge (*gnostikos*). It was originally a word that was used to describe intellectuals; although, as far as the original gnostics were concerned, they were simply philosophical (i.e., Neo-Platonist) Christians. However, this term could also be used to indicate that the gnostics were the Christians who *knew* the secret teachings of the Savior.

According to the gnostic-inspired apocryphal Gospel of Mary, Mary (Magdalene)[337] was someone who understood the secret teachings of the Savior and was actively involved in the ministry by helping to teach the true message. However, Peter did not approve of Mary's influence, which seems to be at least partially due to his traditional views regarding the role of women. Indeed, the schism that occurred between Mary and Peter is recorded in other texts as well.[338] Of course, the other branch of Christianity that arose during that formative time was Peter and Paul's sect. In the early years, the different Christian groups competed against each other. The rivalry

---

[337] There has been some debate regarding which Mary this referred to; however, the gnostics knew that it was Mary Magdalene who was the special beloved one of the Savior.

[338] The Gospel of Thomas; the Pistis Sophia; the (Greek) Gospel of the Egyptians.

even extended to a personal level between Peter and the gnostic-influenced Simon Magus.[339] In the apocryphal book the Acts of Peter, Simon and Peter are recorded directly competing with one another. Of course in this version, which is presented by the followers of the Petrine/Pauline sect, Simon is depicted as a foolish charlatan, and Peter as a righteous victor. (Although the apostle Peter played a role in what has become so-called "orthodox" Christianity, no single person was as influential as Paul. Therefore, Peter and Paul's sect will henceforth be referred to as the "Paulines.") In this account, it is reported that Simon had the ability to levitate high up into the air, and while he was displaying that ability to the people Peter prayed for him to fall to the ground. After Peter's prayers were answered and Simon plummeted to the ground, the crowd proceeded to stone him to death.

In another story that is recorded in the Pauline New Testament (Acts 8.9), Simon is said to have even resorted to the nefarious scheme of attempting to bribe Peter and John into giving him the ability to cure by the laying on of hands—a practice that has come to be known as "Simony." Simon was also given the title "Magus," which would seem to indicate that he was a practitioner of magic; however, the three heralded wise-men who visited the infant Savior in Bethlehem were also said to be magi. (That is, if we were to assume that the account that is reported in Gospel of Matthew is accurate.)

According to the second-century Pauline bishop Irenaeus,[340] among some of the other so-called "heresies" that were committed by Simon the Magus and his gnostic followers (i.e., the Simonians) was their practice of the "magical arts." These practices allegedly included the reciting of incantations—which may have actually been nothing more than liturgical hymns and prayers—as well as the making of love potions, the exorcism of spirits, and the communication with spirits. However, most of these accusations could have been leveled at Yeshua the Savior as well.

---

[339] Simon Magus is actually considered more of a proto or quasi gnostic.
[340] Against Heresies (Book I)

Another one of the alleged heresies of Simon Magus was his teaching that the Old Testament god was not the highest God.[341] Of course, such a proclamation was not well-received by the Paulines, and as a result, Simon was unjustly vilified.

Another unconventional notion that offended the Judeo-Christian[342] "orthodox," is that the gnostics taught that there were other divine beings besides Yeshua—such as Derdekeas, Hermes Trismigestus, and Mirotheos (i.e., Meirothea). However, Yeshua himself is recorded instructing his disciples, in even the canonical gospels, to go forth and do as he did; which meant that they themselves were to help save their world by becoming saviors according to their own abilities. This concept is exemplified in the (most likely symbolic) walking on water story, for instance, in which Yeshua invited the disciples to perform miracles along with him. Of course, to the priests of "Jehovah"—which is the anglicized Latin name for Yahweh—such supernatural ability was considered to be sorcery.

Yeshua urged the people to discover and develop the latent divinity that was within themselves, and to teach and heal others just as he himself had done. Inspired individuals, such as Simon Magus and the gnostic Persian prophet Mani, were only seeking to fulfill this very teaching.

> For this person is no longer a Christian but a Christ.
> —Nag Hammadi codices, The Gospel of Philip

* * *

The second- and third-century CE Judeo-Christian author Tertullian, directed his condemnation against a Christian man named Marcion.

---

[341] Clementine Homilies III (Chapter 10)

[342] The term "Judeo-Christian" will be used (as opposed to Christian) as a way in which to designate the Christians (e.g., Paulines) who accept Yahweh and the scriptures of the Old Testament.

According to Tertullian,[343] Marcion was a devious barbarian from Sinope in Pontus, where residents supposedly engaged in cannibalism and freely partook of "libidinous desires" without shame. Marcion's other purported crime was that he preached the existence of "two gods"—which pertained to the difference between Yahweh and the Father. Although, Marcion did not directly equate Yahweh with the devil, but rather with the "Demiurge." (The subject of the Demiurge will be examined further ahead.)

Despite Tertullian's accusation that Marcion was a barbarian, records[344] indicate that Marcion was actually born into an upper-class Judeo-Christian family (his father was a bishop), and that he himself was a "ship-master." Marcion even highly regarded Paul, and included his writings in his own collection (i.e., the *Apostolikon*). It is also peculiar that Tertullian described Sinope as barbaric, since the city was actually settled by the Greeks. Furthermore, far from being an unrestrained sexual deviant, Marcion preached the value of celibacy.[345]

One of the differences between Marcionite and Pauline Christianity, was that Marcion permitted woman to attain important positions in the church.[346] Other than this, it seems that the only significant difference that set him apart from the Paulines was that he knew that the Old Testament god was not the heavenly Father of Jesus. To him this was not secret knowledge, but rather only common sense. Marcion also compiled the sayings of Jesus and compared

---

[343] Five Books Against Marcion (Book I)

[344] Five Books Against Marcion (Book I); The Panarion (i.e., Against Heresies) (Book I, part 42) by Epiphanius of Salamis.

[345] The Panarion (i.e., Against Heresies) (Book I, part 42) by Epiphanius of Salamis.

[346] The Panarion (i.e., Against Heresies) (Book I, part 42) by Epiphanius of Salamis.

them to the sayings of Jehovah, in order to show the contradictions between these two individuals, in his book Antitheses.[347]

During those early years, the Marcionite branch was so successful that it even rivaled that of the Paulines. In some regions, Marcionism was even the dominant form of Christianity (Ehrman 2003: 109). However, it was Paul's sect that of course reigned supreme when they—among other reasons—received the support of the Roman emperors. It was this consolidation that worked to finally stamp out the competitive influence of the other Christian groups.

It was during that time that the Hebrew name Yeshua was changed to the Greek *Iesous*, and then again to the Latin *Iesus*; which eventually lead to the English rendering, *Jesus*. (Henceforth, Yeshua will be referred to as Jesus—which signifies his revised Judeo-Christian identity.)

* * *

In 1945, an earthenware jar was uncovered by local farmers in Nag Hammadi Egypt. When the jar was broken open, papyrus pages bound together into thirteen leather codices were discovered. The books had clearly been intentionally buried in order to preserve them from Pauline censorship.

Among the writings in the gnostic Nag Hammadi collection that make a reference to the secret teachings of the Savior is the Gospel of Thomas, which is a text that can be dated as far back as the gospels of the New Testament (Valantasis 1997: 9). In this record are further references to "hidden" matters, and a quote that directly refutes Old Testament law:

> His disciples said to him, "Is circumcision useful or not?" He [Jesus] said to them, "If it were useful, their father would produce children already circumcised

---

[347] Antitheses is no longer extant; however, references to it appear in Five Books Against Marcion (Book II) by Tertullian, and Refutation of All Heresies (Book VII) by Hippolytus.

from their mother. Rather, the true circumcision in spirit has become profitable in every respect."
—Nag Hammadi Codices, Gospel of Thomas

In the following passage, Jesus indicates that the underlying truth that he was alluding to would not be easy for the people to accept, since it was such a controversial and astounding revelation:

> Let him who seeks continue seeking until he finds. When he finds, he will become troubled. When he becomes troubled, he will be astonished, and he will rule over all.
> —Jesus, The Nag Hammadi Codices, The Gospel of Thomas

The next passage is especially significant. Based on Thomas's reaction, it is evident that Jesus had told him the difference between Yahweh and the Father:

> And he [Jesus] took him [Thomas], and withdrew, and spoke three sayings to him. When Thomas came back to his friends they asked him, "What did Jesus say to you?" Thomas said to them, "If I tell you one of the sayings he spoke to me, you will pick up rocks and stone me, and fire will come from the rocks and devour you."
> —Nag Hammadi Codices, The Gospel of Thomas

The gnostic account also emphasized a more humanistic mentality; a Christianity that was based not on the authority of a hierarchical institution, but from within every single person:

> If those who lead you say to you, "See, the Kingdom is in the sky," then the birds of the sky will precede you. If they say to you, "It is in the sea," then the fish will precede you. Rather, the Kingdom is inside of you, and

> it is outside of you. When you come to know yourselves, then you will become known, you will realize that it is you who are the sons of the living Father. But if you do not know yourselves, you dwell in poverty and it is you who are that poverty.
> —Jesus, Nag Hammadi Codices, The Gospel of Thomas,

> That which you have will save you if you bring it forth from yourselves.
> —Jesus, Nag Hammadi Codices, The Gospel of Thomas

> Beware that no one lead you astray, saying, "Lo here," or "Lo there!" For the Son of Man is within you.
> —Jesus, The Gospel of Mary

Gnostic teaching, which placed a greater emphasis on spirituality, philosophy, and individuality, presented a subversive threat to the authoritative institution that the Pauline bishops, priests, and deacons were endeavoring to establish.

While Jehovah had condemned the self-empowered individual, according to the gnostics, Jesus had actually praised such a person:

> Blessed are the solitary and elect.
> —Jesus, Nag Hammadi Codices, The Gospel of Thomas

Of course, such a type of Christianity would not have appealed to the Roman emperors, who required a creed that would help bolster their own power by creating a society of workers, warriors, and worshipers; and thus, help to preserve their fractured empire. What they needed was not individual wisdom, spirituality, and self-discovery, but rather a system that turned the people's attention toward service to God, King, and Country.

Indeed, Machiavelli was aware of the advantages of the merging of church and state, just as the Roman emperors before him:

> For though they [ecclesiastical principalities] [...] may be kept [...] because they are sustained by ancient laws rooted in religion that have proved capable of keeping princes in power no matter how they live or rule [...] And since they are sustained by superior causes which transcend human understanding, I will not discuss them because they are supported and exalted by God, it would be an act of presumption and rashness to speak of them.
> —Niccolo Machiavelli, *The Prince*

> Sovereignties, in particular possess strength, unity, stability, only to the degree to which they are sanctified by religion.
> —Niccolo Machiavelli, *The Prince*

* * *

Unlike their Pauline counterparts, the gnostics did not band together to form any singular institutional authoritative council. The result was an abstruse variety of supernatural characters, arcane cosmologies, and philosophic theologies that were not only not always in complete accord but were also mostly unintelligible to the lay-person. This is, most likely, one of the reasons why the Pauline sect was able to take over the interpretation.

Some of the gnostic Christian groups that arose during that early period were the Valentinians and the Sethians; each of which were named after their founding teachers. Most of the information that was recorded about these groups were documented by their Pauline opponents (e.g., Epiphanius and Irenaeus). The result is a distorted interpretation that accuses some of these groups, such as the so-called "Borborites" (although they were more likely originally known as the Barbeloites) and the Carpocratians, of engaging in bloody and sexual sacraments. However, it is unlikely that these descriptions are accurate. In fact, most gnostics saw the physical world as corrupt, and sought to free themselves from captivity to this world by focusing on matters related to the eternal spirit. Furthermore, according to the gnostic Book of Thomas the Contender, engaging in "polluted intercourse" was something that was discouraged. It is also ironic that the Paulines would make such accusations, since similar charges were leveled at them by the pagans (e.g., the Greek philosopher Celsus) (Ehrman 2005: 26).

In the second century of the Common Era, the proto-orthodox bishop Irenaeus of Lyons urged his followers to condemn the gnostics.[348] This hostile polemic set the precedent for the years of strife and persecution that were to follow.

One of the original Christian patriarchs who may have been aware of the secrets of the Savior and who was also accused of heresy was James "the Just." This individual may have not only been one of the original disciples (not to be confused with the apostle James, son of Zebedee) but may have even been Jesus's half-brother as well (Galatians 1.19). Moreover, according to the Gospel of Thomas, Jesus did not name Peter as the future leader of the movement, but rather James the Just.

In the following passage, James admonishes the people for having been deceived by the wrong "Lord":

> The Lord has taken you captive from the Lord, having closed your ears, that they may not hear the sound of my word.
> —James, Nag Hammadi Codices, The Second Apocalypse of James

He then goes on to say that he shall "doom to destruction" the house that they believe that God has made. The Yahwehist crowd reacted to what they perceived as blasphemy by seizing him and burying him halfway into the ground before stoning him to death.

---

[348] Against Heresies (*Adversus Haereses*)

* * *

In the following gnostic tractate, Jesus refers to the Machiavellian "schemes" that were used by the malevolent ruler to assert his selfish will:

> [. . .] his gifts are not blessings. His promises are evil schemes.
> —Jesus, Nag Hammadi Codices, The Second Apocalypse of James

The gnostics knew that Yahweh/Jehovah masqueraded as the heavenly Father. Although they did not refer to him as either Yahweh or Jehovah, but rather as "Yaldabaoth."[349]

> When he [Yaldabaoth] gazed upon his creation surrounding him, he said to his host of demons, the ones who had come forth from out of him: "I am a jealous God and there is no God but me!"
> —Nag Hammadi Codices, The Apocryphon of John

The gnostics also referred to Jehovah as Sakla/Saclas (the fool) and Samael (the blind god/god of the blind).

---

[349] The most prominent etymological interpretation of the theonym Yaldabaoth (i.e., Yaltabaoth, or Ialdabaoth) is that it could mean "child of chaos" (*yalda bahut*) (Kasser, et al. 2006: 37); although this interpretation is contested by some scholars. Gershom Scholem (a prominent twentieth-century scholar of Jewish mysticism) postulated that the name is most likely a composite of the words *Yald* and *sabaoth* (Gruenwald 2014: 113). The Hebrew root *YLD* could indeed mean child; however, it could also be a reference to the verb form, which means "to beget." This not only relates to the Aramaic use of the term but also relates to his description as the "begetter" in the gnostic texts (e.g., the Sophia of Jesus Christ). Furthermore, if *abaoth* derives from the Hebrew word *sabaoth*, this would indicate that Yaldabaoth was the begetter of armies—which is a reference to Yahweh Sabaoth.

> But Ialdabaoth, Saclas, who possesses many forms in order to reveal himself with diverse forms as he pleases [. . .]
> —Nag Hammadi Codices, Apocryphon of John

Lord Saclas Yaldabaoth[350] was not the heavenly Father, but rather the "Demiurge."

The word Demiurge is derived from the Greek word for public or people worker, which signified his role as the creator, or "craftsmen," of the physical world of mortal beings. Although Plato and the Neo-Platonists saw the Demiurge as a supernatural Creator, it was the gnostics who relegated him to the creator of not only the inferior material world but also identified him with the wicked god of the Old Testament.

The Demiurge was able to gain power for himself through the use of fear, violence, and deceptive schemes. The conditioning of mental perspective led to what enlightened sages refer to as the veil of illusion.

> And thus when the world came to be in distraction, it wandered astray throughout time. For all the men who are on earth served the demons from the foundation until the consummation of the Aeon—the angels served justice and the men served injustice. Thus the world came to be in a distraction and an ignorance and a stupor. They all erred until the appearance of the true man.

---

[350] According to the Gospel of Judas, Yaldabaoth was also referred to as "Nebro, which means 'rebel'." In this account, Saklas is described as an assistant to Nebro, which differs from the account that is found in the Apocryphon of John. Furthermore, if Nebro is the same individual who is called "Nebruel" in the gnostic Holy Book of the Great Invisible Spirit text (i.e., the Coptic Gospel of the Egyptians), this would be another differing account. It is likely that these discrepancies represent different understandings that were held by different gnostic individuals and groups.

—Nag Hammadi codices, On the Origin of the World

The "true man" was the Savior, who came at the end of the age to help free the servants of the Lord of Armies.

> That which was revealed to me was hidden from everyone and shall only be revealed through him [. . .] I hasten to make them free and want to take them above him who wants to rule over them.
> —Jesus, Nag Hammadi codices, The Second Apocalypse of James

In the following passage, Jesus encourages the people to discern the difference between the one who rules over them and the Father:

> His promises are evil schemes. For you are not an instrument of his compassion, but it is through you that he does violence [. . .] But understand and know the Father who has compassion.
> —Jesus, Nag Hammadi codices, The Second Apocalypse of James

> Hear and understand—for a multitude, when they hear, will be slow witted. But you, understand as I shall be able to tell you. Your father is not my father.
> —Jesus, Nag Hammadi Codices, The Second Apocalypse of James

* * *

The Christians faced significant persecution during the early years. One especially zealous antagonist was a man named Saul. Saul was a Greek Jew who was not only a Pharisee but was also a Roman citizen.

According to the account that is recorded in the book of Acts, Saul was traveling down a road on his way to Damascus to deliver letters to the high-priests that would petition for the rejection of the Christians, when he was blinded by a white light that shot down on him from out of the sky. Inside the light he heard the voice of the Christian Savior who spoke to him asking, "Why are you persecuting me?" After having witnessed the vision, Saul underwent a conversion of faith; and, after changing his name to Paul, he became a devout follower of Jesus.

Although Paul had undergone a profound experience and transformation, he was still prone to the ideology that his upbringing had instilled in him. By linking the Father with Jehovah, he was better able to make that monumental leap of faith. Unfortunately, no single person played as much of an influence in propagating the mistaken interpretation of Christianity to the world than Paul of Tarsus.

While it is true that Paul was a proponent of "love" (1 Corinthians 13), he also urged his followers to "become slaves of God" (Romans 6.22), and "slaves of Christ" (1 Corinthians 7.22). He also warned against provoking "the Lord's jealous anger" (1 Corinthians 10.22), and to beware of "the wrath of God" (Romans 1.18). He also exhorted his followers to cultivate "obedience of faith" (Romans 1.5), so that their "faith might rest not on human wisdom but on the power of God" (1 Corinthians 2.5). Likewise, he emphasized the importance of following without "questioning" (Philippians 2.14), and to work towards salvation with "fear and trembling" (Philippians 2.12). He also instructed the members of his sect to "follow the faith of Abraham" (Romans 4.16), and to "hold fast to traditions" (1 Corinthians 11.2) so as to "confirm the promises of the patriarchs" (Romans 15.8). He also continually cited passages from the Old Testament/Tanakh in order to justify his mistaken reasoning (Romans 9). Paul was a rigidly devout man. He apparently saw himself not as a humanitarian and enlightened benefactor, but rather as a "prisoner of Christ" (Ephesians 3.1)[351] and an "ambassador in chains" (Ephesians 6.20).

---

[351] Scholars do not believe that the Epistle to the Ephesians was written by Paul. Nevertheless, this book does reflect the Pauline perspective.

> Slaves, obey your earthly masters with respect and fear, and with sincerity of heart, just as you would obey Christ. Obey them not only to win their favor when their eye is on you, but as slaves of Christ, doing the will of God from your heart.
> —Ephesians 6.5

Paul had been conditioned all his life to believe that virtue was to be found in self-sacrifice, suffering, obedience, tradition, and servitude to Lord Yahweh. However, Jesus himself indicated, in even the Pauline New Testament itself, that he thought of his disciples neither as servants nor as slaves, but rather as "friends" (John 15.14). Indeed, other differences can be found between the teachings of Jesus and Paul. For example, Mark 2.15 records that Jesus ate with sinners, but according to 1 Corinthians 5.11, Paul instructed his followers *not* to eat with sinners. Furthermore, Jesus emphasized the value of repentance (Luke 3.3, 13.3), while Paul emphasized the value of faith (Romans 1:17, 3.26-28). It can also be said that Paul inadvertently contradicted the teachings of Jesus when he put "new wine into old wine-skin"—that is, when he conflated the heavenly Father and his son with Yahweh/Jehovah.

According to Paul's own interpretation, Israel needed a Savior because they thought that they could achieve salvation through the works of law rather than by faith (Romans 9.31-32). It was also Paul who helped to emphasis not only servitude to Yahweh/Jehovah but submission to the death of the Christ. Paul also claimed that he received his knowledge through "revelation" (Galatians 1.12); although, according to the account that is recorded in the book of Acts, "Jesus" did not reveal any special information to him— especially information related to the secret matters.

Paul was generally an ethically-minded man who truly wanted to do the right thing, and did so to the best of his ability; although, he was not present during the time when Jesus was referring to the underlying secret truth of his mission. Indeed, Paul even admits to at times being "perplexed" by the situation (2 Corinthians 4.8).

In the following passage, Paul attempts to explain why the people rejected the Savior by referring to a quote from the Old Testament:

> God gave them a spirit of stupor, eyes that could not see and ears that could not hear, to this very day.
> —Paul, Romans 11.8

Unfortunately, Paul did not realize how far the confusion extended. When Jesus was denouncing the Pharisees as "blind guides" (Matthew 15.9-14), these words could have been directed at Paul; a man who had been one of these priests himself.

* * *

Paul was not the only one who misunderstood the true meaning of the mission of Jesus. The other was one of the original disciples himself, Simon Peter.

Roman Catholic Paulines often cite a passage that is found in the Gospel of Matthew (16.18), where Jesus supposedly refers to Peter as the foundational "rock" of the future Christian movement. However, this passage is found only in the Gospel of Matthew, which, as was previously noted, is the least reliable of all the gospels. It is also telling that a similar conversation appears in the gnostic Gospel of Thomas; although, in this version there is no mention of Peter being declared the foundation stone of a future movement. It is evident that this reference was inserted by what scholars refer to as the "M source."

The reason why Jesus *may* have especially regarded Peter was because of the sincerity of his optimistic faith. Indeed, that is the context of the quote, which occurred directly after Peter declared Jesus to be "the Christ, the Son of the living God." Therefore, if this quote is accurate, it could mean that Jesus meant that he would build the church of the Father upon the type of faith in his divinity that was professed by Peter.

It must also be kept in mind that in the very same gospel there are also instances of Jesus actually rebuking Peter for not understanding what he was doing and what was supposed to happen. In Matthew 16.21-24, for instance, Peter protests Jesus's destiny to be harmed by the elders of Jerusalem, which ends with Jesus turning to Peter and saying, "Get behind thee Satan!" Jesus also admonished Peter for not understanding one of his parables (Matthew 15.16), and even asked Peter if he was "dull" (or "without understanding," in other translations). Peter is again admonished when his faith wavered when he attempted to walk on water (Matthew 14.31), and again for falling asleep in Gethsemane (Matthew 26.36-45). When Judas brought the authorities to seize Jesus, it was Peter who took out a sword and cut off the ear of a servant of one of the high-priests (John 18.10-11). During that incident, Jesus reprimanded Peter again for not only acting out in violence but for not understanding what was supposed to take place. This misunderstanding is also reflected in the book of the Acts of the Apostles, in which Peter announces to the crowd that it is the "God of Abraham, the God of Isaac, and the God of Jacob, the God of our ancestors," who has "glorified his servant Jesus" (Acts 3.13).

Although Peter was obviously a devout and well-meaning person, it is evident that he never fully understood the true meaning of the mission of the Savior. This is most likely because Jesus felt that Peter—who was clearly what could be described as a simpleton—would not have been able to handle the astounding truth, and therefore never directly told him.[352] Therefore, if Jesus called Peter the cornerstone rock of the future Christian movement, it was a statement that was clearly based on Peter's optimistic disposition, rather than his mental comprehension.

---

[352] According to the gnostic Apocryphon of James text, James and Peter were both told the sacred secret. However, the Apocryphon of James is not only a second- to fourth-century CE pseudonymous work but the entire text does not accord well with other gnostic accounts and doctrine. The account seems to be the author's—most likely a man known to scholars as Cerinthus—own interpretation of events (Robinson 1996: 29-30). When we apply coherence theory to this problem, this contradiction can be dismissed as an isolated aberration.

* * *

In the early fourth century CE, the Romans were fighting among themselves for control of a declining empire. It was during that time that Emperor Constantine allegedly experienced a revelation (most likely in a dream) concerning a Christian symbol. (A monogramatic Christogram in the form of a *labarum*.)[353] Constantine went on to be victorious in battle. As a result, he converted and issued the Edict of Milan, which officially sanctioned the Pauline sect.

Why then was this particular group chosen, as opposed to the others that were also around during that time, such as the Montanists,[354] the Ebionites,[355] the Marcionites, and gnostic groups such as the Manicheans? There are several reasons for this: (1) The Paulines were likely the most active sect in Rome at that time, and therefore may have been the most familiar to Constantine. (2) The Paulines were organized into a manageable hierarchical regime who were willing to submit themselves to authority systems.

> Remind them to be in subjection to rulers and to authorities, to be obedient, to be ready for every good work.
> —Titus 3.1[356]

(3) The Pauline sect would have also been seen as more legitimate than some of the others because of its connection with a long-standing Old Testament tradition—as opposed to the Marcionites and the gnostics, who rejected the Old Testament. Therefore, the Paulines

---

[353] The labarum (i.e., the *chi rho*) is a symbol that is formed from the first two letters of the Greek word for Christ, which appear as the letter P superimposed over the letter X.

[354] The Montanists were an early Christian sect that engaged in ecstatic prophecy and emphasized a conservative lifestyle.

[355] The Ebionites were an early ascetic Jewish Christian movement who saw Jesus as the Jewish Messiah and rejected Paul and the gentile world.

[356] See also Romans 13.1.

would have been perceived as more of an authentic religious institution, rather than a mere fledgling cult (Ehrman 2003: 111-112). (4) The Ebionites required members to strictly observe Yahwehist law, which included not only not eating certain kinds of meat but circumcision as well. It also seems that the Ebionites were devoted to lives of poverty (the name Ebionite itself most likely derives from the Hebrew word *ebyonim*, which can be translated as "the poor ones"). It is likely that such requirements would not have been appealing to the gentiles—especially Constantine; despite the fact that Jesus himself spoke out against those who hoarded wealth. Moreover, the Ebionites had no intention of integrating with the gentiles. (5) The Roman Pauline orthodox were not as elitist as the gnostics, and therefore were more accessible to the common citizen. Indeed, this is what the word "catholic"[357] refers to.

This turn of events proved to be a crushing setback for other Christian groups, as persecution by the Paulines, who had the enforcement power of the Roman army behind them, increased.

It was also during that formative period that the books of the Bible were being collected and edited, as the records of the Old Testament/Tanakh were combined with the books of the New Testament. It was a mistake that would serve to justify the centuries of human rights abuses that followed.

Due to the misunderstandings of those early Christians, what essentially happened is that the priesthood of Yahweh was carried over to the priesthood of Jehovah. What happened next was the most consequential and ironic twist in all of human history. Instead of a doctrine that was centered on the life and true teachings of the Savior and the compassionate heavenly Father, a dogmatic creed was founded that was based on the death of the Christ and servitude to Lord Jehovah.

---

[357] The word "Catholic" is derived from the Greek word *katholikismos*, which means "all-embracing," "universal," or "according to the whole." It is a word that represents the belief that this Pauline denomination teaches the whole truth and represents all of Christianity.

> Some who do not understand mystery speak of things that which they do not understand, but they will boast that the mystery of the truth is theirs alone [. . .] But immediately they join with one of those who misled them [. . .] They do business in my word, and they propagate harsh fate.
>
> —Jesus, Nag Hammadi codices, The Apocalypse of Peter

What the Paulines did was build up a grand Christology that turned a young rabbi healer and reformer from Galilee into "God" himself. However, what should be understood is that the movement that was begun by Yeshua was never originally intended to be its own separate gentile religion, nor was Yeshua God.

Judeo-Christianity was able to succeed not only because of the support that it received from the Roman empire but because it did indeed fulfill basic human needs related to security and community. Nevertheless, it should be understood that it is an institution that is based on the greatest misunderstanding of all time.

* * *

The next step in the great misconception was when the book of Revelation was added to the New Testament. This final book of the Pauline collection is also known as the "Apocalypse of John."

The book of Revelation was written late in the first century (circa 95 CE) by a man who is referred to as "John the Revelator," or "John of Patmos." These epithets are used to distinguish him from John the disciple, who would have been of an improbable age—especially in that era. It is also telling that, according to Acts 4.13, John the disciple was described as "uneducated" (i.e., illiterate). Indeed, scholars who have examined the original text do not believe that the book of Revelation was written by the original disciple of the same name (Cory 2006: 7). Even the early Judeo-Christian scholar Dionysius did not believe that this book was written by John the disciple (Ehrman 2011: 21). Indeed, the author of Revelation never even made the direct claim that he was John the disciple.

It is also known that the apocryphal Acts of John, as well as the Acts of Peter, Andrew, Paul, and Thomas, were not written by John the disciple, but rather by a man named Leucius Charinus (Hoeller 2002: 98-99; Schneemelcher 1992: 87). This practice of titling a text with the name of one of the original disciples was a way to make the work seem more authentic.[358]

> To give authority to the vision, the apocalyptist (the author of the apocalypse) takes on the name of a great figure of the past, an apostle or a patriarch of the Old Testament. To increase its prophetic value, the work is usually placed in the past; thus the prophecy of future history can be proved correct (because in fact these earlier events have already occurred).
> —Willis Barnstone,[359] *The Other Bible*

There were other eschatological works (i.e., apocalypses) besides the Revelation of John (e.g., the Apocalypses of Peter, Ezra, and Baruch) that were circulating among the Christians during the early years as well. None of these books were written by the individuals that they were attributed to.

Common mythologems among these apocalypses include a resurrected new Jerusalem, a great war, as well as a trial by fire and a final day of judgment. Not only do modern-day scholars believe that these texts are fraudulent but even the original compilers of the Bible as well. Indeed, the book of the Revelation of John was not included

---

[358] New Testament scholar Bart D. Ehrman provides further elucidation on this topic in his books, *Forged* (Harper One, 2011) and *Forgery and Counterforgery* (Oxford University Press, 2012).

[359] Willis Barnstone, Ph.D., is twentieth- and twenty-first-century New Testament and Gnostic scholar. He is a former O'Connor professor of Greek at Colgate University and a distinguished professor emeritus of comparative literature and Spanish and Portuguese at Indiana University.

in the earliest compilations of the New Testament. The inclusion of Revelation appears to have been mostly due to the efforts of one man alone, the Pauline bishop, Athanasius (Pagels 2012: 135, 144-149, 160).

In another text that is referred to as the Apocalypse of Peter (not to be confused with the gnostic text of the same name), Jesus is used in the role of Jehovah, in much the same way that he is represented in the book of Revelation. In this account, Peter is taken up into another world where he is shown the hellish punishments of those who turned away from the Lord's "righteousness." Among other horrors, they are hung by their tongues and hair, and heated iron and "rays of fire" are forced into their eyes. According to the account, Peter is led there by Jesus himself, who tells him that his Father has the power to command the forces of hell.

It is evident that the authors of these pseudo-apocalypses, including the Revelation of John, simply used the Old Testament books of Zechariah, Ezekiel, and Daniel, as models for their own work. In the following passage, for instance, we find the inspiration behind the four horsemen of the Apocalypse that is found the book of Revelation:

> Again I lifted up my eyes, and saw, and behold, four chariots came out from between two mountains; and the mountains were mountains of brass. In the first chariot were red horses; in the second chariot black horses; in the third chariot white horses; and in the fourth chariot dappled horses, all of them powerful.
> —Zechariah 6.1-3

In the Apocalypse of Zechariah, we also find seven "lamp-stands" of the light of the Lord, as well as the same type of "scrolls," "crowns," "prophecies," "Daughter of Babylon," and the hellish punishments that await the disobedient. All of these symbols and afflictions are related to "the curse that is going out over the whole land" that is ordered by Jehovah. Likewise, the leviathan dragon images that appear in the book of Revelation were lifted from Psalm 74.14, Jeremiah 51.34, and Ezekiel 32.2.

In these apocalyptic stories are references to a great Machiavellian war that is to come, in which the Lord will cause everyone to "attack each other." Most of the people will be wiped out with biological weapons in the form of plague.

We are told in the book of Revelation that a man named "John" was approached by a divine being who "was like the Son of man"—that is, someone who was *like* Jesus. Initially, the divine being only identifies himself as the one who "holds the seven stars in his right hand," and the one who holds "the keys of death and hell" (Revelation 1:17-18). We are eventually led to believe that the divine being is Jesus; however, according to the gospels, Jesus himself indicated that he held the keys to the kingdom of the heavenly Father (Matthew 16.19), not to death and hell. It is therefore much more likely that the individual who is described is someone who represents Jehovah.

John is then supposedly told a "sacred secret" that began with the "prophet" "slaves" of the Old Testament. A secret that will be consummated in a new Judeo-Christianized era (Revelation 10.7). He is eventually taken into a throne room, where an elaborate pageant is played out before him. In this symbolic performance, the Savior is represented as a "Lamb." Through the use of sorrow, compassion, remorse, guilt, and fear, the people will feel compelled to surrender their hearts and minds to the institution of the wounded lamb that will rule over the Earth (Revelation 5.9-10).

In the following passage, the Lord praises those who did not figure out the so-called "deep things of Satan." This may be a reference to the gnostic and/or Marcionite Christians, who were aware of the difference between Yahweh/Jehovah and the heavenly Father.

> But the rest of you in Thyatira—all who don't hold on to Jezebel's teaching, who haven't learned what are called the deep things of Satan—I won't burden you with anything else. Just hold on to what you have until I come. I have received authority from my Father. I will

give authority over the nations to everyone who wins the victory and continues to do what I want until the end. Those people will rule the nations with iron scepters and shatter them like pottery.
—Revelation 2.24-27

Of course, Jesus himself never spoke of ruling over the people with an iron rod and breaking them into pieces.

In the following passage, we find more references to the slavery and fear that Jesus had sought to abolish:

Praise our God, all you his servants, you who fear him, both great and small!
—Revelation 19.5

In Revelation 22.16, the author begins speaking as Jesus himself, saying that it is "I, Jesus" who inspired the Apocalypse. However, this simply is not true. Indeed, Jesus warned that the minions of the devil would attempt to use his name and usurp his identity after his departure:

Be careful not to let anyone deceive you. Many will come using my name. They will, 'I am he,' and they will deceive many people.
—Jesus, Mark 13.5-6

When you see "the abomination that causes desolation" standing where it does not belong—let the reader understand [. . .]
—Jesus, Mark 13.14

The "abomination that causes desolation" is the "Almighty" "Lord of Armies." The place where "he does not belong" is on the throne next to the "Lamb." The truth is that the abomination was the same being that Jesus referred to as the "Father of the lie" (John 8.44).

Jesus knew that the forces of Yahweh would try to thwart him, both during and after his ministry. He tried to prepare the people for this inevitable scheme when he warned them about "false prophets" (Mark 13.21-22). In the following passage, Jesus tells the people how to identify false prophets:

> Watch out for false prophets. They come to you in sheep's clothing, but inwardly they are ferocious wolves. By their fruit you will recognize them [. . .]
> —Jesus, Matthew 7.15-16

The fruit of Jehovah in the book of Revelation is war, famine, suffering, fear, corruption, death, and disease. The "Lamb" who deceptively posed as the Christ is the Wolf in "sheep's covering":

> The Lamb who was slain deserves to receive power, wealth, wisdom, strength, honor, glory, and praise.
> —Revelation 5.12

It is not Jesus who sought to acquire wealth and egotistical power and fame, but rather the Wolf who masqueraded as the Lamb. We are also told that it was the so-called "Lamb" who unleashed the four horsemen of the Apocalypse to bring hell on Earth. Therefore, it is likely that the author of the book of Revelation was in-league with the Wolf. In this case, it can be concluded that the author was one of the false prophets that Jesus warned his followers about.

Moreover, the "vision" that "John" claimed to have witnessed never actually happened. What the cunning author did was simply use Old Testament motifs in order to advance a Yahwehist agenda.

Despite Jesus's repeated warnings, the leaders of Paul's sect failed to recognize the signs. One of the ways that this was able to happen was due to the unfortunate inability of the people to read, analyze, and interpret the scriptures for themselves. Because of this lack of education, the parochial laity were forced to trust Pauline officials—officials who never understood the true mission of Jesus to begin with!

Of course, the result of all of this deliberate deception and unintentional misunderstanding is that the forces of Jehovah prevailed.

> [. . .] See, the Lion of the tribe of Judah, the Root of David, has triumphed [. . .]
> —Revelation 5.5

In the following passage, the Wolf commends those who have kept his (Jehovah's) name, and rewards the obedient by offering to spare them from the horrors of the Tribulation:

> I know that you [the church of Philadelphia] have little strength, yet you have kept my word and have not denied my name [. . .] Since you have kept my command to endure patiently, I will also keep you from the hour of trial that is going to come upon the world to test those who live on the earth.
> —Revelation 3.8-10

Realizing what was actually happening, the gnostic Christians referred to this tragedy in the following passage:

> After he [Yaldabaoth] imprisoned those from the Father, he seized them and fashioned them to resemble himself. And it is with him that they exist.
> —Nag Hammadi Codices, The Second Apocalypse of James

A similar sentiment is found in the canonical account:

> When anyone hears the message about the kingdom and does not understand it, the evil one comes and snatches away what was sown in his heart [. . .]
> —Jesus, Matthew 13.19

Peter and Paul were the ones who heard the message but did not fully understand it. Because of their ignorance, the legacy of the Lord of Armies was able to not only survive but to advance. However, it is unlikely that this operation had been orchestrated by Yahweh himself, but rather by agents who had been conditioned to uphold his legacy. One of those agents was the author of the book of Revelation.

What the book of the Revelation of John documents is a massive Machiavellian operation that was intended to extend the power of the Lord of Armies outside the old Semitic realm. This new Machiavellian operation called for the affliction of the people of the world with disease, famine, and war, in order to "kill a third of mankind," before stepping in at a later time to present himself as their glorious benefactor:

> [. . .] And God will wipe away every tear from their eyes.
> —Revelation 7.17

Of course, this act of generous mercy was granted only on the condition that the people submit themselves as slaves that "serve him day and night":

> [. . .] the survivors were terrified and gave glory to the God of heaven.
> —Revelation 11.13

* * *

Many Paulines believe that the Apocalypse will be accompanied by the return of Jesus to the world. One of the sources that is used to justify this belief is found in Luke 21.27. In this passage, Jesus warns his followers about a future calamity. He warns that they may not survive the turmoil, and that they should be ready to meet him in the Kingdom of the Heavenly Father. In this case, Jesus was not literally returning to Earth in physical form, but rather was helping his followers to cross over to the other side upon their deaths from this unfortunate event.

> In my Father's house are many homes. If it weren't so, I would have told you. I am going to prepare a place for you. If I go and prepare a place for you, I will come again, and will receive you to myself; that where I am, you may be there also.
> —Jesus, John 14.2-3

Jesus told his followers that the Tribulation would occur in their own generation:

> Truly I tell you, this generation will certainly not pass away until all these things have happened.
> —Jesus, Mark 13.30

Therefore, the Tribulation was never supposed to occur in some apocalyptic age in the far future, such as the present time—as is commonly believed. For Paulines to disagree with this conclusion they would have to disagree with the words that are actually recorded in the Bible itself. Even the account that is documented in the book of Revelation clearly indicates that the End Times were not thousands of years in the future, but rather "near" and "soon":

> Do not seal up the words of the prophecy of this scroll, because the time is near.
> —Revelation 22.10

> Look, I am coming soon! [. . .]
> —Revelation 22.12

Likewise, according to the New Testament letters, the "antichrist" was predicted to arrive in the age in which it was written:

> Children, it's the end of time. You've heard that an antichrist is coming. Certainly, many antichrists are already here. That's how we know it's the end of time.
> —1 John 2.18

It could be said that the destruction of Jerusalem and the apocalyptic Tribulation were fulfilled in 70 CE, when the Romans attacked Jerusalem and destroyed the Temple of Yahweh. Indeed, in Mark 13.1 Jesus seems to predict the destruction of the Temple—that is, if we assume that this passage was not inserted by the author after the events had already occurred. It is also reported that during the Tribulation event, when Jerusalem was "surrounded by armies," the people would "flee to the mountains" (Luke 21.20-21); which did occur when the Jewish rebels fled to Masada in order to escape the Roman army.

The return of the Savior (i.e., *Parousia*) could either be the resurrection after the crucifixion, or perhaps more likely his spiritual manifestation in heaven—which is what Marcion[360] and the gnostics[361] believed. Indeed, according to Mark 12.25, Jesus indicated that the resurrection was not physical, but rather spiritual; which is also corroborated in 1 Corinthians 15:35-52. Therefore, Judgment Day is the event in which a soul ascends into the world of spirit after the death of the physical body; which would be an event that would be more frequent during the Tribulation. Those whose hearts and minds were pure would be able to ascend into the Kingdom of Heaven. This was also an event that was supposed to occur not thousands of years in the future, but rather during the time of Jesus; hence the following statement:

> [. . .] And you will see the Son of Man sitting at the right hand of the Mighty One and coming on the clouds of heaven.

---

[360] The *Panarion* (i.e., Against Heresies) (Book I, part 42) by Epiphanius of Salamis.

[361] The Sophia of Jesus Christ, Nag Hammadi codices.

—Mark 14.62

In this case, references to a physical resurrection that appear in the New Testament may actually be revised interpretations that that were inserted by well-meaning devotees who were not present during the time of the actual events. Furthermore, it must be understood that it was the scribe's duty to impress the reader and thereby inspire conversion to Judeo-Christianity. Expanding upon the supernatural element was a way in which to do this.

A primary reason why passages that confirm the preterist[362] position have been ignored by many Paulines is because the belief that Armageddon, the Tribulation, the Rapture, the Apocalypse, and the return of the Christ are about to happen, infuses their belief system with an element of exciting faith-bolstering relevance. The other effect of this interpretation is that it compels fearful obedience for those who may be concerned about the horrendous adversity that is foretold, and about the horror of being "left behind" during the "Rapture"—which further acts to benefit the institution of Pauline Judeo-Christianity itself. However, the concept of the Rapture is based on yet another misunderstanding.

* * *

The concept of the Rapture is related to the Tribulation that is believed will occur during the apocalyptic End Times. Many Judeo-Christians believe that those who are faithful to Lord Jehovah will be taken up into heaven, while those who are not will be "left behind" to suffer the prophesized calamity. The biblical precedent for this belief is found in the following passages:

> Two men will be in the field; one will be taken and the other left. Two women will be grinding with a hand mill; one will be taken and the other left. Therefore

---

[362] Preterism is the belief that the events described in the book of Revelation have already happened.

keep watch, because you do not know on what day your Lord will come.
—Jesus, Matthew 24.40-42

But no one know of that day and hour, not even the angels of heaven, but my Father only.
—Jesus, Matthew 24.36

Therefore keep watch, because you do not know the day or the hour.
—Jesus, Matthew 25.13

For the Lord himself will come down from heaven, with a loud command, with the voice of the archangel and with the trumpet call of God, and the dead in Christ will rise first. After that, we who are still alive and are left will be caught up together with them in the clouds to meet the Lord in the air. And so we will be with the Lord forever. Therefore encourage one another with these words.
—1 Thessalonians 4.16-18

According to the gospels, Jesus himself indicated that on that day a person's material possessions will no longer mean anything. The reason for this is because on that day that person will no longer be in the physical world because he or she would have ascended into the world of spirit. In the Gospel of Luke, where some of the alleged references to the Rapture occur, Jesus also makes a reference to a "dead body" (Luke 17:37). The original reason for the references to a dead body and to the end of material possessions is because what he was referring to was the death of the physical body. This is what the Rapture actually is. This is also what the "Day of Judgment" is as well.

The word "rapture" is a term that is found in the Latin Vulgate translation of the Bible. It is derived from the word *raeptis*, which means "taken away" or "caught up." This is a reference to someone's spirit form, or soul, being caught up and taken away into heaven, which is essentially a euphemistic term for death.

> When these people, however, have completed the time of the kingdom and the spirit leaves them, their bodies will die but their souls will be alive, and they will be taken up.
> —Jesus, The Gospel of Judas

When Jesus was referring to people who would be "taken up," it was a reference to those whose lives would be *taken*—that is, taken up out of the physical world and raptured into the realm of "everlasting life" in the spiritual world. What he was doing was advising the people to be mentally and spiritually prepared for that event. This is something that Paul never fully understood; hence his statement in 1 Thessalonians 4.16-17, in which he differentiates between the dead and the living, who he says will both be taken up separately into the clouds with Jesus.

When Jesus talked about what has come to be known as the Rapture, he referred to when the people of Sodom had been killed in the fire of the Lord, which is another reference to death. According to Jesus, that was not the end of their lives, for they were resurrected into the realm of everlasting life in the spiritual world. In this case, the only ones who will be "left behind" are those who do not enter into the light after the expiration of their physical form.

Indeed, one of Jesus's primary aims was to make the people aware of the Kingdom of Heaven in the after-life. This is why he emphasized preparation by cultivating a virtuous psyche in this world. Up until that point, the emphasis in the Old Testament/Tanakh was mostly placed on an underworld state that is referred to as "Sheol," or "Gehenna," which are essentially Hebrew words for the grave, underworld, and hell. Although, there were instances in which a joyous afterlife existence was alluded to (e.g., Isaiah 26:19; Daniel 12:2), these exceptions are both sparse and brief. The word translated as "heaven" or "the heavens" (*shamayim*) originally only referred to the firmament of the sky (e.g., Deuteronomy 28.12), not to the blissful "Kingdom of Heaven" in the after-life (*Basileia tou Ouranou*) that is referred to by Jesus (e.g., Luke 10.15).

* * *

Other misunderstandings in the book of Revelation relate to so-called "beasts," "serpents," leviathan "dragons," and the infamous "antichrist." However, what is commonly misunderstood about the term "antichrist" is that it does not even appear in the book of Revelation. This term actually only appears in the New Testament Epistles of John, who is not the author of Revelation. Of course, the "false prophet" of Revelation could be considered to be an antichrist, just as anyone who denies the significance of the Christ could be— which is the original definition of the term according to "John" himself. Indeed, this is why the term appears in plural form (1 John 2.18; 2 John 1.7). The book of Revelation indicates that the prophesied events would appear in their own generation, and it is in that generation that we will discover the identity of its characters.

In the latter half of the first century of the Common Era, the primary antagonists of the Christians were not only the Yahwehists but the Romans. The Romans were not only originally unreceptive to the religions of Yahweh/Jehovah but were also involved in self exultation and the worship of pagan gods. Therefore, the author of Revelation related Rome to the city of Babylon. There is even a cryptic reference to Rome in the following passage:

> This calls for a mind with wisdom. The seven heads [of the beast] are seven hills on which the woman [the Whore of Babylon] sits.
> —Revelation 17.9

The "seven hills" refers to the seven hills on which the city of Rome was built (Cory 2006: 76; Ehrman 2009: 98; White 1999). The following verse indicates that the "woman" represents the city of Rome itself:

> The woman you saw is the great city that rules over the kings of the earth.
> —Revelation 17.18

The hostile pagan Romans incarcerated the Christians and sentenced them to cruel deaths:

> I saw that the woman was drunk with the blood of God's holy people, the blood of those who bore testimony to Jesus.
> —Revelation 17.6

The reason why Revelation was encrypted with symbolism was to conceal its message from the Roman authorities, who would have destroyed it along with anyone who possessed it.

In the following passage, we find a cryptic reference to a series of kings:

> This call for a mind with wisdom. The seven heads are seven hills on which the woman sits. They are also seven kings. Five have fallen, one is, the other has not yet come; but when he does come, he must remain for only a little while. The beast who once was, and now is not, is an eighth king. He belongs to the seven and is going to his destruction.
> —Revelation 17.9-11

The five who have fallen refers to the five Roman emperors of the Julio-Claudian dynasty who had ruled since the time Jesus was born—those emperors being: Augustus, Tiberius, Gaius (i.e., Caligula), Claudius, and Nero. The emperor "kings" who came after the first five were from the Flavian dynasty of Vespasian (White 1999). It was the eighth "king," Domitian, who sought to establish the formation of an imperial cult that was based on himself and the Flavian family that he descended from (Friesen 2001: 46, 60; White 1999). This bloodline included Vespasian and Titus; the very same

Roman persecutors who were involved in the attack on the rebels in Judea. By establishing an imperial pagan institution with religious connotations that was hostile to Judeo-Christianity, Domitian had established himself as a "false prophet" in the mind of the author of Revelation.

The beast that is from the "earth" is either a high-priest of the imperial cult or a provincial governor, most likely in Asia Minor (White 1999)—the same location where the churches mentioned in Revelation were located. This so-called "beast" made the people worship the image of the beast that is from the "sea." Both the "image" and the "mark" refer to monetary currency. We know this because in Revelation 13.17 we are told that they were linked to "buying" and "selling." Therefore, the mark is most likely a reference to an imperial seal (i.e., *charagma*) or a certificate of sacrifice (i.e., *libellus*). What this means is that the beast from the sea and his cohorts sought to control the wealth of the land. The Old Testament tells us that this was an indulgence that was reserved for Lord Yahweh (and his elite agents) alone. The "image" of the beast was the face of one of the emperors that appeared on the Roman coinage of the time. The "mark" is associated with the identity of the beast who is from the sea. This character is identified by the number "666." Scholars have discovered that by assigning numbers to the Hebrew alphabet using a numerology process known as gematria, that the numbers 666 can be derived from the words "Nero Caesar" (Cory 2006: 61; Pagels 2012: 33). Therefore, the image that appeared on the currency was that of the Emperor Nero, who was an infamous persecutor of the early Christians. In this case, it can be concluded that Nero was the beast that came from the sea—namely, from out of the Mediterranean Sea from Rome. Indeed, it was Nero who initiated the Jewish-Roman Wars when he sent troops to quash the Great Revolt. It is highly likely that the author of Revelation was a refugee of that first war, which explains not only his exile to the island of Patmos but his Jewish Christian perspective.

According to Revelation 13.3, 13.14, the beast will be mortally wounded but will survive. This is another reference to Nero, who allegedly committed suicide but was thought to have survived (Pagels

2012: 32-33; White 1999). Indeed, the return of Nero is known as the *Nero Redivivus* legend.[363] The reality is that the "signs of the times" have always occurred and will continue to occur for years, centuries, and millennia to come. This is because the events and characters that are described in the book of Revelation are unintentionally infused with a fundamental archetypal significance that can be applied to a wide array of individuals and situations, including ones in the present day. However, it is now known that the situation that is referred to in the Apocalypse of John originally only applied to individuals and events that were occurring in the period in which it was written. This includes the character of the so-called "antichrist." This is why, according to 1 John 4.2-3, the spirit of the antichrist is said to be "already in the world."

* * *

According to the book of Revelation, the war between the forces of good and evil will eventually bring about a new theocratic empire. The agents of the Lord of Armies were supposed to bring about this regime by the use of violence, oppression, deception, and fear. It could be said that what Revelation successfully predicted was the subsequent "Dark Ages." Indeed, the alliance that was made between the Paulines, the Roman Empire, and the barbarian tribes (more on this subject to come), put the power structure in place that enabled the forces of the Lord of Armies to inadvertently bring about the apocalyptic scenario that is described in the book of Revelation. Consequently, the age that followed can be associated with the four horsemen of the Apocalypse. It was a time of war, famine, death, disease, and the attainment of power and wealth by an elite and ruthless few.[364] It is no coincidence that the rise of Christendom, and

---

[363] The Nero Redivivus legend is referred to in the book The City of God (Book XX) by Augustine of Hippo, and in the Sibylline Oracles (Book IV, V).

[364] In his book *The Gods of Eden*, William Bramley presents the case that the plagues of the Dark Ages were deliberately inflicted by agents of an other-wordly Judeo-Christian cabal who he refers to as the "Custodians."

the apocalyptic Tribulation that followed, corresponds precisely with the time-frame of the medieval Dark Ages. Therefore, the Apocalypse was not the end of the world, but rather the end of an age.

A trend in recent scholarship is to restrict what was previously referred to as the Dark Ages—which is now usually referred to as the Early Middle Ages—to about the years AD 500 to AD 1000. This period is characterized by a decrease in information and civilization. However, a definition of "darkness" should be expanded to include the increase in repression, wars, famine, plague, and corruption that occurred during that time. Indeed, the brevity of this reduced time-frame leaves out some of the darkest moments in European history—which was the region that was the most effected. Not only did most of the infamous wars of the time—which is the "white horse" of the Apocalypse—not begin until the eleventh century but the Black Death plague—which is the "pale horse" of the Apocalypse—did not fully occur until the fourteenth century. The Paulines were also able to acquire a good deal of wealth from the spoils of conquest—which is the "black horse" of the Apocalypse. All of those wars, plagues, and abuses of power, led to sporadic food shortages due to lost crops. The resulting famine was the "red horse" of the Apocalypse. It was also during that period that the atrocities of the Inquisitions occurred. Not only were the alleged enemies of Jehovah imprisoned in dungeons and executed but many were subjected to horrific acts of torture—in much the same way that the so-called "wicked" will be judged and punished during the Apocalypse.

Despite all of the tribulations that occurred after the year AD 1000, many historians are satisfied with the truncated time-line; however, the region that was most effected by this especially tumultuous period did not actually begin to see the light until the time of the Renaissance (from the French word for "rebirth") in the fifteenth and sixteenth centuries; as well as the following Age of Enlightenment. Indeed, the original Dark Ages coincided with the Middle Ages—which is the period between the fall of Rome and the Renaissance. This is the most accurate time-frame. Therefore, ideally, the medieval Dark Ages should be classified as the Early and Late

Dark Ages. Moreover, it can be contended that the "Dark Ages" moniker still applies.

It is true that Renaissance type movements did occur during the Middle Ages. The Carolingian Renaissance, of the eighth and ninth centuries, was the first of three cultural revivals. However, these revivals were not only limited to the rediscovery of literature from the Roman Pauline empire of the fourth century CE but the effects of the rebirth were mostly limited to the clergy and the literate elite. Likewise, the Ottonian Renaissance of the tenth century was also limited in its scope and impact. The closest movement that could be described as a true cultural revival was perhaps the Twelfth Century Renaissance; but even that was more of a precursor of things to come.

The turning point in this time-line should be marked by the invention of the printing press in 1450 CE. This invention allowed the mass production of information, which made it more possible for the public to read, learn, and think for themselves—which in turn led to subsequent progressive movements. Indeed, one of the books that was published during that time was the Bible. (It was also during that time that Niccolo Machiavelli's books were first printed.) Therefore, the entire Dark Ages actually lasted from the fifth to fifteenth centuries. Perhaps it is only coincidence that this is the thousand-year period of the reign of Jehovah that is predicted in the book of Revelation.[365]

It was also during that time that the pagan Eleusis festival in Greece was shut down. The last remnants of the Eleusian mysteries were eradicated in 396 CE, when barbarian Goths, under the leadership of Alaric, who himself was an Arian Pauline, invaded Greece. The historian and Neo-Platonist philosopher Eunapius reports[366] that Alaric was accompanied by Christian monks "clad in black raiment." According to Eunapius, the monks were "tyrants" who showed "contempt for things divine," and "allowed countless

---

[365] In Revelation 20.3, it is reported that the adversary of Jehovah, who is likened to a "dragon," will be cast into the abyss for one thousand years to keep him from "deceiving" the nations. During that thousand-year period, the theocracy of Jehovah would prevail.

[366] *Lives of the Philosophers*

unspeakable crimes." It was also during that same period that the Pauline emperor Theodosius 1 ordered the ban on the Olympic games, since it was also perceived as a pagan festival.

The Dark Ages are commonly blamed on the barbarians who attacked and destroyed Rome, thus bringing to an end the most advanced civilization of the ancient era—that is, despite its many ethical shortcomings. However, what is not commonly known is that the barbarians themselves were Judeo-Christians! The Visigoths, the Ostrogoths, the Vandals, as well as other smaller tribes, were all Pauline Judeo-Christians[367] (Kelly 2010: 102, 115; Le Goff 1991: 14; O'Donnell 2009: 91). The Germanic barbarians (as opposed to the Huns, who rejected the Paulines) had been converted to Judeo-Christianity by missionaries[368] even before the raid of Rome (LeGoff 1991: 14; Wolfram 1997: 76).

* * *

Between the years 325 and 451 CE, four Catholic Church councils were convened. It was these councils that developed a doctrine by which all others were to be considered heresy. One of the ways that they were able to control mental perspective was to destroy books that documented different viewpoints.

The book burning of the Pauline era began when the followers of Paul burned scrolls that were said to deal with "sorcery" (Acts 19.19). Likewise, Emperor Constantine ordered the burning of the book titled Against the Christians (written by the third-century Greek scholar and Neo-Platonist Philosopher Porphyry of Tyre). The same book was also burned by the Pauline emperors Valentinian II and Theodosius II (Bosmajian 2006: 38). (Porphyry himself was also physically assaulted by the Paulines.) The Catholic emperor Jovian ordered the burning of an entire library in Antioch (Jackson 2001: 407). Also in

---

[367] Most were Arian Judeo-Christians. Arianism (attributed to the presbyter, Arius) is the belief that Jesus was not God himself, but rather the son of God (which is affirmed in John 14.28).

[368] The most influential of which was the bishop Ulfilas.

the fourth century, the theologian Arius was denounced by Constantine's first Council of Nicaea because of his views pertaining to Jesus and the Father; and as a result, his books were destroyed (Barry 1907; Bosmajian 2006: 38-39). The decree against Arian Christian writings was also made by the Catholic Visigoth king Reccared, in the sixth century (McMillan, et al. 1984: 137). In the thirteenth century, the Roman Catholic Paulines burned the books of the Christian Cathars (Bosmajian 2006: 52). It was in that same century that the library of Constantinople was destroyed by Roman Catholic crusaders (Miller 2007). In the thirteenth and fourteenth centuries, the Roman Catholics burned thousands of Jewish books (e.g., the works of Maimonides and the Talmud, etc.) (Bosmajian 2006: 44-46). In the fifteenth century, the writings of the Judeo-Christian reformist John Wycliffe were banned and burned by order of the Roman Catholic Church (Lechler 1904: 502). Also in the fifteenth century, the Spanish inquisitor Tomas de Torquemada ordered the burning of non-Roman Catholic approved literature (Bosmajian 2006: 63; Cooke 2006: 499). It was around this same time that the Roman Catholic cardinal Cisneros ordered that the books of the Nasrid library in Granada Spain be removed and burned (Irwin 2005: 95). In the sixteenth century, the Spanish bishop Diego de Landa and his accomplices burned the codices of the Maya in the Yucatan (Bosmajian 2006: 28). It was around this same time that missionaries destroyed Buddhist texts in Japan (2006: 105-106, 128). Also in the sixteenth century, all books written in the native Sanskrit and Marathi languages were burned by the agents of the Inquisition in the Portuguese colony of Goa, no matter what the subject matter (2006: 28). One of the first known cases of book burning committed by Protestant Paulines occurred in 1520, when Martin Luther burned the theologian Angelo Carletti di Chivasso's book *Summa de Casibus Conscientiae*, along with other papal literature (Luther 1970: 225). Likewise, Protestants burned the Italian theologian Francesco Stancaro's book *Collatio doctrinae Arii* (etc.) (Bosmajian 2006: 88; Sher Tinsley 2001: 289). These are only a few examples of the extensive destruction and repression that occurred.[369] Although it is true that non-Judeo-Christians were also guilty of committing similar transgressions, it should be understood that Judeo-Christendom was not always the great guardian benefactor of civilization that is commonly proclaimed by the Paulines.

During the Dark Ages, the Paulines ruled with an iron hand. If a supposed "heretic" was repentant he would only have his books burned and he would be forced to publicly recant. The unfortunate

---

[369] The following is a list of some prominent intellectuals whose work was also banned by the Paulines: The theologian Ratramnus of Corbu (1050); The philosopher Berengar of Tours (1050); The philosopher, theologian, and logician Peter Abelard (1121, 1140); The philosopher Almaric (Amaury) of Bene (1210); The philosopher and theologian David of Dinant (1210); The philosopher Johannes Scotus Erigena (John the Scot) (1225); The rabbi and philosopher Maimonides (1233); The Judeo-Christian reformist Gerard Segarelli (1300); The theologian Petrus Johannes Oliva (Peter Olivi) (1328); The poet Dante Alighieri (1329); The physician, encyclopaedist, and professor Cecco d'Ascoli (1328); The philosopher, priest, and university rector John Hus (1415); The poet, astrologer/astronomer, and historian Don Enrique de Aragon (Marquis de Villa) (1434); The professor Pedro Martinez de Osma (1481); The physician and astrologer/astronomer Simon de Phares (1494); The humanist and lawyer Johannes Reuchlin (1514); The humanist and translator Louis de Berquin (1523); The French scholar and printer Etienne Dolet (1546); The Protestant reformer Martin Bucer (1555); The political philosopher Jean Bodin (1591); The philosopher, mathematician, and astrologer/astronomer Giordano Bruno (1599); The poet Agrippa d'Aubigne (1620); The poet Theophile de Viau (1623); The theologian Ferninando de las Infantes (1605); The scientist Marco Antonio de Dominis (1624); The physician Francisco Maldonado de Silva (1639); The theologian Jonas Schlichting (1646); The Unitarian reformer John Biddle (1647, 1655); The mathematician, physicist, inventor, and philosopher Blaise Pascal (1657, 1660); The political philosopher Thomas Hobbes (1683); The theologian and rector of Exeter College at Oxford Arthur Bury (1690); The philosopher John Toland (1697); The historian Pietro Giannone (1723); The philosopher and writer Voltaire (1734, 1759 etc.); The philosopher and writer Denis Diderot (1746, 1759); The philosopher, musician, and writer Jean Jacques Rousseau (1762, 1764 etc.). Note: even more cases are cited in Haig Bosmajian's *Burning Books* (2006).

ones were either imprisoned for life or sentenced to death. Some who received the death penalty were strangled before their bodies were burned at the stake. Others were slowly burned alive (Bosmajian 2006: 9, 53, 77). Indeed, many were burned at the stake together with their books. Furthermore, not only were authors arrested and killed but booksellers and printers as well (2006: 77-78). A papal bull issued by Pope Leo X stated that books could not be printed until they passed a rigorous inspection process by ecclesiastical censors (2006: 68). During the Spanish Inquisition, manuscripts were also required to be reviewed by censors. (i.e., the Censorship Law of 1558), and all printing had to be licensed by a bishop (Kamen 1999: 104-105). In England, no printing was allowed outside of Cambridge, Oxford, and London (Bosmajian 2006: 93). Thousands of books were also confiscated and destroyed at Oxford in 1535 and 1550 (2006: 66). Because church and state were linked, the book burning was extended to not only heretical writings but anything that could be interpreted to be "seditious" (2006: 8-9). Consequently, private homes and private libraries were also raided. For example, in Spain, the books in the library of the Marquis of Villena were seized and burned in 1434 (2006: 56). Moreover, in the sixteenth century the List of Prohibited Books (*Index Librorum Prohibitorum*) was created in order to suppress free speech and thought.

The following quote that is attributed to Jesus could have been addressed to those censors of free speech:

> Woe to you experts in the law, because you have taken away the key to knowledge. You yourselves have not entered, and you have hindered those who were entering.
> —Yeshua, Luke 11:52

Although some classical works of literature were collected and preserved in Catholic monastery libraries, in some cases these documents were erased in order to make way for Pauline literature. This was a practice known as palimpsesting. An example of this is the Archimedes palimpsest, in which the works of one of the greatest mathematic minds of the ancient world was erased and written over by a Catholic scribe who needed a writing surface (i.e., vellum) on which he could record liturgical text (Miller 2007).

In the sixteenth century, the English scholar William Tyndale translated the Bible directly from early Greek and Hebrew texts. The result was not only an edition that was more accurate than the official Latin Vulgate edition that had been sanctioned by the Catholic Church but a Bible that was more accessible to the public, since it was written in the common language of the day. However, his translations were not considered favorable to the Catholic Church, since, among other reasons, he used words such as "congregation" instead of church, and "senior" instead of priest. Therefore, his Bible was confiscated and he was strangled to death before his corpse was burned at the stake (Bosmajian 2006: 73-74). Despite his demise, his ground-breaking translations became the foundation of the popular King James Bible that is still in use to this day—although, it should be understood that even this edition cannot be considered to be entirely accurate (Ehrman 2005: 209).

* * *

The Pauline Roman emperors saw themselves as existing above the common morality of man, since their authority was thought to be ordained by God. It was a position that was recommended by Machiavelli in the following passage:

> And in fact, there was never anyone who ordained new and unusual laws among the people without having recourse to God, for they would not otherwise have been accepted. This is so because prudent men know of many beneficial things which, having no persuasive evidence for them, they cannot get others to accept. Consequently, wise men who wish to avoid this difficulty resort to divine authority.
> —Niccolo Machiavelli, *Discourses on Livy*

One of the advantages of combining the affairs of church and state is that it can be used to justify war. Soldier martyrs were promised eternal rewards for those who gave up their lives in service to the "Creator." As a result, a great many pious followers died while following the orders of rulers who they were led to believe were inspired by "God." During the Crusades, for example, the Roman Catholic soldiers believed that they were fighting not only for the liberation of the Holy Land but for the remission of sin (Phillips 2014: 35, 71, 166; Riley Smith 2005: 133-134); even though the "Holy Wars" themselves were brutally violent events in which looting and rape occurred (e.g., the sack of Constantinople) (Armstrong 1988: 386). Eventually, the lure of acquiring wealth also became a factor as new markets opened up and donations and pillaging helped to enrich the Roman Catholic Church (Armstrong 1988: 387; Phillips 2014: 144, 147, 166).

The Pauline sect developed a dogmatic creed that was not only based on the death of the Christ and servitude to Jehovah but emphasized the fear-based threat of eternal damnation. The laity were also led to believe that to follow Jesus was to embrace the meek and mild mentality of a suffering martyr, which was an interpretation that further acted to weaken and subdue the people.

> If our religion claims of us fortitude of soul, it is more to enable us to suffer than to achieve great deeds. These principles seem to me to have made men feeble, and caused them to become an easy prey to evil-minded men, who can control them more securely, seeing that the great body of men, for the sake of gaining paradise, are more disposed to endure injuries than to avenge them.
> —Niccolo Machiavelli, *Discourses on Livy*

Church officials were also able to deflect criticism directed at themselves and the situations that the suffering laity found themselves in by redirecting blame onto an adversarial devil scapegoat; even though the book of Revelation tells us that it was the will of Jehovah himself to inflict the world with tribulation.

> So simple-minded are men and so controlled by immediate necessities, that a prince who deceives always finds men who let themselves be deceived.
> —Niccolo Machiavelli, *The Prince*

During the apocalyptic age, widespread corruption in the highest levels of the church took place as various Roman Catholic officials jockeyed among themselves for power, prestige, and wealth. Some high-ranking positions could even be bought for a nominal sum—a practice known as "Simony."[370] Even though Simony was officially condemned by the Roman Catholic Church, it was a widespread practice in the ninth and tenth centuries (Doniger 1999: 1012).

Another problem that resulted was the continuation of a sexist male-dominated power system (see also 1 Corinthians 14.34-35).

> A woman should learn in quietness and full submission.
> —1Timothy 2.11

To this day, women are forbidden from holding any high positions—especially in the Catholic Church.

> Thus we must conclude, that a husband is meant to rule over his wife as the spirit rules over the flesh.
> —Augustine of Hippo,[371] *City of God*

* * *

---

[370] A reference to Simon Magus, who, according to Acts 8.18, attempted to purchase the ability to heal from the apostles.

[371] Augustine was an influential fifth-century Pauline theologian, philosopher, and bishop of Hippo.

It is commonly believed that Judeo-Christendom was a beacon of light in the Dark Ages. While it is true that some Paulines did actually follow the example that was established by Jesus, such as Saint Francis of Assisi and the other great lesser-known altruistic saints of the time, the fact is that the rise of that oppressive and tumultuous age in human history coincides with the rise of Judeo-Christendom.

It is true that hospitals and universities were constructed by the Paulines; however, during that time hospitals were more devoted to care rather than to cures. Even though limited attempts were made into the study of medicine, diseases were also often thought to be the punishment inflicted on sinners by God—especially before the emergence of the natural philosophers, which did not occur until the late Dark Ages. Holy relics and icons were also sometimes used in an attempt to ward off illnesses as a greater emphasis was placed on faith rather than reason (Freeman 2003: 320-321). Philosophical inquiry and rational thought were discouraged by such prominent Catholic archbishops as Gregory Nazianzus and John Chrysostom (2003: 309-311). It was not until the end of the Dark Ages that breakthroughs in health care were made by scientists, such as the microbiologist Antonie van Leeuwenhoek, and the polymath Michael Servetus—who nevertheless was sentenced to death by the Paulines, both Catholic and Protestant, due to differences that pertained to theological interpretations (Encyclopedia Britannica Online, s.v. "Michael Servetus").

The truth is that hospitals existed even before the rise of Judeo-Christendom. Some of the first hospitals were built by the Greeks, as well as the Indian and Sri Lankan Buddhists in the third and fourth centuries BCE (Rannan-Eliya and De Mel 1997: 19; World Book Encyclopedia, s.v. "Hospital" 2015). The ancient Egyptians also created medical treatises, such as the Ebers Papyrus, circa 1550 BCE (Encyclopedia Britannica Online, s.v. "Ebers Papyrus") and what has come to be referred to as the Edwin Smith Papyrus, circa 1600 BCE—although there is evidence that this document had been copied from older texts that date back to the Old Kingdom period, circa 3000 BCE (Encyclopedia Britannica Online, s.v. "Edwin Smith Papyrus"). Furthermore, the earliest attempt at a pharmacopoeia, in the form of a

Sumerian written tablet, also dates back to the third millennium BCE (Kramer 1981: 60-63).

Although monasteries, convents, and cathedral schools provided educational services during the Dark Ages, the curriculum was significantly limited. The Catholic cathedral schools in particular were focused primarily on religious instruction. Indeed, most of the students were male children of nobility who were preparing for careers in the church.

It is also true that the works of the philosopher Aristotle were preserved—that is, preserved from destruction by the Paulines themselves. It should also be remembered that it was a Pauline emperor (i.e., Justinian I) who initially shut down the last philosophic Academy. It is also disingenuous for Pauline traditionalists to claim that the Judeo-Christians were the leading scholars of the age when they forcefully silenced other voices. Although the Judeo-Christian Paulines are usually credited with preserving the works of Aristotle during the Middle Ages, there were also times when the Paulines (e.g., the Dominicans)[372] condemned Aristotle and ordered his writings to be burned (Bosmajian 2006: 45). Furthermore, in 1210 an edict was issued by a Catholic synod at the University of Paris that forbid lecturing on the subject of Aristotelian metaphysics and natural science (Perry, et al. 2008: 263; Spade, et al., 2013). In 1231, Pope Gregory IX ordered that Aristotle's writings could not be reinstated until all offending errors in his works were removed (Spade, et al. 2013). In 1245, Pope Innocent IV extended the ban to the University of Toulouse (2013). Likewise, acting on a request by Pope John XXI, the bishop of Paris, Stephen Tempier, issued an edict in 1277 that prohibited the teaching of 219 theological and philosophical works at the University of Paris that he deemed to be heretical (Thijssen 2013).

The sixteenth-century astronomer Nicolaus Copernicus was at first unable to publish the discovery that the Earth revolved around the Sun, because the findings contradicted the Pauline belief in an Earth-centered universe. Therefore, he did not release his findings until he

---

[372] The Dominicans are a Roman Catholic organization that was founded in the thirteenth century.

was near the end of his life. Indeed, Copernicus was condemned by the Dominican theologian and astrologer Giovanni Maria Tolosani, who believed—among other false assumptions—that mathematics could not be used to calculate activities in the natural world (Feldhay 1995: 205); although mathematics is now the primary tool that is used by physicists.

However, it was not only the Catholics who were impeding progress. The Protestant theologian Philipp Melanchthon—who was Martin Luther's primary collaborator, also called for Copernicus's theory to be repressed by governmental force (Rosen 1995: 198).

> [. . .] certain people believe it is a marvelous achievement to extol so crazy a thing, like that Polish astronomer who makes the earth move and the sun stand still. Really, wise governments ought to repress impudence of mind.
> —Philipp Melanchthon, Letter to Mithobius

The Protestant theologian John Owen also condemned the works of Copernicus. He declared that the heliocentric theory was based on "fallible phenomena" that contradicted the "testimonies of Scripture" (1995: 166-167). Likewise, the following statement was written by the Protestant-influenced Pauline theologian, John Calvin:

> We indeed are not ignorant that the circuit of the heavens is finite, and that the earth, like a little globe, is placed in the center.
> —John Calvin, *Commentaries of the First Book of Moses called Genesis*

Although Copernicus managed to avoid any serious consequences, in 1663 Galileo Galilei was convicted of heresy for similar findings and was placed under house arrest by the Roman Catholic Paulines for the rest of his life.

* * *

During the Pauline Dark Ages, free thought was repressed and the academic curriculum was restricted. It was the humanists that helped to unrestrict it. Humanism was a movement that promoted rationalism, empiricism, ethics, and science. Although the first humanists were a small progressive faction within the Pauline institution, over the centuries the movement became increasingly secular. It was the Humanists who transformed the universities into the institutions of higher learning that we know today.

> Humanism drove much of the curricular and research innovation. After joining university faculties, humanists transformed the study of grammar and rhetoric into the *studia humanitatis*. Once humanistic studies and humanists became established in the university, scholars in other disciplines acquired their philological and linguistic expertise, along with the humanistic ideology that viewed ancient texts as the true source of learning, and medieval scholarship as a barrier to them. Humanistically inclined scholars changed the content of instruction and research in all disciplines except theology.
> —Paul F. Grendler,[373] *The Universities of the Italian Renaissance*

What is less commonly known is that some of the first universities were initially independent from Judeo-Christendom. Indeed, the world's first university, the University of Bologna (founded in 1088), was initially a student-run institution that was independent of both kings and popes.

It is true that the founding of these institutions, or simply allowing the secular universities to exist, was a step in the right direction; however, both students and teachers were required to operate under

---

[373] P. F. Grendler, Ph.D., is a professor emeritus in the history department at the University of Toronto.

limiting regulations. Indeed, the pioneering seventeenth-century philosopher Rene Descartes cautiously remarked that the people of his time were more influenced by custom than by genuine knowledge of the truth.[374]

It should also be understood that the Pauline wars, such as the Thirty Years' War, disrupted the progress that was being made by the universities.

* * *

It was also during that period that the Roman Catholics engaged in the selling of "indulgences," which was money paid to the church for the remission of the punishments for sin. A percentage of what little money, or grain, animals, etc., that the peasant class did possess, was forced back into the church coffers in the form of "tithes." It was also during that time that the Roman Catholic armies would sweep through the lands and either kill or heavily tax its inhabitants. Of course, all of this would have appalled Jesus.

> For what does it profit a man if he gains the whole world, and loses or forfeits his own self?
> —Jesus, Luke 9.25

In the thirteenth century, a series of battles that were directed against a tribe of peasants in northern Germany (i.e., the Stedingers) were launched by both the Archbishop of Bremen and Pope Gregory IX. The war began after the peasants refused to perform forced labor and pay tithes to the archbishop. Despite some early success, the peasants were eventually violently suppressed (Kirsch 1907).

Jewish people were targeted because they denied that Jesus was the Mashiach. In 1391, the Spanish cleric and Archdeacon of Ecija in Seville, Ferrand Martinez, instigated a "Holy War" against the Jews for blasphemy. Those who did not convert were slaughtered (Poliakov 2003: 156-157). The persecution of the Jews (i.e., "pogroms")

---

[374] *Meditations on First Philosophy*

throughout the centuries included not only expulsions and forced conversions but outright massacres (e.g., the Rhineland Massacre of 1096; the Strasbourg Massacre of 1349; the Lisbon Massacre of 1506; etc.).

Likewise, over a millennium earlier, the Roman Catholic emperor Theodosius I sentenced the gnostic Manichaens, along with other smaller Christian sects (i.e., Encratites; Saccophores; Hydroparastates), to death (Arendzen 1910). Other Pauline emperors (i.e., Justin and Justinian) initiated similar campaigns, until the Manichaens were also struck down (Arendzen 1910).

The Medieval Inquisitions (e.g., the Episcopal Inquisition, etc.) were initiated in 1184 by Pope Lucius III. Likewise, Pope Gregory IX authorized the Papal Inquisition in 1233. In 1478, the atrocities of the Spanish Inquisition began. It was during that period that censorship and the burning of books intensified. The result was a decline not only in freedom but in intellectual achievement (Roth 1996: 21. 274). Other Inquisitions included the Roman Inquisition, the Portugese Inquisition, and the Goa Inquisition, etc.

Just as Jesus had been convicted of heresy by the priests of Jehovah, so did Pauline officials sentence "heretics" to death for the crime of disobeying the will of Jehovah.

> [. . .] Beware of the yeast of the Pharisees, which is hypocrisy.
> —Jesus, Luke 12.1

It can therefore be concluded that the form of Christianity that the world has become familiar with was not successful because it was the most truthful, but rather because it was the most oppressive and violent.

* * *

Throughout the Inquisitions, torture occurred in dark underground dungeons. During the torture sessions, an inquisitor, or some other commissioner of the Holy Office, would be present to act as overseers. The torture itself was carried out by men clad in black who often wore masks.

Torture was first authorized by Pope Innocent IV in 1252, and amended in 1256 by Pope Alexander IV. Alexander's addition authorized inquisitors to absolve one another of transgressions (i.e., "irregularities") that might occur while carrying out their duties (Peters 1985: 65).

Torture instruments that were used by the inquisitors included the *strappado*, which was a rope and pulley device that hoisted victims who were suspended by their hands behind their backs up into the air before letting them down in quick jerking movements that inflicted horrible pain and damage (Kirsch 2008: 105-106; Peters 1985: 68, 167). Some were choked with water (2008: 104-105; 1985: 167), while others suffered upon the rack; which was a device that separated bones from their joints by pulling the feet and arms in opposite directions (2008: 106; 1985: 68, 167). Another practice was the ordeal by fire, in which the victim was strapped down and fire was held to his or her feet (2008: 105). The torture was not supposed to last more than the time it took to recite a prayer, and was not supposed to last more than a single session; however, these rules were not always heeded (Blotzer 1910). If the victim confessed to heresy, the confession had to be repeated outside the torture chamber. If the victim refused to confirm the confession, he or she might be sent back and tortured again (Kirsch 2008: 16, 111; Peters 1985: 69). Children were not spared and were subjected to painful interrogation practices as well (2008: 76). There were also sadists among the torturers, who stripped their victims naked before abusing them (2008: 109). One of the primary aims was not only to inflict pain and terror but to denigrate and humiliate their victims until the ordeal could no longer be tolerated and the desired confession was obtained (2008: 108). The Roman Catholic Paulines were also able to profit from their activities by seizing the property of alleged "heretics" (2008: 120).

It is unknown for sure how many people were imprisoned, injured, and killed by the Inquisitions; just as it is also unknown how many innocent victims falsely confessed to heresy only as a means of ending their ordeal. It is known, however, that many victims were condemned based only on circumstantial evidence (2008: 103).

In recent times, some individuals, who are often affiliated with the Catholic Church, have attempted to soften the atrocities that were committed during that period. Despite such blatant revisionism, it should be understood that it is indeed evidential fact that crimes against humanity were indeed perpetrated by the servants of Jehovah during the Dark Ages.

Some of the physical persecution lasted all the way up until the 1700s, in the form of witch hunts—many of which were committed by the Protestant Paulines. The peak witch-hunting years occurred in Germany in the sixteenth and seventeenth centuries. A conservative estimate of the number of people who were killed for alleged witchcraft by the Paulines during that period is 45,000 (Levack 2006: 23).

The Paulines not only attacked anything related to paganism but mysticism as well, since it was believed that anything pertaining to the supernatural that was not in the Bible was the devil's sorcery. Of course, these were the same type of charges that were directed at Jesus when he was performing miracles as well:

> And the teachers of the law who came down from Jerusalem said, "He is possessed by Beelzebul! By the prince of demons he is driving out demons."
> —Mark 3.22

The irony and the error of the situation went unrealized by the Paulines, and as a result, their oppressive law prevailed.

> Our first debt to the Church and her priests is that, thanks to them, we Italians have become irreligious and wicked. But we owe it a still greater debt—the second cause of our ruin: that is, that the Church has kept and still keeps this country divided.
> —Machiavelli, *Discourses on Livy*

The Paulines brought about more adversity when missionaries entered into other sovereign regions around the world. They brought with them not only their mistaken interpretations of the Bible but their physical diseases as well. Although it is true that horrible indigenous practices, such as human sacrifice, were eradicated, in many cases the missionaries also banned harmless activities and set up a repressive imperialist church system that instilled "the fear of God" into the people.

* * *

In the twelfth century, gnostic Christianity reemerged in the form of the Cathars (or *Kathari,* which means "Pure Ones"). According to the Cistercian monk Raynaldus,[375] one of the Cathari blasphemies was that they believed that the Roman Catholic Church was corrupt.

The Cathars (also known as the Albigenses) were related to the Bogomiles in Bulgaria. Both of these movements appear to have been inspired by the Manichean gnostics of third-century Persia. Another dualist[376] group that arose during that period were the Paulicians, who were also violently persecuted.

The Roman Catholic Paulines eventually recommenced its war on the resurgent gnostic Christians. One of the ways that they were able to do this was by offering freedom from eternal damnation, as well as the material incentive of land, to the loyal servants of Jehovah who fought against the Cathars (Nigg 1990: 190). In the war that followed (i.e., the Albigensian Crusades), hundreds of thousands of people, including woman and children, were slaughtered.

These types of uprisings continued sporadically throughout the years. Each time they were struck down by the Roman Catholic army. As a result, the Pauline faction continued to dominate as the gnostic Christians were once again forcefully suppressed.

---

[375] *On the Accusations Against the Albigensians*

[376] Dualism, in this context, is the belief in the existence of two different deities. The first is the spiritual true God of light, which can be associated with the heavenly Father of Jesus. The other is the lower level Demiurge, which is Yahweh.

* * *

In 1302, Pope Boniface VIII declared his absolute authority over all human beings:

> Furthermore, we declare, state, define, and pronounce that it is altogether necessary to salvation for every human creature to be subject to the Roman pontiff.
> —Pope Boniface VIII, *Unum Sanctum*

In the twelfth century, the Augustinian reformer and monastery prior Arnold of Brescia was hanged and burned, and his writings were destroyed after he denounced the Catholic Church's materialism and advocated for greater liberty. Likewise, in 1600 the brilliant hermetic philosopher and astrologer Giordano Bruno was burned alive for, among other reasons, believing that the universe was populated with planetary worlds other than our own (i.e., the "plurality of worlds"). The only person who barely managed to escape with his life for disagreeing with the Roman Catholics was the Pauline reformist, Martin Luther.

Although the intention of the Protestant movement was certainly well-meaning, it was also a Christian movement that had nothing to do with dualistic gnostic knowledge—that is, in regard to the difference between Yahweh and the Father. The result was a creed that was about as equally flawed as its predecessor. Like the Catholic Paulines before him, Luther also did not understand the underlying truth of the mission of the Savior; and as a result, the mistake that was committed by the Catholics was carried over into the Protestant Reformation.

Luther may have been inspired by the Old Testament Lord of Armies when he exhorted his followers to kill thousands of rebellious peasants who were fighting for equality and human rights in feudalistic Europe during the German Peasants' War (Edwards 1975: 66). Although it is commonly believed that Luther deplored violence, he actually did condone such actions when he believed that it was necessary to crush those who were operating "outside the law of God and Empire."

> Therefore let everyone who can, smite, slay, and stab, secretly or openly, remembering that nothing can be more poisonous, hurtful, or devilish than a rebel [. . .]
> —Martin Luther, *Against the Murderous, Thieving Hordes of Peasants*

Luther also approved of the executions of the members of a Judeo-Christian reformist group known as the Anabaptists (Nigg 1990: 304), whose only major difference with the Lutherans, besides their pacifism, was that they believed in adult baptism, as opposed to infant baptism. Luther referred to the Anabaptists as "brainsick" "bastards," whose "parents were all adulterers and whoremongers."[377] Luther not only advocated for a policy of submission (i.e., serfdom) to the masters of the world, both religious and secular, but was also severe in his attitude toward the Jews, who he described as "poisonous bitter worms."[378] Luther even advocated for setting synagogues on fire and destroying Jewish homes. In his book *On the Jews and their Lies*, he stated that Jews should be prevented from teaching "on pain of life and limb." His hateful diatribes helped to instigate the violent persecution of Jews in Germany.

In the sixteenth century, the Protestant-influenced theologian John Calvin began his career in Switzerland when he spear-headed a movement that maligned free will and reserved salvation solely for an elite predestined "elect" (i.e., "unconditional election"). Those who were not fortunate enough to be born with such a status were destined for damnation (i.e., "reprobation"). Calvin based his interpretation on the premise that God controls every aspect of his creation and therefore does not leave anything to free will or chance. What is especially peculiar about this deterministic interpretation is that it not only contradicts scientific findings that prove the existence of

---

[377] *A Commentary on Saint Paul's Epistle to the Galatians*

[378] *On the Jews and Their Lies*

randomness and uncertainty in nature but also passages in the Bible itself, where Jehovah clearly did not have foreknowledge of events (e.g., Genesis 3.9, etc.).

Calvin also played a part in the death of the theologian, physician, and Renaissance humanist Michael Servetus, who he condemned as a heretic. Likewise, he also ordered that Servetus's book (*Christianismi Restitutio*) be burned. Indeed, the Calvinists sentenced other books to the flames as well (Bosmajian 2006: 88, 95).

The Puritans were an Old Testament-based Calvinist offshoot sect who believed that humanity was inherently contaminated and in need of strict Judeo-Christian guidance. Sexuality was repressed, attendance at church was mandatory, and questioning the Bible was prohibited. Punishments included: fines, public shaming, imprisonment, whippings, and executions. In 1684, Puritan ministers published a book titled *An Arrow Against Profane and Promiscuous Dancing, drawn out of the Quiver of the Scriptures*. Of course, all of these oppressive rules actually contradicted the original spirit of the Savior. Jesus himself would have been appalled by such heartless small-mindedness.

> We played music for you, but you did not dance [. . .]
> —Jesus, Matthew 11.17

> When you strip without being ashamed, and you take your clothes and put them under your feet like little children and trample them, then [you] will see the son of the living one and you will not be afraid.
> —Jesus, The Gospel of Thomas, Nag Hammadi Codices

The Puritans were also the same sect that engaged in the witch hunts of the 1690s, in which over a hundred people were arrested and imprisoned—some of whom were tortured and killed. The Puritan theocracy that was established in the New World of America justified torture and persecuted not only people who they thought were pagans but non-Puritan Christians, such as the Quakers, as well.

The European "Wars of Religion" began in the sixteenth century and lasted for over a hundred years (e.g., the German Peasants' War; the War at Kappel; the Schmalkaldic War; the Eighty Years' War; the French Wars of Religion; the Thirty Years' War; the Wars of the Three Kingdoms; etc.). Some of these wars, such as the French Wars of Religion, were known for their brutal massacres (e.g., the Massacre of Vassy; the Saint Bartholomew Day Massacre; the Massacre of Merindol).

As a result of all of the Pauline infighting that occurred in Europe during the Thirty Year's War, many peasant farmers who were caught in the middle lost their crops, and after decades of warfare, plague, and starvation, the population of Germany dropped by approximately 25% to 40% (Encyclopedia Britannica Online, s.v. "History of Europe").

It was during that time that the Puritan "New Model Army" arose in England, Ireland, and Scotland. Its leader, Oliver Cromwell, cited Old Testament passages in order to justify his ruthless campaigns.

* * *

The Roman Catholic Paulines were aware of the power that the arts (i.e., aesthetics) had on the hearts and the minds of the people. They sought to control this power by establishing rules related to permissible artistic expression, and by soliciting the most talented artists of the day to promote their own interpretation of Christianity[379]—especially during the Counter Reformation (Paoletti and Radke 2005: 27, 51).

The word aesthetic is derived from the Greek word *aisthetikos* (i.e., *aisthanomai*), which essentially means "to perceive." The definition of the word relates to the ability of beauty to affect the perception of the observer.

The Roman Catholic Paulines commissioned monumental works of architecture in the form of majestic cathedrals that were intended

---

[379] The need for an orthodox artistic standard was affirmed at the Council of Trent in the sixteenth century.

to impress upon the observer the majesty and glory of not only Jehovah but of the institution of the Catholic Church itself. However, not even the power of Pauline Judeo-Christendom could stop the inevitable rise of the Renaissance that appeared at the end of the Dark Ages.

The Renaissance began in Florence Italy in the fifteenth century. It was a movement that not only promoted the humanistic genius of man but drew inspiration from pre-Judeo-Christian art and knowledge as well. The Renaissance also helped lead the way for the following Scientific Revolution, the Enlightenment, and the Romantic period; which promoted not only beauty, nature, and the imagination but the emotional spirit of freedom as well. Major reforms took place as a greater emphasis was placed on human rights, civility, reason, liberty, and respect for diversity. Consequently, the Paulines made the necessary adjustments in order to ensure their relevance; although, at the same time also doing everything they could to silence the rising voices of the modern age—such as confiscating the literary works of the luminary intellectuals of the Enlightenment (e.g., Voltaire; Rousseau; Diderot; etc.) (Bosmajian 2006: 118).

Therefore, the Enlightenment and the modern-day world that we enjoy today, which places a greater value on human life, liberty, democracy, science, and ethics, did not arise because of Judeo-Christianity, but rather despite Judeo-Christianity.

* * *

The rediscovery of ancient Greek culture and ideas also helped to revive an interest in democracy. This rebirth helped to spark the historic revolutions that occurred in the eighteenth century. It was also during that time that an emphasis was placed on the need for the separation of church and state. This statute was intended to not only prevent a government from restricting personal liberty, and not only to prevent the infighting that was occurring between the Judeo-Christian denominations, but to prevent a leader from claiming some type of holy mandate that is above the common ethical standard and law of man—as Pauline rulers had done for centuries in the Old

World. This safeguard was put into place because the humanists of the Enlightenment knew that when the affairs of church and state are intertwined, irrational and inhumane activities can be justified when the people are led to believe that there is a divine significance behind it. This is the situation behind the so-called "Holy War."

> Those who believe in absurdities will commit atrocities.
> —Voltaire, *Questions sur les miracles*

The humanistic principles that were advanced during the post Dark Age era is especially exemplified by the movements that occurred in the American New World. Although it is widely believed by many American Paulines that the United States of America was founded by Judeo-Christians, this is not entirely true. The laws of Jehovah actually contradict the original American values of liberty and democracy. Jehovah himself would have preferred a totalitarian theocracy. In fact, when Jehovah was on the Earth he actually embodied the persona of a Middle Eastern dictator with weapons of mass destruction!

American Paulines will sometimes cite the words "In God We Trust," which is the motto that is printed on federal property, in order to justify their position; however, this Judeo-Christian maxim is not the original motto of the United States. In fact, it did not even appear until almost a hundred years after the founding of the nation—although, even then it was used only on coinage. It did not become an officially recognized motto until 1956. The original motto that was used by the Founding Fathers was actually *E Pluribus Unum* (One Out of Many). Moreover, it does not say "In God We Trust" in the U.S. Constitution, but rather "We the People"; which is another reference to a democracy, not a theocracy.

American Paulines cite examples of the Founding Fathers referring to God and the principles of the Judeo-Christian faith; however, while it is true that many of the Founding Fathers of America did respect some of the basic principles of the Judeo-Christian religion—perhaps out of practical necessity—they were also students of philosophy and Enlightenment thinking. Some were also Deists, which included such prominent statesmen as Thomas Jefferson and Thomas Paine. Deism was a progressive Christian philosophy that embraced the value of reason over dogmatic faith. Some of the Founding Fathers, such as James Madison and Thomas Jefferson, even condemned traditional Judeo-Christianity as being corrupt.

> Paul was the [. . .] corrupter of the doctrines of Jesus.
> —Thomas Jefferson, Letter to William Short

> His [Jesus's] object was the reformation of some articles in the religion of the Jews, as taught by Moses. That sect had presented for the object of their worship, a being of terrific[380] character, cruel, vindictive, capricious and unjust.
> —Thomas Jefferson, Letter to William Short

> Experience witnesseth that ecclesiastical establishments, instead of maintaining the purity and efficacy of religion, have had a contrary operation. During almost fifteen centuries has the legal establishment of Christianity been on trial. What has been its fruits? More or less, in all places, pride and indolence in the clergy; ignorance and servility in the laity; in both, superstition, bigotry and persecution. [. . .] What influence, in fact, have ecclesiastical establishments had on society? In some instances they have been seen to erect a spiritual tyranny on the ruins of the civil authority; on many instances they have been seen upholding the thrones of political tyranny; in no instance have they been the guardians of the liberties of the people. Rulers who wish to subvert the

---

[380] Jefferson uses the word "terrific" here in the original sense of the word, which meant terrifying.

> public liberty may have found an established clergy convenient auxiliaries. A just government, instituted to secure and perpetuate it, needs them not.
>
> —James Madison, A Memorial and Remonstrance (Addressed to the General Assembly of the Commonwealth of Virginia.)

The other influential group that was active during those years were the Freemasons, which included such esteemed members as Benjamin Franklin, John Hancock, Paul Revere, Andrew Jackson, and George Washington. The following passage is taken from a classic Masonic manuscript:

> The two great motors are Truth and Love. When all these Forces are combined, and guided by the Intellect, and regulated by the Rule of Right, and Justice, and of combined and systematic movement and effort, the great revolution prepared for by the ages will begin to march.
>
> —Albert Pike,[381] *Morals and Dogma*

Although the origin of Freemasonry is not known for certain, it is believed that its roots may be traced back to mysterious "brotherhoods" (e.g., the Rosicrucians and the Knights Templar of the Middle Ages, etc.) who were thought to possess secret knowledge. This mystery has inspired many nefarious conspiracy theories over the years. Many present-day investigators who are influenced by the Pauline creed have linked these groups to a world-wide satanic conspiracy involving the antichrist, the so-called "New World Order," the Illuminati, and the apocalyptic end of the world.

Although its history may be open to debate, Freemasonry itself can only really be traced back to the stone-masons of the early European Renaissance period. It was originally a local trade

---

[381] Albert Pike was an eminent nineteenth-century Scottish-rite Freemason, poet, and attorney.

fraternity. It was not until the early eighteenth century that some more inquisitive individuals began to delve into other more esoteric topics. These more philosophical members were called "speculative," as opposed to "operative," Masons.

It is very likely that one of the reasons for all the secrecy had to do with the continuing threat that was posed by the officials of the Pauline institutions and the laity who they influenced. Indeed, the Paulines have had a tumultuous relationship with the esoteric brotherhoods over the centuries. Even though by the time the Freemasons arose the Inquisitions were coming to an end, the Paulines still had the ability to blacklist those who did not conform to its own tradition. Free-thinking and progressive ideas continued to be discouraged. Such condemnation and "excommunication" were enough to effect public reputations and careers. Indeed, many Freemasons were, and are, eminent public figures.

The reason for the secrecy does not have anything to do with a sinister plot, as is commonly believed, but rather with what was referred to by Jesus:

> It is those who are worthy of my mysteries that I tell my mysteries.
> —Jesus, The Nag Hammadi Codices, The Gospel of Thomas

> Do not give dogs what is sacred; do not throw your pearls to pigs. If you do, they may trample them under their feet, and turn and tear you to pieces.
> —Jesus, Matthew 7.6

The unfortunate fact is that not everyone in the world is capable of understanding and accepting higher truths. Therefore, retaining special information may be necessary in some instances due to the imperfect condition of a world that is still in the midst of recovering from a less fortunate time. The truth is just about every informed individual hopes for the day when the truth can be made public to the entire world.

> Truth is not for those who are unworthy or unable to receive it, or would pervert it. The Teachers, even of Christianity, are, in general, the most ignorant of the true meaning of that which they teach. There is no book of which so little is known as the Bible.
>
> —Albert Pike, *Morals and Dogma*

Although it is debatable whether or not at least some of the Freemasons were (and are) aware of the difference between Yahweh/Jehovah and the heavenly Father of the universal Godhead source, the official Masonic stance regarding the Supreme Being is that "The Great Architect of the Universe" is based upon one's own understanding of the Creator. This is why members of different faiths can all come together as fellow Masons.

Some present-day conspiratorialists believe that the Freemasons, along with the Bilderberger Group, the Trilateral Commission, the Council on Foreign Relations, as well as the mysterious Illuminati, are all scheming to create a tyrannical "one-world government" that will be ruled over by the biblical antichrist. However, it should be understood that most of these theories are influenced by the mistaken precepts of the Pauline creed.

An inscription that is found on the U.S. dollar bill reads *Novus Ordo Seclorum*; however, this is not a reference to some sinister "New World Order," but rather to a "new order of the ages." This motto actually pertains to the new era that the great thinkers of that time were ushering in. The "great revolution prepared for by the ages" that was referred to by the Freemasons, as well as the New Order of the Ages that is printed on the U.S. dollar bill, actually have a much more benign meaning. After all, it was Masonic types who helped to pave the way for the great revolution that founded the free country of the United States of America, which became a modern-day model for freedom, prosperity, and democracy throughout the world.

Conspiracy theorists will undoubtedly counter this thesis by citing events in which some confirmed member of a "brotherhood organization" was exposed committing some unethical transgression.

My response to that is that, statistically speaking, among any large group of people there are likely to be a minority who succumb to their shortcomings. However, I contend that burning down an entire orchard because of a few bad apples is both unnecessary and unwise.

Nevertheless, it is also acknowledged that the presence of powerful special-interest groups and individuals who are actively pursuing an unethical agenda that would work to undermine democracy is indeed a realistic threat. However, in most cases these factions have more to do with the wealthy corporate elite (i.e., oligarchs/plutocrats) than with the defunct Illuminati. The primary aim of the wealthy corporate elite is to make money for themselves, which was not the primary aim of groups such as the Illuminati. The objective of the Illuminati was to oppose the irrational and dysfunctional dominance of Pauline Judeo-Christendom. What the Illuminati sought to establish in its place was a more humanistic regime that place a greater emphasis on secular philosophy and science. Therefore, the wealthy corporate elite and the so-called "brotherhood" organizations, such as the Illuminati, should not be conflated.

What the Founding Masonic Fathers of America established was not specifically a Judeo-Christian nation, but rather a free and independent country that more highly regards personal liberty. The United States of America was founded upon the humanistic principle that all people should have the right to practice their religious beliefs, or not to at all, and not be persecuted one way or the other.

Proponents of the Pauline tradition, who believe that America was founded by Judeo-Christians, also often cite Paulines such as Christopher Columbus and the Puritans as examples; however, such people did not actually found the independent republic of the United States of America.

If there was no separation of church and state, various Pauline institutions might still be physically fighting for control of the government, since each of them would claim to know the-one-and-only "truth," and might even be willing to fight to the death in order to establish and preserve it. This is what the Founding Fathers were clearly hoping to prevent.

* * *

In the eighteenth century, a new Protestant movement began in Britain, and especially in the American colonies, that came to be referred to as the "Great Awakening." One of the primary instigators of that revival was Jonathan Edwards, a Calvinist preacher and theologian who was most known for his sermon "Sinners in the Hands of an Angry God."

In the early twentieth century, fundamentalism was officially reborn in America when sixty-four American and British preachers and theologians wrote a collection of essays titled *The Fundamentals*. This Old-World Calvinist doctrine preceded the extremist "Christian Reconstructionism" movement, as well as other Old Testament-influenced ideologies that eschew science and higher forms of education, which they believe are being forced upon them by a corrupted secular materialist world. Consequently, an anti-intellectual backlash developed around the Puritanical flag of Old World tradition. The fundamentalists proceeded to establish their own academic institutions; institutions that not only asserted the mistaken Pauline interpretation but promoted a revisionist view of world history.

Many of these errors can be found in the present-day Evangelical movement.[382] Among other misconceptions, Evangelicals believe in the literal interpretation of the Bible; its absolute inerrancy; that salvation is attained through the death of the Christ (i.e., Substitionary Atonement/Crucicentrism), and that Jesus will return to Earth in a future Second Coming event. All of which are misinterpretations.

Many present-day Paulines cite tribulations in our own time, such as murder, drug use, sexual immorality, and the general decline in moral behavior, as signs of a fallen society that has strayed from God. However, it should be understood that just because a minority of

---

[382] There has been some debate whether the Evangelical movement can be equated with fundamentalism or not. After examination, I maintain that the two movements are indeed closely related.

individuals have not handled their freedom in a responsible manner does not mean that freedom or humanity itself is bad; nor is immorality something that is new to the human experience. Indeed, the actions of Yahweh and some of the Yahwehists themselves were what could be described as immoral.

The unfortunate and ironic reality is that if Jesus were to literally return, as many Paulines believe that he will, it is very likely that he would not conform to the stereotype that they have constructed of him; and it would be they themselves, like the Pharisees and the laity who they influenced before them, who would condemn him—most likely as "the antichrist"—and would seek to have him put to death. Despite significant progress in recent centuries, the unfortunate fact is that, in many cases, the primary mentality of the contentious and affected mind of the devoted Yahwehist has remained unchanged.

* * *

The Roman Pauline Papacy is the last of the ancient autocracies with a following larger than ever. Over one-billion people acknowledge the authority of the pope. Fortunately, the Catholic Church has taken some progressive steps in the past century. For example, despite objections by some in the Vatican council, Pope John Paul II apologized for the mistakes that were committed by members of the Catholic Church through the centuries (Caroll 2000). However, as long as "Jehovah" is continued to be misidentified as the heavenly Father of "Jesus," the very core foundation of this historic problem will continue to exist. In this case, the Paulines—including the Protestants—must not only acknowledge the dark side of their history but also the great misinterpretation that was committed by the early church patriarchs as well.

One of the success stories of the Pauline denominations in recent times, on the other hand, has been the commitment by its members to emphasize the beneficial power of the Holy Spirit, which is the energy of love, rejuvenation, and of healing. Unfortunately, this emotional truth can be used to mislead the faithful into believing the mistaken concepts that its leaders attach onto it.

CHAPTER XII

EPILOGUE

It is commonly believed that subjects related to mind control and so-called "brain-washing" only apply to members of cults and totalitarian political regimes; however, this is not true. There are two basic types of mind control: direct and indirect. Direct cognitive manipulation pertains to victims who are forced into a situation in which they are overtly coerced into surrendering their mental autonomy (i.e., menticide) and adopting the beliefs of the dominating agent. Indirect mind control occurs on a subtle level. In these instances, the subject is not usually aware that the effect is occurring. Indirect thought reform can be accomplished through common sources: e.g., radio, television, books, etc. Through the use of these non-threatening everyday mediums, the agent asserts a particular conviction by presenting information that not only advances only one side of the issue, and not only presents their interpretations as irrefutable fact, and not only evokes emotional triggers, but uses fallacious disinformation to mislead their subjects. In a great majority of these instances, the agents themselves are not even aware that the information that they are asserting is either unethical and/or not true. This is because they themselves have been subjected to this very same conditioning. The most common aim of the dominating agent is to convince the subject to replace their own self interest with that of the agent and the institution that they are affiliated with.

The development and proliferation of false consciousness[383] also occurs in some conventional religious institutions. Those who resist are made to feel as if they are refusing "God's will." The price of disobeying the will of God is said to be eternal punishment in hell. The subject is also told that being a good person is not good enough, and that to make it into heaven one must become a devoted money-tithing, church leader-exalting, doctrine-adhering, member—and in some cases: soldier—of the organization.

The indoctrination process is most effective when it begins during childhood. When this occurs, the ideology becomes infused into the subject's fundamental cognitive orientation (i.e., *weltanschauung*).

> Common men find themselves inheriting their beliefs, they know not how.
> —William James,[384] *A Pluralistic Universe*

When confronted with the fact that this worldview is untrue many years later, the subject may be unable to come to terms with such an extreme schism. This type of cognitive dissonance upsets the comforts of the type of regulated normalcy that is represented in the Freudian concept of the super-ego. This resistance is especially salient when one's religious beliefs form the very bedrock of one's mores. When confronted with the facts, the subject may also be made to feel that the devil is trying to tempt them away from God. For others, the humiliation of having been misled is intolerable, and they will resist reform due to this reason alone. Similarly, others who have high opinions of themselves may also be under the impression that so-called "brain-washing" could never happen to them, and therefore resist reform due to an arrogant sense of certainty.

---

[383] This philosophical term is defined as a lack of awareness of the true source of one's beliefs. It can be related to *avidya*, which is the Hindu concept of ignorance, and *bi* (or *pi*), which is the Chinese Confucian term (from the philosopher Xunzi) for blindness of the mind.

[384] William James was a nineteenth- and twentieth-century professor of philosophy and psychology at Harvard University.

In some countries, most notably in the United States of America, the link between Pauline Judeo-Christianity, politics, big business, and the military-industrial-congressional-complex[385] is especially consequential. Some at the top, who, in previous times were referred to as the "Robber Barons" and "the masters of mankind,"[386] and who are now more commonly referred to as plutocrats and oligarchs, have been able to use this conglomerate for their own advantage. Through the use of "think tanks," they have devised ways in which to not only foment an environment of moral skepticism (i.e., moral particularism)—especially in regard to honesty and avarice—but convince the general populace to replace their own self-interest, as well as the welfare of the environment, with that of the wealthy corporate elite. By disseminating disinformation (i.e., social constructivism) that exploit the fears, the prejudices, and the ignorance of the people, this faction has been able to take advantage of low-information voters—many of which themselves are abject members of the precariat.[387] The plutocrats/oligarchs execute this reckless and unethical campaign while simultaneously aligning themselves with the cherished symbols of tradition (e.g., Bible; Constitution; flag; etc.) (i.e., the hyperreal simulation), which they use to conceal the sociopathic greed that underlies their patriotic pretense. This stratagem has been, for the most part, extremely effective.[388]

* * *

---

[385] The "military industrial complex" most often refers to the American military and the industry that supplies it. The term was popularized by President Dwight D. Eisenhower in his 1961 farewell address. In more recent times, the term has been extended to include the military industrial congressional complex, due to the role that politicians play in this operation.

[386] From Adam Smith's book *Wealth of Nations* (1776).

[387] The term "precariat" denotes the precarious state of the laborer class: i.e., the proletariat.

[388] For more on this subject, I recommend the documentary *Requiem for the American Dream* (2016).

The contradictions, the duplicity, and the mistaken interpretations can finally be remedied by beginning with the fundamental fact that the heavenly Father of Jesus was not the Demiurge of the Old Testament.

Of course, the members of the institutions of Yahweh/Jehovah will adamantly fight to preserve the ways that they have become accustomed to.

> In short, the social, as opposed to the mystical function of a mythology, is not to open the mind, but to enclose it: to bind a local people together in mutual support by offering images that awaken the heart to recognitions of commonality, without allowing these to escape the monadic compound.
> —Joseph Campbell, *The Inner Reaches of Outer Space*

> New opinions are always suspected, and usually opposed, without any other reason but because they are not already common.
> —John Locke,[389] *An Enquiry Concerning Human Understanding, Dedicatory Epistle*

Some hardened Paulines have even gone so far as to not only disparage the findings of science and the principles of the Enlightenment but even liberty and intellectualism itself. This type of behavior is exemplified in their condemnation of other belief systems, such as the present-day New Age movement; which many Paulines believe is related to sorcery, false gods, and the antichrist. However, according to Jesus himself the impostors will be exposed by their "fruit." The fruit of the New Age movement is peace, love, and enlightenment, which are the very same values that Jesus himself sought to encourage. The condemnation of movements that are

---

[389] John Locke was seventeenth- and eighteenth-century English philosopher. He was an influential proponent of liberalism and the Enlightenment movement.

related to peace, love, and enlightenment, is the residual influence imposed upon the world by the deceptive Lord of Armies.

It must be remembered that it was the Old World Paulines who also rejected the fact that the Earth orbits the Sun. We have also seen this type of reaction manifested in our own time in their attack on the scientific theory of evolution. Proponents of creationism, under the vanguard banner of "intelligent design," will only accept scientific theories that they can use to bolster their interpretations. However, it must be understood that the author of Genesis was not present during the creation of the world, and therefore used metaphorical descriptions in order to convey primary points. Scientific examination has revealed that our planet has developed over the course of billions of years. The author of the book of Genesis used the description of "days" in order to describe the different *stages* that the planet went through as it was formed. The description of days is, of course, not to be taken literally; just as the character of the serpent in the Garden of Eden story was not intended to represent a literal talking snake, but rather was intended to convey the description of a character who was thought to be dangerous.

Some Paulines have been using the "Junkyard Tornado" argument to support their claims about evolution (which is based on an example that was first proposed by the astronomer Fred Hoyle, and is related to the "watchmaker" analogy). According to this analogy, Darwin's theory of evolution is the equivalent of a tornado whirling up spare metal fragments in a junkyard into a fully-formed working 747 jet airplane. However, it must be understood that metal parts are not organic, and therefore do not adapt and mutate over the course of millennia the same way that biological organisms do. Therefore, this analogy does not disprove evolution, but rather only indicates that the Pauline opponents of science who are using this example do not understand biology.

The enlightened Christian should be open to a type of evolution that is influenced by the energy of the universal source, and should not reject the science of evolution altogether. In other words, it could be more accurately said that God (i.e., the Godhead of the heavenly Father) created evolution.

* * *

It is also necessary to clarify the Garden of Eden story by putting it into context with the findings. What must be understood is that the account that was described in the Old Testament/Tanakh is reported from the perspective of a follower of Yahweh/Jehovah, and therefore is incorrect. In other words, it was wrong for the Lord of Adam and Eve to deny their servants access to "life" and "knowledge"—i.e., the Tree of Life and the Tree of Knowledge of Good and Evil. This truth is referred to in the following gnostic Christian tractate:

> But what sort is this God? First he maliciously refused Adam from eating of the tree of knowledge, and, secondly, he said "Adam, where are you?" God does not have foreknowledge? Would he not know from the beginning? And afterwards, he said, "Let us cast him out of this place, lest he eat of the tree of life and live forever." Surely, he has shown himself to be a malicious grudger! And what kind of God is this? For great is the blindness of those who read, and they did not know him. And he said, "I am the jealous God; I will bring the sins of the fathers upon the children until three (and) four generations." And he said, "I will make their heart thick, and I will cause their mind to become blind, that they might not know nor comprehend the things that are said."
> —The Nag Hammadi Codices, The Testimony of Truth

The Old Testament record itself informs us that the so-called "serpent" character did not actually intend to harm Adam and Eve, but rather had compassion for human-beings who had been forced to live life as nothing more than lowly servants in a so-called "garden." However, it must be understood that the garden was not some heavenly realm of recreation and relaxation. The record tells us that the Lord put Adam in the garden to work in it (Genesis 2.15), not to enjoy a permanent holiday in paradise. Therefore, what was referred to as a garden was actually a farm on the steppe plains of the Sumerian *Edinnu* (i.e., Eden). The truth is that it was only a relaxing paradise for the Elohim themselves.

The description of the serpent as the devil is also incorrect. This was actually a hypocritical accusation that was made by the agents of the jealous and deceptive Lord of Armies. The truth is that this individual who was compared to a snake was actually compassionate toward human-beings, and sought to lift them from out of an oppressive system that had made them into nothing more than workers and worshipers. (The warriors of the Lord of Armies came later.) Indeed, in the biblical record itself references to the beneficial symbol of snake can be found (John 3.14; Matthew 10.16). Of course, it was easy for the agents of Yahweh/Jehovah to turn this symbol into something hideous and intimidating, even though this is not the original connotation.

One of the gnostic Christian groups that were active in the early years were what has come to be known as the Naassenes. This designation was applied to this group by one of its proto-orthodox detractors, Hippolytus (Gaffney 2004: 6). According to this third-century Pauline theologian,[390] the Naasenes were a heretical faction that celebrated the serpent. Hippolytus derived the name from the word *naas*, which he believed was the Hebrew word for serpent (i.e., *nehash/nachash*). This group may be related to, or are the same group who are known as, the Ophites—which derives from the Greek word *ophis*, which also means serpent (Gaffney 2004: 6).

* * *

Another step that must be taken in the reformation of the Catholic Church is the abolition of institutionalized clerical celibacy. This obligatory rejection of sexuality that the priests of the Roman Catholic Church (and Eastern Orthodox Church) have been led to practice has led to widespread dysfunction; a dysfunction which has

---

[390] *Refutation of all Heresies* (Book V)

manifested itself in the form of the perversion of pedophilia. This is very likely one of the results of the misdirected libido when it is unnaturally repressed. Celibacy can be a great way of establishing independence and emotional stability. However, this decision must be a personal one; not the life-long result of an institutional decree.

In regard to the child abuse cases, investigations have revealed that priests who engaged in pedophilia was far more widespread than had been previously thought. The official estimate (according to the John Jay report)[391] of the total number of cases involving pedophiliac priest is 4,392, and the number of children who were abused is 10,667! In many cases, even after officials in the Catholic Church were made aware of the allegations law enforcement investigators were never contacted, and the offenders were instead transferred to other parishes where the abuse continued (Clark and Clark 2007: 78). In some cases, compensation payments were made on the condition that the victims remain silent (Robinson, et al. 2002). Investigations revealed that church leaders had not done enough to stop the abuse (Goodstein, et al. 2014). This is apparently because there were some who believed that the welfare of the Catholic Church superseded that of the children.

Dysfunctional behavior related to sexuality and Pauline denomination misunderstandings is also exemplified in the "God Hates Fags" movement of the Kansas Westboro Baptist Church. The excuse that is used to justify this policy of intolerance toward homosexuals is based not only on the New Testament letters of Paul (Romans 1.26; 1 Corinthians 6.9) but from Jehovah himself. The reference is found in Leviticus 18.22, in which it is reported that Jehovah considered homosexuality to be an "abomination." Jesus himself did not directly comment on this issue (that we know of); however, we can draw a conclusion based on actions and teachings of Jesus himself, that he would have not shared the same hostile and

---

[391] The John Jay Report, or "The Nature and Scope of the Problem of Sexual Abuse of Minors by Catholic Priests and Deacons in the United States," is a 2004 study commissioned by the U.S. Conference of Catholic Bishops that was conducted by the John Jay College of Criminal Justice.

narrow-minded mentality that Jehovah and Paul did. In fact, in the very same book of the Bible Jehovah not only ordered a blasphemer to be stoned to death (Leviticus 24.23) but affirmed the law "an eye for an eye, a tooth for a tooth" (Leviticus 24.19); both of which are practices that Jesus directly overturned (Matthew 5.28; John 8.7). Also, in the very same book Jehovah decreed that his followers must not cut the hair at the sides of their head, nor could they trim the edges of their beard (Leviticus 19.27), nor could they "wear clothing woven of two kinds of material," nor were they allowed to plant their fields with "two kinds of seed" (Leviticus 19.19). They were also instructed not to eat pigs (which would include bacon and ham) and seafood that does not have "fins and scales," such as crabs and lobsters (Leviticus 11.7-10). In this case, one must wonder how many hardened Pauline zealots who are discriminating against homosexual people are guilty of committing such erroneous transgressions. Furthermore, according to the very same book of the Old Testament, slavery was permitted! (Leviticus 25.44)

Fundamentalist Paulines also cite Matthew 5.17, in which Jesus appears to affirm the regulations of the Old Testament; however, as was previously noted, the Gospel of Matthew is the least accurate of the accounts. Indeed, this passage contradicts the actions of Jesus elsewhere in the gospels, where it is reported that he endeavored to nullify Old Testament thinking. Therefore, this passage, which only appears in the Gospel of Matthew, and does not appear in the more authentic source material, such as Mark and Q, cannot be considered to be an accurate representation of what he actually said.

Another passage that is cited by the Paulines is the following:

> Some Pharisees came to him to test him. They asked, "Is it lawful for a man to divorce his wife for any and every reason?" "Haven't you read," he replied, "that at the beginning the Creator 'made them male and female,' and said, 'For this reason a man will leave his father and mother and be united to his wife, and the two will become one flesh'? So they are no longer two,

but one flesh. Therefore what God has joined together, let no one separate."
—Matthew 19.3-6

However, in this passage Jesus is actually only referring to the issue of divorce and the institution of marriage itself. The words "man" and "women" were used only because that is the most common form of marriage, not because he was making a statement about homosexuality. Indeed, homosexuality is never specifically mentioned in the gospels. Therefore, the relationship between "sexual immorality" (Matthew 15.19) and homosexuality is only a forced conflation.

It is also disingenuous to cite the words of Paul to support an antihomosexual argument; not only because of the previously cited discrepancies between the teachings of Paul and Jesus but because Paul himself stated that Old Testament law was no longer required for salvation! (Romans 7.6, 6.24; 2 Corinthians 3.14; Galatians 2.16, 3.13, 23-25, 2.21, 5.4) Furthermore, according to Paul's own interpretation the people will not be saved by the old covenant but by the "new covenant" (2 Corinthians 3.6, 3.14), which simply required "faith" in Christ (Galatians 3.11). Therefore, when Paul is condemning homosexuality he is actually contradicting himself!

The enlightened viewpoint of this issue is that the discrimination, the persecution, the hatred, and the violence that is being directed at homosexuals is the real sin. This is the type of blasphemy against the Holy Spirit that the Savior warned about (Mark 3.29). Consenting adults who engage in intimate and loving relationships are not evil. The real issue has to do with the problem of mental conditioning that results in bigotry, hatred, and in some cases, even violence.

* * *

Throughout this book the word "enlightenment" has been referred to. What is enlightenment? How can it be defined?

According to the eighteenth-century German philosopher Immanuel Kant, enlightenment is the successful emergence of an individual from out of a state of mental "immaturity." Kant believed that this state of immaturity is the result of cowardice and mental laziness—I would also add arrogance. He proposed that the enlightened individual, on the other hand, is a bold and free-spirited type of person. A person who dares to be wise.

In ancient India, enlightenment was described in comparable but slightly different terms. Although the higher state of consciousness that was attained by the Buddha is commonly translated as "enlightenment," a more accurate rendering is "awakened" (*bodhi*). To be awakened is to understand the true nature of things, which in turn leads to liberation (*moksha*)[392] from the suffering that is caused by delusion.

It would seem that enlightenment can be defined as both a state of clarity as well as a state of awareness. The clarity aspect refers to the realization of the difference between the artificial (i.e., socially constructed) Demiurge/man-made world and its related military, industrial, religious, economic, and political perspectives (i.e., the hyperreal simulation), and the natural world.

> Follow Nature! Follow Nature! As she works so will I work!
> —Rosicrucian Motto[393]

The awareness aspect may refer to educational knowledge; although it might also relate to spiritual awareness. This may pertain to the realization that there is more occurring in this world than our standard everyday work and material-based experience would influence us to believe. It is possible that the mystical adept (e.g., Yeshua the Savior) is one who is able to attune to a higher level of consciousness, in which the realm of supernatural energy is experienced. However, mental enlightenment may not always be coincident with spiritual awareness and virtue. Indeed, there is a physical life, a mental life,

---

[392] This is the Hindu term for enlightenment.

[393] Schrodter, Willy. *A Rosicrucian Notebook: The Secret Sciences Used by Members of the Order* (Weiser Books, 1992).

and a spiritual life. Spiritual enlightenment, as opposed to mental enlightenment, may refer not only to spiritual awareness but to internal rectification. This may relate to the meditative process by which one removes his or herself from the maladies that might cause one to deviate from a more life-supporting ethical standard. To live the spiritually enlightened way is to be personally empowered, without being narcissistic, and to see one's self as being connected and in peaceful harmony with the "way" (i.e., the *Tao*) of the natural spirit of the universe.

Some people may choose not to embrace the belief in the divinity of one's self, and instead may turn to the aid of a protective and inspirational religious figure, which can also be beneficial—that is, as long as those instructing individuals do not abuse their positions.

> The Supreme Personality of the Godhead said: O sinless Arjuna, I have already explained that there are two classes of men who try to realize the self. Some are inclined to understand it by empirical, philosophical speculation, and others by devotional service.
> —Krishna, The Bhagavad Gita (Chapter III)

People who have achieved both mental and spiritual enlightenment have conquered issues, such as fear and hatred, by not only establishing a strong personal connection with the divine but by understanding and feeling compassion for those who do not. Indeed, many people are stuck in negative habit cycles because no one has helped to guide them from out of the ignorant and hostile world that they have been effected by.

However, an enlightened adept also knows when strength and confrontation is necessary. Yeshua knew that there were times to both "turn the other cheek"—in order to maintain the moral high ground and to appeal to the antagonist's intrinsic ethical nature—and to stand up and challenge the ignorant, the arrogant, and the malicious, in an honorable manner.

There are indeed living beings who can be considered to be "angels," but they do not wear robes or have wings, and their halos are not commonly visible. Indeed, they are unassuming everyday people who do not seek to fit into the stereotypes of man.

* * *

When the intelligent, enlightened, peaceful, and ethical mentality is more regarded than the domineering ways of the selfish, the hateful, the ignorant, the fearful, and the arrogantly close-minded, the world will move a little closer to a Heaven on Earth—which should be the ultimate paragon goal of every virtuous human-being.

Of course, many of those who are still effected by the past will seek to suppress such an ideal. Such hard-hearted and mentally-affected individuals are bound to preserve the institutions that not only have they become accustomed to, or that they have been influenced to preserve, but that they have been able to derive some kind of personal benefit from as well. Of course, these types of benefits usually manifest in the political and economic world as some type of materialistic form of prosperity. In the present-day world, the problem of greed, deceit, ignorance, arrogance, corruption, sociopathic malice, and the lust for power, wealth, and fame, is just as prevalent as it was in biblical times. This is the same type of immature and destructive behavior that fueled the internal fire of the wicked god of war. Indeed, this is the same type of iniquity that Yeshua spoke out against (Mark 7.22).

The traditional practice of white-washing the abominations of Yahweh/Jehovah must also be brought to an end. The cover-up of the atrocities that were committed by the Machiavellian Lord of Armies and his associates continues to this very day by devoted priests, reverends, rabbis, etc., who often tell the people what they want to hear, instead of what is actually documented in the Bible itself.

Due to the influence of not only pious sermons but Renaissance artworks and Hollywood movies, the world has been effected by a magical view of the biblical past. This phantasia has caused even further confusion. The truth is that before this modern age life was primitive and harsh, and before this Age of Information that we enjoy today, legitimate information was not as available. The truth is that the fantastical Golden Age that many who are immersed in the Judeo-Christian tradition believe happened, never actually happened.

It should also be understood that just because an institution is large, or an interpretation is old, does not mean that it is correct. The fact is that the time for discovering and understanding the truth is not in the primitive past, but now. Indeed, just because some choose not to believe in the reality of evolution does not mean that it is not actually happening.

The unfortunate fact is that many people, if not most, are not sincere seekers of the truth. Such people adhere to beliefs that either they themselves feel the most comfortable with, or that others have imposed upon them. For many, the shock that the belief system that has been instilled into them since childhood is incorrect will be too much to bear, and they will resist change due to this reason alone. Nevertheless, the truth must be made available to those who are ready to receive it.

Those who are influenced by mistaken Pauline interpretations often use the threat of an antichrist or devil figure to frighten the laity away from progress and back toward the superficial comfort of established doctrine. Although, in regard to such a deceptive character, it must be understood that when people become informed they become empowered, and when they become empowered they are less likely to be misled by "Almighty" "Lord" or "Master" types. Indeed, the enlightened individual is more likely to be inspired by the least expected of life's little miracles. This is the principle that Jesus himself promoted when he said that the very least among us will be the greatest in heaven (Luke 9.48).

Of course, this epic story is far from over; rather, this post-apocalyptic age is still developing. Indeed, Yeshua knew that the historic revolution that he initiated would not occur in only a single generation. Instead, he stated that his teachings were like a seed that would one day sprout and flourish; and it is under this same type of hopeful spirit that this work is submitted.

Then you will know the truth, and the truth will set you free.
—Yeshua, John 8.32

ILLUSTRATIONS

Figure 1. Babylonian god (possibly Shamash). Painted fire clay statue from a shrine in Ur. 2000 –1750 BCE. British Museum #122934.

Figure 2. Winged disc. From the Stela of Pebeh. Egypt. Painted plaster. Ptolemaic Period. British Museum EA8466.

Figure 3. Winged disc (from a male figure grasping a tree). Ivory Panel. 8th century BCE, Neo-Assyrian. Metropolitan Museum of Art #59.107.6.

Figure 4. Ashur/Shamash in his Flying Disc. Part of an alabaster wall relief from the throne room of Ashurnasirpal II at Nimrud in Norther Iraq. 870 – 860 BCE, Neo-Assyrian. British Museum ME 124531.

Figure 5. Winged disc and levitating figure. Chalcedony cylinder seal of Nabu-Nasi, Son of Amel-Resh. 8th – 9th century BCE, Neo-Assyrian. British Museum #89082.

Figure 6. Top Section of a Stele Decorated With a Libation Scene Before a Seated God. White limestone, late 3rd millennium BCE, Mesopotamia. Louvre, Department of Near Eastern Antiquities. J. de Morgan Excavations sb 7.

Figure 7. King Nabonidus. Basalt stele. 6th century BCE, Neo Babylonian. British Museum #90837.

Figure 8. Cult scene: Worship of the sun god Shamash. Limestone cylinder seal. Mesopotamia. Louvre, Department of Oriental Antiquities (AO 9132). Public domain image adapted from commons.wikipeia.org. Original photograph by Jastrow.

Figure 9. The Standard of Ur. Sumerian, 2500 BCE. Bitumen, lapis lazuli, limestone, shell. British Museum #121201.

Figure 10. Tablet of Shamash. Babylonian, early 9th century BCE. British Museum. ME 91000-91004.

Figure 11. Procession of giants. Cylinder seal. 2300 BCE, Akkadian. British Museum #89137.

Figure 12. Stela of Nakhtefmut. Egypt. Third Intermediate Period. British Museum. EA37899.

Figure 13. Mushroom couple. Molded fired clay relief plaque. Old Babylonian, 1000 BCE–1750 BCE. British Museum #116812.

Figure 14. The royal seal of the Hittite king Tudhalija IV, from a message to the King of Ugarit, Ras Shamra. Seal imprint on a clay tablet. 1250-1220 BCE. Louvre AO 21091.

Figure 15. Stone Mushroom Figure. From Guatemala. Circa 110/200–1000/1100 BCE. American Museum of Natural History, #30.0/6360.

Figure 16. *Genius* (Spirit) holding a poppy flower. Relief from Sargon IIs palace at Khorsabad. Louvre AO 19869.

Figure 17. A seated figure holding a lotus flower. Limestone relief. Circa 9th century BCE, Neo-Hittite. The Metropolitan Museum of Art #43.135.1.

Figure 18. Ivory horse frontlet with a nude goddess holding lotus flowers. Circa 8th – 9th century BCE, Neo-Assyrian. Metropolitan Museum of Art #61.197.5.

Figure 19. Chalcedony cylinder seal from Hillah (near Babylon). Achaemenid Persian Empire, circa 6th – 4th century BCE. British Museum ME 89352.

Figure 20. Stone relief from the throne room of Ashurnasirpal II at Nimrud in northern Iraq. 870–860 BCE, Neo-Assyrian. British Museum #ME124531.

Figure 21. Winged disc (from a male figure grasping a tree) Ivory Panel. 8th century BCE, Neo-Assyrian. Metropolitan Museum of Art #59.107.6.

Figure 22. Master of the Exaltation of the flower. (*L'Exaltation de la Fleur*) Fragment from a grave stele. Parian marble. Two woman (perhaps Demeter and Persephone). Pharsalos, Thesaly. Circa 470–460 BCE. Louvre Museum, Department of Greek, Etruscan, Roman Antiquities. Ma 701.

Figure 23. Diagram

Figure 24. Buddha Sheltered by a Naga. Bronze. 12th century CE, Angkor period. Metropolitan Museum of Art #1987.424.19a, b.

Figure 25. Caduceus. Illustration

Figure 26. Snake god. Cylinder Seal. 2400BC–2200 BCE, Akkadian. British Museum #102511.

Figure 27. Pazuzu. Bronze. Assyrian. Early 1st millenium BCE. Louvre MNB 46

Figure 28. Bearded Underworld God. Molded ceramic plaque. Circa 2000 – 1600 BCE, Old Babylonian. Metropolitan Museum of Art #48.104.1.

BIBLIOGRAPHY

"'Abducted': The Myth of Alien Kidnappings." *www.npr.org*. November 9, 2005 (Updated July 17, 2011). Accessed December 23, 2015.

Abraham, Karl. *Nervous and Mental Disease Monograph Series: Dreams and Myths: A Study in Race Psychology*, Vol. 15. William A. White (trans.), The Journal of Nervous and Mental Disease, 1913.

Abraham, Ralph, et al. *Trialogues at the Edge of the West: Chaos, Creativity, and the Resacralization of the World*. Bear & Co., 1992.

Achtemeier, Paul J., (ed). *The Harper Collins Bible Dictionary*. Harper, 1985.

Achterberg, Jeanne, et al. "Evidence for Correlations Between Distant Intentionality and Brain Function in Recipients: a Functional Magnetic Resonance Imaging Analysis." *The Journal of Alternative and Complementary Medicine*. Vol. 11, No. 6, 2005.

Adamnana, William Reeves, (ed). *Life of Saint Columba, founder of Hy*. Edmonston & Douglas, 1874.

Aguilar-Moreno, Manuel. *Handbook to Life in the Aztec World*. Oxford University Press, 2006.

Alexander, Eben. *Proof of Heaven: A Neurosurgeon's Journey into the Afterlife*. Simon & Schuster, 2012.

Allen, James P. *The Ancient Egyptian Pyramid Texts*. Society of Biblical Literature, 2005.
———. *Middle Egyptian, An Introduction to the Language and Culture of Hieroglyphics*. 3rd ed. Cambridge University Press, 2000 [reprint 2014].

Almeder, Robert. *Death & Personal Survival: The Evidence for Life After Death*. Littlefield Adams, 1992.

Andrews, Tamra. *Dictionary of Nature Myths: Legends of the Earth, Sea, and Sky*. Oxford University Press, 2000.

Arendzen, J. "Manichaenism." *The Catholic Encyclopedia*. Vol 9. Appleton Co., *Newadvent.org*, 1910. Accessed February 2, 2015.

Armstrong, Karen. *Holy War: The Crusades and Their Impact on Today's World*. Doubleday, 1988.

Aronson, Martin (ed.). *Jesus and Lao Tzu: The Parallel Sayings, with Commentaries*. Seastone, 2000.

Ascalone, Enrico. *Mesopotamia*. Rosanna M. Giammanco Frongia (trans). University of California Press, 2005 [trans. 2007].

Assmann, Jan. "Amun," *Dictionary of Deities and Demons in the Bible*: 2nd Edition. van der Toorn, Karel, (ed.) et al. Brill, W. B. Eerdman's Publishing, 1995 [reprint 1999].

Audi, Robert (general ed.), Paul Audi (associate ed.). *The Cambridge Dictionary of Philosophy: Third Edition*. Cambridge University Press, 1995 [reprint 2015].

Aune, David E. "Heracles," *Dictionary of Deities and Demons in the Bible*: 2nd Edition. van der Toorn, Karel, (ed.) et al. Brill, W. B. Eerdman's Publishing, 1995 [reprint 1999].

Aveni, Anthony F. *Skywatchers: A Revised and Updated Version of Skywatchers of Ancient Mexico*. University of Texas Press, 2001.

Avise, John C. *The Genetic Gods: Evolution and Belief in Human Affairs*. Harvard University Press, 1998.

Baden, Joel S. *J,E, and the Redaction of the Pentateuch*. Mohr Siebeck, 2009.

Bader, Christopher D. "UFO Abduction Support Groups: Who are the Members?" Tumminia, Diana G.(ed.) *Alien Worlds: Social and Religious Dimensions of Extraterrestrial Contact*. Syracuse University Press, 2007.

Ballingrud, David. "Underwater World: Man's Doing or Nature's? *St. Petersburg Times*. Sptimes.com. November 17, 2002. Accessed July 13, 2015.

Barber, Richard. *The Holy Grail: Imagination and Belief*. Harvard University Press, 2004.

Barker, S.A., et al. "LC/MS/MS analysis of the endogenous dimethyltryptamine hallucinogens, their precursors, and major metabolites in rat pineal gland microdialysate." *Biomed Chromatogr*. Vol. 27 (12), July 2013.

Barnstone, Willis (ed). *The Other Bible: Ancient Alternative Scriptures*. Harper, 1984.
———. Marvin Meyer, ed. *The Gnostic Bible*, Shambhala (Rev. Ed.), 2009.

Barr, Stephen M. *Modern Physics and Ancient Faith*. University of Notre Dame Press, 2003.

Barry, J. "General and comparative study of the psychokinetic effect on a fungus culture," *Journal of Parapsychology*, Vol. 32 (94), 1968.

Barry, William. "Arius." *The Catholic Encyclopedia*. Vol. 1. New York: Robert Appleton Co., 1907.

Bede. *Ecclesiastical History of England*, (Book IV, chapter 7). Henry G Bohm, [reprint 1867].

Beischel, Julie, et. al. "Anomalous Information Reception by Research Mediums Under Blinded Conditions II: Replication and Extension." *Explore: The Journal of Science and Healing*. Vol. 11, No.2, March/April 2015.

Bem, Daryl J. Charles Honorton. "Does Psi Exist? Replicable Evidence for an Anomalous Process of Information Transfer." *Psychological Bulletin*. Vol. 115, No. 1, 1994.

Benard, Elisabeth, Beverly Moon. *Goddesses Who Rule*. Oxford University Press, 2000.

Benet, Sula. "Early Diffusion and Folk Uses of Hemp." Rubin, Vera (ed.). *Cannabis and Culture*. Mouton & Co., 1975.

Benor, Daniel J. "Distant Healing." *Subtle Energies & Energy Medicine*, Vol.11, No. 3.

Berlin, Adele (ed.). *The Oxford Dictionary of the Jewish Religion*. 2nd Ed., Oxford University Press, 2011.

Betz, Hans Dieter. "Dynamis," *Dictionary of Deities and Demons in the Bible*: 2nd Edition. van der Toorn, Karel, (ed.) et al. Brill, W. B. Eerdman's Publishing, 1995 [reprint 1999].

Black, Jeremy (ed.). Et al. *A Concise Dictionary of Akkadian*. 2nd Corrected Printing. Harrassowitz Verlag, 1999 [reprint 2007].

Black, Matthew (ed.). Rowley, Harold Henry (ed.). *Peake's Commentary on the Bible*. Routledge, 1962 [reprint 2001].

Blake, I. M., Cynthia J. Weber. "Radioactivity of Jericho Bones." Balliol College, Oxford University, Research Laboratory for Archeology and the History of Art, Oxford University. *Onlinelibrary.wiley.com*. August 2007. Accessed August 2013.

Blenkinsopp, Joseph. *A History of Prophecy in Israel*. Westminster John Knox Press, 1996.
———. *The Pentateuch: An Introduction to the First Five Books of the Bible*. Anchor Bible Reference Library/Doubleday, 1992.

Blotzer, Joseph. "Inquisition," *The Catholic Encyclopdia*, Vol. 8 (Robert Appleton Co., 1910). Accessed July 16, 2015. www.newadvent.org/cathen/08026a.htm

Boomer, Megan. "Priestess and King: Representations of Power and Gender in the Akkadian Royal Family." ARCH 1600: Archaeologies of the Near East, Joukowski Institute for Archaeology and the Ancient World, Brown University, 2008.

Borenstein, Seth. "Red Dwarf is Mother to an Earth-like Planet. *SMH.com*, Associated Press. April 2007. Accessed May 2013.

Borg, Marcus, (ed.) *Jesus and the Buddha*: The Parallel Sayings. Ulysses Press,1997.
———. (ed.), and Ray Riegert, ed. *The Lost Gospel of Q: The Original Sayings of Jesus.* Ulysses Press, 1999.

Borucki, William J., et al. "Kepler-62: A Five Planet System with Planets of 1.4 and 1.6 Earth Radii in the Habitable Zone." *Sciencemag.org*, April 18, 2013. Accessed March 29, 2015.

Bosmajian, Haig. *Burning Books*. McFarland, 2006.

Botterweck, G. Johannes (ed.), et al. David A. Green (trans.). *Theological Dictionary of the Old Testament.* Vol. VII, William B. Eerdmans Publishing Co. 1982 [reprint 1995].
———. *Theological Dictionary of the Old Testament.* Vol. XIV, William B. Eerdmans Publishing Co. 1992 [reprint 2004].

Bramley, William. *The Gods of Eden*. Avon Books, 1989.

Breasted, J. H. *Ancient Records of Egypt*, University of Chicago Press, 1906.

Bremmer, Jan N. "Hades," *Dictionary of Deities and Demons in the Bible*: 2nd Edition. van der Toorn, Karel, (ed.) et al. Brill, W. B. Eerdman's Publishing, 1995 [reprint 1999].

Brier, Bob. *Egyptian Mummies: Unraveling the Secrets of an Ancient Art.* William Morrow & Co, Inc., 1994.

Brockington, J. *The Sanskrit Epics.* Brill, 1998.

Brockman, John (ed.). *My Einstein: Essays by Twenty-four of the World's Leading Thinkers on the Man, his Work, and his Legacy.* Vintage Books, 2007.

Broderick, Robert, C. (ed.). *The Catholic Encyclopedia.* Thomas Nelson Inc., 1976.

Brown, David T. (ed.) *Cannabis: The Genus Cannabis.* CRC Press, 1998.

Brown, Dwayne, et al. "NASA Spacecraft Confirms Martian Water, Mission Extended." *NASA.com.* July 2008. Accessed April 2013.

Brown, Francis (ed.), et al. *The Hebrew and English Lexicon of the Old Testament*, Oxford University Press, 1906.

Bruins, Hendrik J., Johannes Van der Plicht. "Tell Es-Sultan (Jericho): Radiocarbon Results of Short-Lived Cereal and Multiyear Charcoal Samples From the End of the Middle Bronze Age." Vol. 37, No.2, *Journals.uair.arizona.edu*, 1995. Accessed November 2013.

Budge, Ernest A. Wallis. *Legends of the Gods: The Egyptian Texts.* Cosimo Classics, 1912 [reprint 2010].
———. *Papyrus of Ani: The Egyptian Book of the Dead.* Dover Publications, 1967.
———. *Tutankhamen: Amenism, Atenism and Egyptian Monotheism with Hieroglyphic Texts of Hymns to Amen and Aten.* Courier Dover Publications, 2012.

Burke, John. Kaj Halberg. *Seed of Knowledge, Stone of Plenty: Understanding the Lost Technology of the Ancient Megalith-Builders.* Council Oak Book, 2005.

Burkett, Delbert. *An Introduction to the New Testament and the Origins of Christianity.* Cambridge University Press, 2002.

Burkhart, Kienast. "Igigu und Anunnaku nach den Akkadischen Quellen" (Igigu and Anunnaku in Accordance with the Akkadian Sources) (Kalene Barry, trans., 2010), Assyriological Studies 16, 1965.

Cameron, Averil (ed.), and Peter Garnsey (ed). *The Cambridge Ancient History Vol. 13: The Late Empire, AD 337-425.* Cambridge University Press, 1998.

Campbell, Joseph. *The Hero with a Thousand Faces.* Bollingen Series, XVII, Princeton University Press, 1949 [reprint 2008].
———. *The Masks of God: Occidental Mythology.* Penguin, 1964.
The Mysteries: Papers from the Eranos Yearbooks. Joseph Campell (ed.), Bollingen Series XXX, 2, Princeton University Press, 1955 [reprint 1990].
———. *The Mythic Image.* Princeton University Press, Bollingen Series C, 1974.
———. with Bill Moyers. *The Power of Myth.* Doubleday, 1988.

Cantril, Hadley. *The Invasion of Mars: A Study in the Psychology of Panic.* Transaction Publishers, 2005.

Caroll, Rory. "Pope says sorry for sins of church." *Theguardian.com.* Mar 13, 2000. Accessed Jan 2, 2015.

Cassan, A., et al. "One or more bound planets per Milky Way star from microlensing observations." *Nature*, Vol. 481, No. 7380, January 12, 2012.

Cavalli, Thom F. *Alchemical Psychology: Old Recipes for Living in a New World.* Tarcher/Putnam/Penquin, 2002.

CBS News. "Ex-Air Force Personnel: UFOs Deactivated Nukes." *Cbsnews.com*, September 28, 2010. Accessed December 23, 2016.

Charlesworth, James H. (ed.) *The Old Testament Pseudepigraphia, Vol.1, Apocalyptic Literature and Testaments.* Doubleday, 1983.

Chu, Jennifer. "Scientists Discover Potentially Habitable Planets Just 40 Light Years from Earth, Planets are best Targets so far for Search for Extraterrestrial Life." *News.mit.edu.*, May 2, 2016. Accessed May 3, 2016.

Clancy, Susan A. *Abducted: How People Came to Believe that they were Kidnapped by Aliens.* Harvard University Press, 2005.

Clark, Robin E., Judith Freeman Clark. *The Encyclopedia of Child Abuse.* 3rd Ed. Facts on File, 2007.

Cleary, Thomas (trans.). *The Secret of the Golden Flower: The Classic Chinese Book of Life.* Harper One, 1991.
———. *The Taoist Classics, Vol. III: The Secret of the Golden Flower.* Shambhala, 2000.

Collins, Steven. "Sodom and the Cities of the Plain." Trinity Southwest University Press, 2013. Accessed August 30, 2013.
———. Hamden, Khalil. "The Tall El-Hammam Excavation Project. Season Activity Report. Season 5: 2010 Excavation, Exploration, and Survey." Trinity Southwest University Press, 2010. Accessed August 2013.

Coogan, Michael David (trans., ed.). *Stories from Ancient Canaan.* The Westminster Press, 1978.

Cook, Emily Williams. "Do Any Near-Death Experiences Provide Evidence for the Survival of Human Personality after Death? Relevant Features and Illustrative Case Reports." *Journal of Scientific Exploration.* Vol. 12, No. 3, 1998.

Cook, Stanley A. *The Religion of Ancient Palestine in the Second Millennium B.C.: In the Light of Archaeology and the Inscriptions.* Constable & Company, 1908 [reprint 1921].

Cooke, Bill. *Dictionary of Atheism, Skepticism, and Humanism.* Prometheus Books, 2006.

Cortizas, Antonio Martinez, et al. "Early Atmospheric Metal Pollution Provides Evidence for Chalcolithic/Bronze Age Mining and Metallurgy in Southwestern Europe." *Science of the Total Environment*, 2015.

Cory, Catherine A. *The Book of Revelation,* New Collegeville Bible Commentary. Liturgical Press, 2006.

Coulter, Charles Russell. Turner, Patricia. *Encyclopedia of Ancient Deities.* McFarland and Company, 2000: Routledge, [reprint 2014].

Coxon, Peter W. "Nephilim," *Dictionary of Deities and Demons in the Bible*: 2nd Edition. van der Toorn, Karel, (ed.) et al. Brill, W. B. Eerdman's Publishing, 1995 [reprint 1999].

Crawford, Thomas. *An Account of Terrible Apparitions and Prodigies Which Hath Been Seen Both Upon Earth and Sea, In the End of Last, and Beginning of this Present Year, 1721.* Thomas Crawford, (Date not given [1721?]).

Cremo, Michael L., Richard L. Thompson. *Forbidden Archeology: The Hidden History of the Human Race.* Bhaktivedanta Publishing, 1998.

Crick, F. H. C., L.E. Orgel. "Directed Panspermia," *Icarus*, Vol. 19, 1973.

Cromie, William J. "Alien Abduction Claims Explained: Sleep Paralysis, False Memories Involved." *news.harvard.edu/gazette.* September 22, 2005. Accessed December 23, 2015.

Cross, Frank Moore. *Canaanite Myth and Hebrew Epic: Essays in the History of the Religion of Israel*. Harvard University Press, 1973 [reprint 1997].

Cross, Samuel H. (ed., trans.), and Olgerd P. Sherbowitz-Wetzor (trans., ed.). *Russian Primary Chronicle: Laurentian Text*. Mediaeval Academy of America (date not given).

Curry, Andrew. "Gobekli Tepe: The World's First Temple?" *Smithsonianmag.com*, November 2008. Accessed December 2016.

Dalley, Stephanie (ed., trans.). *Myths of Mesopotamia: Creation, the Flood, Gilgamesh, and Others*. Oxford University Press, 1989.

Davies, Paul. *Superforce*. Touchstone, 1985.
———. John Gribbin. *The Matter Myth: Dramatic Discoveries that Challenge Our Understanding of Physical Reality*. Touchstone, Simon & Schuster, 1992.

Davis, Robert. *The UFO Phenomenon: Should I believe?* Schiffer Publishing, 2014.

Day, John. *Yahweh and the Gods and Goddesses of Canaan*. Sheffield Academic Press, 2002.

Dean, Jodi. *Aliens in America: Conspiracy Cultures from Outerspace to Cyberspace*. Cornell University Press, 1998.

Denzler, Brenda. *The Lure of the Edge: Scientific Passions, Religious Beliefs, and the Pursuit of UFOs*. University of California Press, 2001.

De Sahagun, Bernardino. *Florentine Codex, General History of the Things of New Spain*. The School of American Research and the University of Utah Press, 1982.

Descartes, Rene. *Meditations on First Philosophy*. Pearson, 1960.

Deutsch, David. *The Fabric of Reality: The Science of Parallel Universes—and Its Implications*. Penguin Books, 1997.

Dewhurst, Richard J. *The Ancient Giants Who Ruled America: The Missing Skeletons and the Smithsonian Cover-Up*. Bear & Company, 2014.

Dobson, Roger. Abul Taher. "Cavegirls Were the First Blondes to Have Fun." *The Times* (London), February 2006. Accessed September 28, 2013.

Doniger, Wendy. *Merriam-Webster's Encyclopedia of World Religions*. Merriam-Webster, 1999.

Druyan, Ann, et al. (writer and director). *Cosmos: A Spacetime Odyssey*. Cosmos Productions, Fuzzy Door Productions, National Geographic Channel, 20th Century Fox, 2014.

Dukstra, Meindert. "Mother," *Dictionary of Deities and Demons in the Bible*: 2nd Edition. van der Toorn, Karel, (ed.) et al. Brill, W. B. Eerdman's Publishing, 1995 [reprint 1999].

Dunbar, Brian. "NASA Confirms Evidence that Liquid Water Flows on Today's Mars." *www.nasa.gov*, September 28, 2015. Accessed October. 7, 2015.

Edwards, Mark U. *Luther and the False Brethren*. Stanford University Press, 1975.

Ehrman, Bart D. *Forged: Writing in the Name of God – Why the Bible's Authors Are Not Who They Think They Are*. Harper One, 2011.
———. *Jesus: Apocalyptic Prophet of the New Millennium*. Oxford University Press, 2001.
———. *Jesus Interrupted: Revealing the Hidden Contradictions in the Bible (and Why We Don't Know About Them)*. Harper One, 2009.
———. *Lost Christianities: The Battles for Scripture and the Faiths We Never Knew*. Oxford University Press, 2003.

———. *Misquoting Jesus: The Story Behind Who Changed the Bible and Why*. Harper One, 2005.

Eiddon, Iowerth, Stephen Edwards. *The Cambridge Ancient History*. Cambridge University Press, 1998.

Eisen, Robert. *The Book of Job in Medieval Jewish Philosophy*. Oxford University Press, 2004.

Eisenberg, Howard, D.C. Donderi. "Telepathic Transfer of Emotional Information in Humans." *The Journal of Psychology*, Vol. 103, 1979.

Elnes, Eric E., Patrick D. Miller. "Elyon," *Dictionary of Deities and Demons in the Bible*: 2nd Edition. van der Toorn, Karel, (ed.) et al. Brill, W. B. Eerdman's Publishing, 1995 [reprint 1999].

Emboden, William A. "The Sacred Narcotic Lilly of the Nile: Nymphia Caerula." *Economic Botany*, Vol. 32, Iss. 4, Oct. 1978.

Emerson, Ralph Waldo. Ziff, Larzer (ed.). *Selected Essays*. Penguin Books, 1982.

Eschenbach, Wolfram Von. *Parzival*. Penguin Books, 1980.

Escude, Guillem Anglada, et al. "A Terrestrial Planet Candidate in a Temperate Orbit around Proxima Centauri." *Nature*, 536, August 24, 2016.

Etlinger, Susan. "What do we do with all this Big Data?" *ted.com*. Sept. 2014. Accessed Sept 18, 2016. www.ted.com/talks/susan_etlinger_what_do_we_do_with_all_this_big_data#t-120773

Eunapius, Philostratus. *The Lives the Sophists. The Lives of the Philosophers*. Wilmer C. Wright (trans.), Loeb Classical Library. No. 134, 1921.

Facco, Enrico. Christian Agrillo. "Near Death Experiences Between Science and Prejudice." *Frontiers in Human Neuroscience*. Vol. 6, Art. 209, July 2012.

Fagen, Brian M. (ed.) *The Oxford Companion to Archaeology*. Oxford University Press, 1996.

Fawcett, Lawrence, Barry J. Greenwood. *UFO Cover-up: What the Government Won't Say*. Simon & Schuster, 1990. [Fireside edition 1992].

Feder, Kenneth L. *Frauds, Myths, and Mysteries: Scientific and Pseudoscience in Archaeology*. 4th ed. McGraw Hill Mayfield, 2002.

Fee, Gordon D. (ed.), Robert L. Hubbard Jr. (ed.) *The Eerdmans Companion to the Bible*. William B. Eerdmans Publishing Company, 2011.

Feldhay, Rivka. *Galileo and the Church*. Cambridge University Press, 1995.

Feliu, Lluis. *The God Dagan in Bronze Age Syria*. Wilfred G.E. Watson (trans.), Brill, 2003.

Filoramo, Giovanni. *A History of Gnosticism*. Anthony Alcock (trans.), Basil Blackwell, 1990.

Finkelstein, Israel. Neil Asher Silberman. *The Bible Unearthed: Archaeology's New Vision of Ancient Israel and the Origins of Its Sacred Texts*. The Free Press, 2001.
———. Amihai Mazar. *The Quest for the Historical Israel: Debating Archaeology and the History of Early Israel*. Brain B. Schmidt (ed.). Society of Biblical Literature, 2007.

Fiore, Silvestro. *Voices of the Clay: The Development of Assryo-Babylonian Literature*. University of Oklahoma Press: Norman, 1965.

Fish, M.S., et al. "Piptadenia Alkaloids. Indole Bases of P. peregrina (L.) Benth. And Related Species." *Journal of the American Chemical Society*. Vol. 77 (22), November 1, 1955.

Forbes, Robert J. *A Short History of the Art of Distillation*. Brill, 1970.

Foster, Benjamin. *Before the Muses: An Anthology of Akkadian Literature*. CDL Press, 1993, [reprint 2005].
———. *Distant Days: Myths and Poetry from Ancient Mesopotamia*. CDL Press, 1995.

Fowke, Gerard. *Archaeological History of Ohio: The Mound Builders and Later Indians*. Ohio State Archaeological and Historical Society, 1902.

Freeman, Charles. *The Closing of the Western Mind: The Rise of Faith and the Fall of Reason*. Alfred A. Knopf, 2003.

Friedman, Richard Elliott. *The Bible with Sources Revealed: A New View into the Five Books of Moses*. Harper Collins, 2003.
———. *Who Wrote the Bible?* Harper San Francisco, 1987 [reprint 1997].

Friesen, Steven J. *Imperial Cults and the Apocalypse of John: Reading Revelation in the Ruins*. New York: Oxford University Press, 2001.
———. *Twice Neokoros: Ephesus, Asia, and the Cult of the Flavian Imperial Family*. Religions in the Graeco-Roman World, Leiden: E.J. Brill, 1993.

Furnivall, Frederick (ed.). *The Book of Quinte Essence or The Fifth Being; That is to Say, Man's Heaven*. Trubner & Co.

Gaffney, Mark H. *Gnostic Secrets of the Naassenes: The Initiatory Teachings of the Last Supper*. Inner Traditions, 2004.

Gamboa, Pedro Sarmiento. *Viajes Al Estrecho de Magallanes Por el Capitan Pedro Sarmiento de Gamboa, En los Annos de 1579. y 1580. Y Noticia de la Expedicion Que despues hizo para poblarle*, 1768.

Gaskill, Melissa. "Space Station Research Shows that Hardy Little Space Travelers Could Colonize Mars," *nasa.gov*, May 2, 2014. Accessed November 26, 2016.

Gaustad, Edwin S. *Church and State in America*: 2nd Ed. Oxford University Press, 2003.

Ghose, Tia. "'Genetic Adam & Eve': Chromosome Study Traces All Men to Man Who Lived 135,000 Years Ago." *Huffingtonpost.com*. August 1, 2013. Accessed August 2, 2013.

Giovino, Mariana. *The Assyrian Sacred Tree: A History of Interpretations*. Academic Press Fribourg Vandenhoek & Ruprecht Gottingen, 2007.

Goble, Phillip E. *The Orthodox Jewish Bible*. Afi International Publishers, 2002.

Goldsmith, Donald. *The Hunt for Life on Mars*. Dutton, 1997.

Good, Timothy. *Above Top Secret: The Worldwide UFO Coverup*. Quill, 1989.

Goodall, Doming (ed. trans.). *Hindu Scriptures*. University of California Press, 1996.

Goodspeed, Edgar J. (trans.) *The Apocrypha: an American Translation*. Vintage Books, A Division of Random House, Inc., 1938.

Goodstein, Laurie, et al. "U.N. Panel Criticizes the Vatican Over Sexual Abuse." *Nytimes.com*, February 5, 2014. Accessed Jan 17, 2015.

Graf, Fritz. *Apollo*. Routledge, 2009.
———. "Zeus," *Dictionary of Deities and Demons in the Bible*: 2nd Edition. van der Toorn, Karel, (ed.) et al. Brill, W. B. Eerdman's Publishing, 1995 [reprint 1999].

Grant, Michael. *Herod the Great*. American Heritage Press, 1971.

Graves, Robert. *The Greek Myths*: The Complete and Unabridged version. Acorn Alliance, 1955-08.

Gray, John. *Near Eastern Mythology: Library of the World's Myths and Legends*. Peter Bedwick Books, 1969.

Greene, Brian. *The Hidden Reality: Parallel Universes and the Deep Laws of the Cosmos*. Alfred A. Knopf, 2011.

Greenfield, Jonas Carl. *Al Kanfei Yonah: Collected Studies of Jonas C. Greenfield on Semitic Philology*, Vol. 1. Brill, 2001.
———. "Apkallu," *Dictionary of Deities and Demons in the Bible*: 2nd Edition. van der Toorn, Karel, (ed.) et al. Brill, W. B. Eerdman's Publishing, 1995 [reprint 1999].

Greenstein, Edward L. "Sages with a Sense of Humor: The Babylonian Dialogue Between a Master and His Servant and the Book of Qohelet." Clifford, Richard J. (ed.) *Wisdom Literature in Mesopotamia and Israel*. Society of Biblical Literature, No.36, 2007.

Greer, Steven M. *Hidden Truth Forbidden Knowledge: It is Time for You to Know*. ZTT Consulting, 2013.

Grendler, P. F. *The Universities of the Italian Renaissance*. Johns Hopkins University Press, 2002.

Greyson, Bruce. "Seeing Dead People Not Known to Have Died: 'Peak in Darien' Experiences." *Anthropology and Humanism*. Vol. 35, Iss. 2, 2010.
———. "Western Scientific Approaches to Near-Death Experiences." *Humanities*, Vol. 4, 2015.

Grof, Stanislov. *Realms of the Human Unconscious: Observations from LSD Research*. Souvenir Press, 1975.

Grossman, Maxine. Adele Berlin. *The Oxford Dictionary of the Jewish Religion*. Oxford University Press, 2011.

Gruenwald, Ithamar. *Apocalyptic and Merkavah Mysticism*: 2nd Revised Ed. Brill, 2014.

———. *Rituals and Ritual Theory in Ancient Israel*. Brill, 2012.

Guarini, Filippo. *I Terremoti a Forli: In Varie Epoche*. Stabilimento, 1880.

Guirand, Felix (ed.). *New Larousse Encyclopedia of Mythology*. Aldinton, Richard (trans.), Ames, Delano (trans.). Crescent Books, 1987.

Gwendolyn, Leick. *A Dictionary of Ancient Near Eastern Mythology*. Routledge, 1998.

"Hacker Fears 'UFO cover-up'." *bbc.co.uk*. May 5, 2006. Accessed September 15, 2016.

Haines, Gerald K. "CIA's Role in the Study of UFOs, 1947-1990: A Die Hard Issue" *Cia.gov.* April 14, 2007 (Updated June 27, 2008). Accessed December 24, 2013.

Hale, Chris (writer and producer), *Atlantis Reborn Again*. BBC Horizon, December 14, 2000.

Hamblin, Dora Jane. "Has the Garden of Eden been located at last?" *Smithsonian Magazine*, Vol. 18, No. 2, May, 1987.

Hancock, Graham. *Fingerprints of the Gods*. Three Rivers Press, 1995.

Handwerk, Brian. "Scientist Replicated 100 Psychology Studies, and Fewer than Half Got the Same Results: The Massive Project Shows that Reproducibility Problems Plague even Top Scientific Journals." *Smithsonianmag.com*, August 27, 2015. Accessed August 14, 2016.

Harrington, Daniel J. *The Gospel of Matthew*. Liturgical Press, 1991.

Harris, Richard. *Rigor Mortis: How Sloppy Science Creates Worthless Cures, Crushes Hope, and Wastes Billions*. Basic Books, 2017.

Hassig, Ross. *Time, History, and Belief in Aztec and Colonial Mexico*. University of Texas Press, 2013.

Hastings, James, et al. *Encyclopedia of Religion and Ethics*, Vol. 6, Charles Scribner & Sons, T&T Clark, 1914.

Healey, John F. "Dagon," "Mot," *Dictionary of Deities and Demons in the Bible*: 2nd Edition. van der Toorn, Karel, (ed.) et al. Brill, W. B. Eerdman's Publishing, 1995 [reprint 1999].

Hearn, Kelly. "Who Built the Great City of Teotihuacan?" *science.nationalgeographic.com*, 2015. Accessed January 8, 2017.

Heil, John F. "mental causation." Audi, Robert (general ed.), Paul Audi (associate ed.). *The Cambridge Dictionary of Philosophy: Third Edition*. Cambridge University Press, 1995 [reprint 2015].

Helmolt, Hans Ferdinand (ed.). *The World's History: A Survey of Man's Record*. Vol. 1. William Heinemann, 1901.

Hendel, Ronald S. "Serpent," *Dictionary of Deities and Demons in the Bible*: 2nd Edition. van der Toorn, Karel, (ed.) et al. Brill, W. B. Eerdman's Publishing, 1995 [reprint 1999].

Herrmann, Wolfgang. "El," *Dictionary of Deities and Demons in the Bible*: 2nd Edition. van der Toorn, Karel, (ed.) et al. Brill, W. B. Eerdman's Publishing, 1995 [reprint 1999].

Hinnels, John R. (ed.) *Dictionary of Religions*. Facts on File, 1984.

———. *Mithraic Studies: Proceedings of the First International Congress of Mithraic Studies*, Vol. I. Manchester University Press, 1975.

"History of Europe." *Encyclopedia Britannica. Britannica.com*. 2015. Accessed Jan 27, 2015.

Hoeller, Stephan A. *Gnosticism: New Light on the Ancient Tradition of Inner Knowing*. Quest Books, 2002.

Holmyard, E.J. *Alchemy*. Dover Publications, 1957, [reprint 1990].

Holye, Fred. *The Intelligent Universe: A New View of Creation and Evolution*. Holt, Rinehart and Winston, 1984.

Hopkins, Budd. *Intruders*. Ballantine Books, 1987.

Huntley, Noel. *ETs and Aliens: Who Are They? And Why Are They Here?* 2002.

Hutter, Manfred. "Heaven," *Dictionary of Deities and Demons in the Bible*: 2nd Edition. van der Toorn, Karel, (ed.) et al. Brill, W. B. Eerdman's Publishing, 1995 [reprint 1999].

Huxley, Aldous. *The Perennial Philosophy*. Perennial Library, Harper & Row Publishers, 1944.
———. *The Doors of Perception: Heaven and Hell*. Fontal Lobe, 2011.

Hyman, Arthur, Alfred L. Ivry. *Maimonidean Studies, Vol. 5*. KTAV Publishing House, Inc. 2008.

Ioannidis, John P.A. "An Epidemic of False Claims: Competition and Conflicts of Interest Distort too many Medical Findings." *Scientificamerican.com*, June 1, 2011. Accessed August 14, 2016.

Irwin, Robert. *The Alhambra*. Profile Books, 2005.

Jackson, Holbrook. *The Anatomy of Bibliomania*. University of Illinois Press, 2001.

Jacobs, David M. *Secret Life: Firsthand Accounts of UFO Abductions.* Simon & Schuster, 1992.

Jacobsen, Thorkild. *The Harps That Once... : Sumerian Poetry in Translation.* Yale University Press, 1987.

———. *The Treasures of Darkness: A History of Mesopotamian Religion.* Yale University Press, 1976/1978.

James, Edwin Oliver. *The Tree of Life: An Archaeological Study.* Brill, 1966.

James, M.R., (trans.). *The Apocryphal New Testament.* Oxford, Carendon Press, 1924.

Jannot, Jean-Rene. *Religion in Ancient Etruria.* The University of Wisconsin Press, 2005.

Janowski, Bernd. "Azazel," *Dictionary of Deities and Demons in the Bible*: 2nd Edition. van der Toorn, Karel, (ed.) et al. Brill, W. B. Eerdman's Publishing, 1995 [reprint 1999].

Jastrow, Morris Jr. *Aspects of Religious Belief and Practice in Babylonia and Assyria.* The Knickerbocker Press, 1911.
———. *Civilization of Babylonia and Assyria.* J.B. Lippincott Co., 1915.
———. *Dictionary of the Targumim, the Talmud Babli and Yerushalmi, and the Midrashic Literature.* Vol. II. W.C.: Luzac & Co., G.P. Putnam's Sons, 1903.

"Jewish Encyclopedia: cherub," *Jewishencyclopedia.com*, 2002-2011, original publication 1906. Accessed March 18, 2014.

Kaku, Michio. *Hyperspace: A Scientific Odyssey Through Parallel Universes, Time Warps, and the Tenth Dimension.* Oxford University Press, 1994.
———. *The Future of the Mind: The Scientific Quest to Understand Enhance, and Empower the Mind.* Doubleday, 2014.

———. *Visions: How Science Will Revolutionize the 21st Century*. Anchor, 1998.

Kamen, Henry. *The Spanish Inquisition: A Historical Revision*. Yale University Press, 1999.

Karmin, Monika, et al. "A Recent Bottleneck of Y Chromosome Diversity Coincides with a Global Change in Culture." *Genome Research*, Vol. 25, Cold Spring Harbor Press, 2016.

Kasser, Rodolphe (ed.), et al. *The Gospel of Judas: from Codex Tchacos*. National Geographic Society, 2006.

Kean, Leslie. *UFOs: Generals, Pilots, and Government Officials Go on the Record*. Harmony Books, 2010.

Kellens, Jean. "Haoma," *Dictionary of Deities and Demons in the Bible*: 2nd Edition. van der Toorn, Karel, (ed.) et al. Brill, W. B. Eerdman's Publishing, 1995 [reprint 1999].

Kelly, Christopher. *The End of Empire: Attila the Hun & the Fall of Rome*. W.W. Norton & Co., 2010.

Kelly, Edward, et al. *Irreducible Mind: Toward a Psychology for the 21st century*. Rowman & Littlefield Publishers, 2007

Kelly, Emily Williams, et al. "Can Experiences Near Death Furnish Evidence of Life After Death?" *Omega: Journal of Death and Dying*. Vol. 40, No. 4, 1999-2000.

Kerenyi, C. Karl. Ralph Manheim (trans.). *Eleusis: Archetypal Image of Mother and Daughter*. Bollingen Series LXV, Vol. 4, Pantheon Books, 1967; Princeton University Press, 1991.

Khan, Amina. "Is Kepler-452b an Earth Twin? More Like a Bigger Older Cousin." *Latimes.com*, July 23, 2015. Accessed November 11, 2015.

"Kidnapped by UFOs: Interview with John Mack, Psychiatrist Harvard University." *www.pbs.org*. 1996. Accessed December 23, 2015.

King, Robert C., et al. *A Dictionary of Genetics*. Oxford University Press, 2006.

Kirsch, Johann Peter. "Stedingers." *The Catholic Encyclopedia*. Vol. 14. New York: Robert Appleton Co., 1912.

Kirsch, Jonathan. *The Grand Inquisitor's Manual: A History of Terror in the Name of God*. Harper One, 2008.

Kiviat, Robert C. (producer). *Aliens on the Moon: The Truth Exposed*. Robert Kiviat Productions, Inc., 2014.

Klotz, Irene. "Subatomic calculations indicate finite lifespan of the universe." www.reuters.com. February 18, 2013. Accessed February 15, 2017.

Knauf, Ernst. A. "Shadday," *Dictionary of Deities and Demons in the Bible*: 2nd Edition. van der Toorn, Karel, (ed.) et al. Brill, W. B. Eerdman's Publishing, 1995 [reprint 1999].

Kormendy, John. Ralf Bender. "Correlations Between Supermassive Black Holes, Velocity Dispersions, and Mass Deficits in Elliptical Galaxies with Cores." *The Astrophysical Journal*, Vol. 691, No. 2. February 1, 2009.

Kornfield, Jack (ed.). *Teachings of the Buddha: Revised and Expanded Edition*. Barnes and Noble Books, 1993.

Kramer, Samuel Noah. *Cradle of Civilization: Great Ages of Man, A History of the World's Cultures*. Time-Life Books, 1978.
———. *History Begins at Sumer: Thirty-Nine Firsts in Man's Recorded History*. 3rd Ed. University Press, 1981.
———. "Reflections of the Mesopotamian Flood: The Cuneiform Data New and Old." *Expedition Magazine*, Vol 9, Iss. 4, Penn Museum, July 1967.

———. *The Sumerians: Their History, Culture and Character.* University of Chicago Press. 1963.

Kramrisch, Stella. *The Presence of Siva.* Princeton University Press, 1981. Motilal Banarsidass, [reprint 1988].

Krauss, Lawrence M. "Lawrence M. Krauss; A Universe from Nothing; Radcliffe Institute," Uploaded by Harvard University, *Youtube.com* (July 17, 2013). Accessed January 9, 2017.

Lagrange, Pierre. "Close Encounters of the French Kind: The Saucerian Construction of "Contacts" and the Controversy over Its Reality in France." Tumminia, Diana G. (ed.) *Alien Worlds: Social and Religious Dimensions of Extraterrestrial Contact.* Syracuse University Press, 2007.

Lambert, W.G. Millard, A.R. *Atra Hasis: the Babylonian Story of the Flood.* Eisenbrauns, 1999.

Landis, Geoffrey A. "Colonization of Venus."American Institute of Physics, Conference. Proc. 654,1193, February 2003. *Scitation.aip.org,* January 28, 2003. Accessed February 23, 2013.

Layton, Bentley. *The Gnostic Scriptures: A New Translation with Annotations and Introductions by Bentley Layton.* The Anchor Bible Reference Library. Doubleday, 1987.

Leary, Timothy. *The Politics of Ecstasy.* Ronin Publishing, 1998.

Lechler, Gotthard Victor. *John Wycliffe and his English Precursors.* The Religious Tract Society, 1904.

Leeming, David. *The Oxford Companion to World Mythology.* Oxford University Press, 2005.

Le Goff, Jacques. *Medieval Civilization.* Blackwell Publishing, 1991.

Leick, Gwendolyn. *A Dictionary of Ancient Near Eastern Mythology.* Routledge, 1991.

Levack, Brian P. *The Witch Hunt in Early Modern Europe,* 3rd Ed. Longman, 2006.

Lewis, Theodore J. "Dead," "First Born of Death," *Dictionary of Deities and Demons in the Bible*: 2nd Edition. van der Toorn, Karel, (ed.) et al. Brill, W. B. Eerdman's Publishing, 1995 [reprint 1999].

Livingstone, Alasdair. "Nergal," *Dictionary of Deities and Demons in the Bible*: 2nd Edition. van der Toorn, Karel, (ed.) et al. Brill, W. B. Eerdman's Publishing, 1995 [reprint 1999].

Lock, Peter. *Routledge Companion to the Crusades.* Routledge, 2006.

Loud, Llewellyn Lemont. Mark Raymond Harrington. *Lovelock Cave*, Vol. 25, University of California Press, 1929.

Luckenbill, Daniel David. *Ancient Records of Assyria and Babylonia: Vol. II.* University of Chicago Press, 1972.

Luther, Martin. *Three Treatises: From the American Edition of Luther's Works.* 2nd Revised Edition, Fortress Press, 1970.

Macchi, M. Mila. Jeffery M. Bruce. "Human pineal physiology and functional significance of melatonin." *Front Neuroendocrinol.* Vol. 25 (3–4), 2004.

MacDonald, David. "The Flood: Mesopotamian Archaeological Evidence." Vol. 8, No. 2, Spring, 1988. *Hide Creation Evolution Journal. Ncse.com.* Accessed February 15, 2014.

Machiavelli, Niccolo. *The Prince and Selected Discourses*: *A New Translation by Daniel Donno.* Bantam Books, 1966.

———. *Selected Political Writings*. David Woofton (ed., trans.), Hacker Publishing Company, Inc., 1994.

Mack, Burton L. *The Lost Gospel: The Book of Q and Christian Origins*. Harper, 1993.

Mack. John E. *Abduction: Human Encounters with Aliens*.Scribner, 2007 [Reprint 1994].
———. *Passport to the Cosmos: Human Transformation and Alien Encounters*. Crown Publishers, 1999.

Marciniak, Barbara. *Bringers of the Dawn: Teachings from the Pleiadians*. Bear & Co., 1992.

Maslow, Abraham H. *Religion, Values, and Peak Experiences*. Penguin Books, 1964.

Matthews, Victor H. Don C. Benjamin. *Old Testament Parallels: Laws and Stories from the Ancient Near East*. Paulist Press, 2006.

Mayer, Jerry. John P. Holms (ed.). *Quotations on just about Everything from the Greatest Minds of the Twentieth Century.* St. Martin's Press, 1996.

Mayor, Adrienne. *Fossil Legends of the First Americans*, Princeton University Press, 2005.
———. William A.S. Sarjeant. "The Folklore of Footprints in Stone: from Classical Antiquity to the Present." *Ichnos: An International Journal for Plant and Animal Traces*, Vol. 8, Pt. 2, 2001.

Mcintosh, Gregory. *The Piri Reis map of 1513.* University of Georgia Press, 2000.

McKenna, Terence. *Food of the Gods: The Search for the Original Tree of Knowledge. A Radical History of Plants, Drugs, and Human Evolution.* Bantam, 1993a.

———. *The Invisible Landscape: Mind, Hallucinations, and the I Ching.* Harper One, 1975.

———. *True Hallucinations: Being an Account of the Author's Extraordinary Adventures in the Devil's Paradise.* Harper One, 1993b.

McMillan, Duncan, et al. *Société Rencesvals, Guillaume d'Orange and the chanson de geste: essays presented to Duncan McMillan in celebration of his seventieth birthday by his friends and colleagues of the Société Rencesvals*, University of Reading, 1984.

McNally, Richard J., et al. "Psychophysiological Responding During Script-Driven Imagery in People Reporting Abduction by Space Aliens." *Psychological Science*, Vol. 15 (7), 2004.

Meador, Betty De Shong. *Inanna Lady of Largest Heart: Poems of the Sumerian High Priestess Enheduanna.* University of Texas Press, 2000.

Meier, Samuel A. "Angels." *Dictionary of Deities and Demons in the Bible*: 2nd Edition. van der Toorn, Karel, (ed.) et al. Brill, W. B. Eerdman's Publishing, 1995 [reprint 1999].

Meissner, Bruno, et al. *Reallexikon der Assyriologie.* Walter de Gruyter, 1983.

Mellink, Machteld J. *Troy and the Trojan War: A Symposium Held at Bryn Mawr College, October 1984.* Bryn Mawr College, 1986 [reprint 1999].

Mettinger, T.N.D. "Cherubim," "Seraphim," "Yahweh Zebaoth," *Dictionary of Deities and Demons in the Bible*: 2nd Edition. van der Toorn, Karel, (ed.) et al. Brill, W. B. Eerdman's Publishing, 1995 [reprint 1999].

"Michael Servetus." *Encyclopedia Britannica, Britannica.com*, 2015. Accessed January 21, 2015.

Midelfort, H.C. Erik. *Witch Hunting in Southwestern Germany 1562–1684.* Stanford University Press, 1972.

Milbrath, Susan. *Heaven and Earth in Ancient Mexico: Astronomy and Seasonal Cycles in the Codex Borgia*. University of Texas Press, 2013.

———. *Star Gods of the Maya: Astronomy in Art, Folklore and Calendars*. University of Texas Press, 1999.

Miles, Jack. *God a Biography*. Alfred A. Knopf, 1995.

Miller, Mary K. "Reading Between the Lines: Scientists with High-Tech Tools are Deciphering Lost Writings of the Ancient Greek Mathematician Archimedes." *Smithsonianmag.com*. March 2007. Accessed January 2, 2015.

Mirabilis Annus Secundus (The Second Year of Prodigies), 1662.

Missler, Chuck. Mark Eastman. *Alien Encounters*. Koinonia House, 1997.

Montgomery, David R. *The Rocks Don't Lie: A Geologist Investigates Noah's Flood*. W. W. Norton & Co., 2012.

Moody, Raymond A. Jr. *Life After Life: The Investigation of a Phenomenon—Survival of Bodily Death*. Harper One, 1975.

Moran, William L. (ed., trans.) *The Amarna Letters*. Johns Hopkins University Press, 1992.

Morton, John. *Natural History of Northamptonshire: With Some Account of the Antiquities*. 1712.

Moskowitz, Clara. "Hundreds of New Exoplanets Validated by Kepler Telescope Team." *Scientificamerican.com*. February 26, 2014. Accessed February 27, 2014.

Moussaieff, Arieh, et al. "Incensole acetate, and incense component, elicits psychoactivity by activating TRPV3 channels in the brain." The FASB Journal. August 2008. Accessed April 15, 2015.

Munnich, Maciej. *The God Resheph in the Ancient Near East*. Tubingen: Mohr Siebeck, 2013.

Mussies, Gerard. "Giants," *Dictionary of Deities and Demons in the Bible*: 2nd Edition. van der Toorn, Karel, (ed.) et al. Brill, W. B. Eerdman's Publishing, 1995 [reprint 1999].

Nash, C. B. "Test of Psychokinetic Control of Bacterial Mutation," *Journal of the American Society for Psychical Research*, Vol. 78, 1984.

National Science Foundation. "Astronomers Discover Link Between Supermassive Black Holes and Galaxy Formation." *Nsf.gov*, February 2009. Accessed November 2013.

Newman, William R. (ed.). Grafton, Anthony (ed.). *Secrets of Nature: Astrology and Alchemy in Early Modern Europe*. The MIT Press, 2006.

Nicholson, H. B. *Topiltzin Quetzalcoatl: The Once and Future Lord of the Toltecs*. University Press of Colorado, 2001.

Niehr, Herbert. "Host of Heaven," "God of Heaven," *Dictionary of Deities and Demons in the Bible*: 2nd Edition. van der Toorn, Karel, (ed.) et al. Brill, W. B. Eerdman's Publishing, 1995 [reprint 1999].

Niemi, Tina M., et al. *The Dead Sea: The Lake and Its Setting*. Oxford Monographs on Geology and Geophysics (Book 36). Oxford University Press, 1997.

Nigg, Walter. *The Heretics: Heresy Through the Ages*. Dorset Press, 1990.

Nolland, John. *The Gospel of Matthew: A Commentary on the Greek Text*. Eerdmans, 2005.

O'Donnell, James J. *The Ruin of the Roman Empire: A New History*. Ecco, 2009.

O' Flaherty, Wendy Doninger (trans.). *The Rig Veda: An Anthology of One Hundred Eight Hymns*. Penguin Books, 1981.

Olivelle, Patrick (trans.). *Upanishads*. Oxford University Press, 1996.

Orr, James (ed.). *The International Standard Bible Encyclopaedia, Volume 4*. Howard Severance Company, 1915.

Overbye, Dennis. "Two Promising Places to Live, 1,200 Light-Years From Earth. *Nytimes.com*, Space and Cosmos. April 2013. Accessed May 2013.

Pagels, Elaine. *Revelations: Visions, Prophecy, & Politics in the Book of Revelation*. Viking, 2012.
———. *The Gnostic Gospels*. Vintage Books, 1979 [reprint 1989].

Paoletti, John T., Gary M. Radke. *Art in Renaissance Italy*. 3rd Ed. Laurence King Publishing, 2005.

Parker, Simon B. "Sons of (the) God(s)," *Dictionary of Deities and Demons in the Bible*: 2nd Edition. van der Toorn, Karel, (ed.) et al. Brill, W. B. Eerdman's Publishing, 1995 [reprint 1999].

Parpola, Asko. "The Problem of the Aryans and the Soma: Textual-linguistic and archaeological evidence." *The Indeo-Aryans of Ancient South Asia: Language, Material Culture and Ethnicity*. Erdosy, George (ed.). Walter de Gruyter, 1995.

Parsche, Franz, Andreas Nerlich. "Presence of Drugs in Tissues of an Egyptian Mummy." *Fresenius Journal of Analytical Chemistry*. Vol. 352, 1995.

Patai, Raphael. *The Jewish Alchemists*. Princeton University Press, 1995.

Peet, Stephen D. (ed.) *The American Antiquarian and Oriental Journal*, Vol. 7, F.H. Revell, 1885.

Penchansky, David. *Twilight of the Gods: Polytheism in the Hebrew Bible.* Westminster John Knox Press, 2005.

Penrose, Roger. *The Emperor's New Mind: Concerning Computers, Minds, and the Laws of Physics.* Penguin, 1989.

Perry, Marvin, et al. *Western Civilization: Ideas, Politics, and Society.* 9th Ed. Houghton Mifflin Harcourt Publishing Co., 2008.

Peters, Edwards. *Torture.* Basil Blackwell, 1985.

Phillips, Johnathan. *The Crusades: 1095-1197.* 2nd Ed. Routledge, 2014.

Pike, Albert. *Morals and Dogma of the Ancient and Accepted Rite of Freemasonry: Prepared for the Supreme Council of the Thirty Third Degree for the Southern Jurisdiction of the United States.* L.H. Jenkins, Inc., 1950.

Poliakov, Leon. *The History of Anti-Semitism, Vol. 2.* University of Pennsylvania, 2003.

Pope, Marvin H. *El in the Ugaritic Texts. Vol. 2.* Brill Archive, 1955.

Powell, J.W. *Report on the Mound Explorations of the Bureau of Ethnology. The Twelfth Annual Report of the Bureau of Ethnology to the Secretary of the Smithsonian Institution,* 1890-'91. 1894.

Poznik David G., et al. "Sequencing Y Chromosomes Resolved Discrepancy in Time to Common Ancestor of Males Versus Females." *Science*, Vol. 341, No. 6145, August 2013.

Prance, Sir Ghillean, Mark Nesbitt. *The Cultural History of Plants.* Routledge, 2012.

Principe, Lawrence M. *The Aspiring Adept: Robert Boyle and His Alchemical Quest.* Princeton University Press, 1998.

Pritchard, James B. (ed). *Ancient Near Eastern Texts Relating to the Old Testament*. Princeton University Press, 3rd Ed., 1969.

Targ, Russell. *The Reality of ESP: A Physicist's Proof of Psychic Abilities*. Quest Books, 2012.

Radin, Dean. "Beyond Belief: Exploring Interaction Among Body and Environment," *Subtle Energies*, 2 (3), 1992.
———. *Entanglement: Extrasensory Experiences in a Quantum Reality*. Paraview Pocket Books, 2006.
———. *Supernormal: Science, Yoga, and the Evidence for Extraordinary Psychic Abilities*. Deepak Chopra Books, 2013.

Rainey, Anson F. *Stones for Bread: Archaeology versus History in Near Eastern Archaeology*. Vol. 64, No. 3, *The American Schools of Oriental Research*, 2001.

Randle, Kevin D. *Conspiracy of Silence: From Roswell to Project Bluebook – What the Government Doesn't Want You to Know About UFOs*. Harper Perennial, 1997.

Rannan-Eliya, Ravi P., Nishan De Mel. "Resource Mobilization in Sri Lanka's Health Sector." Department of Population & International Health, Harvard School of Public Health & Health Policy Programme, Institute of Policy Studies. *Hsph.harvard.edu*, 1997. Accessed December 28, 2014.

Rao, Nathan. "Is this picture a 'seed' sent to Earth by aliens? Scientists discover mysterious organism." *express.co.uk*, January 24, 2015. Accessed July 11, 2015.

Rassam, Hormuzd. *Ashur and the Land of Nimrod*. Curtz and Jennings, Eaton & Mains, 1897.
———. *Babylonian Cities: A Paper, with Comments by Professor Delitzsch and an Appendix by Mr. W. St. Chad Boscawen*. Nabu Press, (no date given).

Ratsch, Christian. *Marijuana Medicine: A World Tour of the Healing and Visionary Powers of Cannabis*. Inner Traditions/Bear & Co., 2001.

Regardie, Israel. *The Tree of Life: An Illustrated Study in Magic*. Llewellyn Publications, 2001.

Reiling, Jannes. "Holy Spirit," *Dictionary of Deities and Demons in the Bible*: 2nd Edition. van der Toorn, Karel, (ed.) et al. Brill, W. B. Eerdman's Publishing, 1995 [reprint 1999].

Report of the Thirtieth Meeting of the British Association for the Advancement of Science. John Murray, Albemarle Street, 1850.

Rhea, Karen, Nemet Nejat. *Daily Life in Mesopotamia*. Greenwood Press "Daily Life through History" Series, 1998.

Rhine, J.B. J.G. Pratt. *Parapsychology: Frontier Science of the Mind*. Charles C Thomas, 1957.

Ribichini, Sergio. "Melqart," *Dictionary of Deities and Demons in the Bible*: 2nd Edition. van der Toorn, Karel, (ed.) et al. Brill, W. B. Eerdman's Publishing, 1995 [reprint 1999].

Richards, William A. *Sacred Knowledge: Psychedelics and Religious Experiences*. Columbia University Press, 2016.

Riley Smith, Johnathan. *The Crusades: A History*. Yale University Press, 2005.
———. *The Crusades, Christianity, and Islam*. Columbia University Press, 2013.

Ringgren, Helmer. *Religions of the Ancient Near East*. John Sturdy (trans.), The Westminster Press, 1973.

Roberts, J.J.M. "Erra: the Scorched Earth." *Journal of Cuneiform Studies*. Vol. 24, 1971.

Robinson, James M. (ed.) *The Nag Hammadi Library in English*. Harper, 1988. [reprint: 4th Ed. Brill, 1996].

Robinson, Walter V., et al. "Scores of Priests Involved in Sex Abuse Cases: Settlements Kept Scope of Issue Out of Public Eye." *Boston.com*. January 31, 2002. Accessed January 15, 2015.

Ronnevig, Georg M. "Toward an Explanation of the 'Abduction Epidemic': The Ritualization of Alien Abduction Mythology in Therapeutic Settings." Tumminia, Diana G. (ed.) *Alien Worlds: Social and Religious Dimensions of Extraterrestrial Contact*. Syracuse University Press, 2007.

Rosen, Edward. Erna Hilfstein (ed.) *Copernicus and his Successors*. Hambledon Press, 1995.

Rosenburg, David, Harold Bloom. *The Book of J*. Grove Weidenfeld, 1990.

Roth, Cecil. *The Spanish Inquisition*. W.W. Norton & Co., 1996.

Rouillard, Hedwige. "Rephaim," *Dictionary of Deities and Demons in the Bible*: 2nd Edition. van der Toorn, Karel, (ed.) et al. Brill, W. B. Eerdman's Publishing, 1995 [reprint 1999].

Ruiz, Ana. *The Spirit of Ancient Egypt*. Algora Publishing, 2001.

Rutersworden, Udo. "King of Terrors," *Dictionary of Deities and Demons in the Bible*: 2nd Edition. van der Toorn, Karel, (ed.) et al. Brill, W. B. Eerdman's Publishing, 1995 [reprint 1999].

Sagan, Carl. *The Dragons of Eden*. Random House, 1977.

Saler, Benson. "Secondary Beliefs and the Alien Abduction Phenomenon." Tumminia, Diana G. (ed.) *Alien Worlds: Social and Religious Dimensions of Extraterrestrial Contact*. Syracuse University Press, 2007.

Schiller, Ronald. "The African Cradle of the Human Race." St. Petersburg Times, August 1973.

———. "New Findings on the Origin of Man." Reader's Digest, August 1973.

Schneemelcher, Wilhelm. *The New Testament Apocrypha: Volume Two: Writings Related to the Apostles; Apocalypses and Related Subjects.* Rev. Ed. James Clarke & Co. Ltd. and Westminster/John Knox Press, 1989 [English translation 1992].

Scholz, Bernhard Walter (trans.). *Carolingian Chronicles: Royal Frankish Annals and Nithard's Histories.* The University of Michigan Press, 1972.

Schott, Siegfried. *Bücher und Sprüche gegen den Gott Seth, Urkunden des ägyptischen Altertums, sechste Abteilung.* Heft 1, 1929.

Schrodter, Willy. *A Rosicrucian Notebook: The Secret Sciences Used by Members of the Order.* Weiser Books, 1992.

Schultes, Richard E. "Antiquity of the Use of New World Hallucinogens." *The Heffter Review of Psychedelic Research*, Vol. 1, 1998.
———. *Hallucinogenic Plants.* Golden Press, 1976.
———. *Man and Marijuana.* Natural History Magazine, Vol. 82, 1973.
———. Albert Hofmann, Christian Ratsch. *Plant of the Gods: Their Sacred, Healing, and Hallucinogenic Powers.* Healing Arts Press, 1979 [reprint 2001].

Schwartz, Stephan A. "Six Protocols, Neuroscience, and Near Death: An Emerging Paradigm Incorporating Nonlocal Consciousness," *Explore: The Journal of Science and Healing.* Vol. 11, No. 4, 2015.

Scott, George Ryley. *The History of Torture Throughout the Ages.* Oakley Press, 2010.

Scribner, Scott R. "Alien Abduction Narratives and Religious Contexts." Tumminia, Diana G.(ed.) *Alien Worlds: Social and Religious Dimensions of Extraterrestrial Contact*. Syracuse University Press, 2007.

Sheldrake, Rupert. *The Sense of Being Stared At: And Other Unexplained Powers of Human Minds*. Park Street Press, 2003 [reprint 2015].

Sher Tinsley, Barbara. *Pierre Bayle's Reformation: Conscience and Criticism on the Eve of the Enlightenment*. Susquehanna University Press, 2001.

Shiga, David. "Could Black Holes be Portals to other Universes?" *Newscientist.com*, April 2007. Accessed November 15, 2013.

Shklovskii, I.S., Carl Sagan. *Intelligent Life in the Universe*. Dell, 1966.

Shuker, Karl. *Unexplained*. Carlton Books, 2003.

Silverberg, Robert. *Mound Builders of Ancient America: Archaeology of a Myth*. New York Graphic Society, 1968.

"Simony" *Encyclopedia Britannica, Britannica.com*, 2015. Accessed January 21, 2015.

Singer, Isidore (ed.), et al. *The Jewish Encyclopedia*, Funk and Wagnalls Co. 1901-1905.

Sinha, Indra. *The Great Book of Tantra: Translations and Images from the Classic Indian Texts*. Park Street Press, 1993.

Sitchin, Zecheria. *The Cosmic Code: Book VI of the Earth Chronicles*. Avon Books, 1998.
———. *Divine Encounters: A Guide to Visions, Angels, and Other Emissaries*. Avon Books, 1995.
———. *Genesis Revisited: Is Modern Science Catching Up with Ancient Knowledge?* Avon Books, 1990.

———. *The Lost Realms: The Fourth Book of the Earth Chronicles Series*. Avon Books, 1990.

———. *Stairway to Heaven: Book II of the Earth Chronicles*. Avon, 1980.

———. *The 12th Planet*. Avon Books, 1978.

———. *The Wars of Gods and Men: Book Three of the Earth Chronicles Series*. Avon Books, 1985.

———. *When Time Began: Book Five of the Earth Chronicles*. Avon Books, 1993.

Sleigh, Robert C. "Leibniz." Audi, Robert (general ed.). *The Cambridge Dictionary of Philosophy: Third Edition*. Cambridge University Press, 1995 [reprint 2015].

Slezak, Michael. "Ghost Universes Kill Schrodinger's Quantum Cat." *www.newscientist.com*. November 5, 2014. Accessed October 9, 2015.

Smith, Mark S. *The Early History of God: Yahweh and the Other Deities in Ancient Israel*. 2nd Ed. Wm. B. Eerdman's Publishing Company, 1990 [reprint 2002].

———. *The Origins of Biblical Monotheism: Israel's Polytheistic Background and the Ugaritic Texts*. Oxford University Press, 2001.

Spade, Paul Vincent, et al. "Medieval Philosophy." *The Stanford Encyclopedia of Philosophy*, *Plato.stanford.edu.*, 2013. Accessed January 22, 2015.

Speigal, Lee. "WATCH: Cloud-Like UFO Glides Through Philadelphia (Not New York)," Huffingtonpost.com. July 10, 2015. Accessed September 6, 2015.

Speirs, Kayleigh. North American Mounds. *UMASA Journal*, Vol. 32, 2014.

Steiger, Brad. *The Fellowship*. Dolphin Doubleday, 1988.
———. *Worlds Before Our Own*. G.P. Putnam and Sons, 1978.

Steiner, Rudolf. *Christianity and the Occult Mysteries of Antiquity*. Steiner books, 1978.

———. *Christianity as Mystical Fact and the Mysteries of Antiquity*. Steiner Books: Affiliate of Multimedia Publishing Corp., 1961.

Stone, Joshua David. *Hidden Mysteries: ETs, Ancient Mystery Schools and Ascension*. Light Technology Publishing, 1995.

Stonehill, Paul. *The Soviet UFO Files*. Quadrillion Publishing, 1998.

Storm, Lance, et al. "Meta-Analysis of Free-Response Studies, 1992-2008: Assessing the Noise Reduction Model in Parapsychology." *Psychological Bulletin*, Vol. 136, No. 4., 2010.

Strange, James F. "Observations from Archaeology and Religious Studies on First Contact and ETI Evidence." Tumminia, Diana G. (ed.) *Alien Worlds: Social and Religious Dimensions of Extraterrestrial Contact*. Syracuse University Press, 2007.

Strassman, Rick. *DMT and the Soul of Prophecy: A New Science of Spiritual Revelation in the Hebrew Bible*. Park Street Press, 2014.

———. *DMT: The Spirit Molecule: A Doctor's Revolutionary Research into the Biology of Near-death and Mystical Experiences*. Park Street Press, 2001.

Sturm, Allan. *ULOs: Unidentified Lunar Objects Revealed in NASA Photography*. Lunomaly Research Group, 2008.

Sturrock, Peter A. *The UFO Enigma: The First Major Scientific Inquiry Since the Condon Report*. Warner Book, 1999.

Suzuki, Daisetz Teitaro. *On Indian Mahayana Buddhism*. Harper & Row, 1968.

Szalavitz, Maia. "Magic Mushrooms Expand the Mind By Dampening Brain Activity." *Healthland.time.com*. January 2012. Accessed July 2013.

Targ, Russell. *The Reality of ESP: A Physicist's Proof of Psychic Abilities.* Quest Books, 2012.

Tart, Charles T. "Towards an Evidence-based Spirituality: Some Glimpses of an Evolving Vision." *Subtle Energies & Energy Medicine*, Vol. 20, No.2.

Teddler, W.H., M.L. Mont. "Exploration of a long-distance PK: a conceptual replication of the influence on a biological system," *Research in Parapsychology*, 1980.

Than, Ker. "Every Black Hole Contains Another Universe?: And our Universe May sit in Another Universe's Black Hole, Equations Predict." *Nationalgeographic.com*, April 2010. Accessed November 2013.
———. "Major Discovery: New Planet Could Harbor Water and Life." *Space.co.* April 2007. Accessed April 2013.

"The Extraordinary Equation of George Van Tassel." *KVOS Channel 12 Films. Center for Pacific Northwest Studies, Western Washington University.* June 18, 1964. Film. You Tube video, posted by Western Washington University.

Thijssen, Hans. "Condemnation of 1277." Edward N. Zalta, ed. *The Stanford Encyclopedia of Philosophy, plato.stanford.edu.*, 2013. Accessed Jan 23, 2015.

Thompson, C.J.S. *Lure and Romance of Alchemy.* Kessinger Publishing, 2003.

Thorndike, Lynn. "Renaissance or Prenaissance?" *Journal of the History of Ideas,* 4, 1943.

Torrey, Charles C. "A Hebrew Seal from the Reign of Ahaz." Bulletin of the American Schools of Oriental Research, No. 79, 1940.

Trench, Brinsley Le Poer. *The Flying Saucer Story.* Ace Star Book, 1966.

Tucker J.B. *Life Before Life: A Scientific Investigation of Children's Memories of Previous Lives.* St. Martin's Press, 2005.

Tumminia, Diana G. (ed.) *Alien Worlds: Social and Religious Dimensions of Extraterrestrial Contact.* Syracuse University Press, 2007.

Valantasis, Richard. *The Gospel of Thomas.* Routledge, 1997.

Vallee, Jacques. *Forbidden Science: Journals 1957-1967.* North Atlantic Books, 1993.
———. Vallee, Jacques. Aubeck, Chris. *Wonders in the Sky: Unexplained Aerial Objects from Antiquity to Modern Times: and Their Impact on Human Culture, History, and Beliefs.* Jeremy P. Tarcher/Penguin, 2009.

van den Broek, Roelof. "Apollo," *Dictionary of Deities and Demons in the Bible*: 2nd Edition. van der Toorn, Karel, (ed.) et al. Brill, W. B. Eerdman's Publishing, 1995 [reprint 1999].

van der Toorn, Karel. *Family Religion in Babylonia, Syria, and Israel: Continuity and Change in the Forms of Religious Life.* Brill, 1996.
———. *Scribal Culture: and the Making of the Hebrew Bible.* Harvard University Press, 2007.
———. "Yahweh," "God (I)," *Dictionary of Deities and Demons in the Bible*: 2nd Edition. Brill, W. B. Eerdman's Publishing, 1995 [reprint 1999].

Van Seters, John. *The Pentateuch: A Social Science Commentary.* Sheffield Academic Press, 1999.
———. *Prologue to History: The Yahwehist as Historian in Genesis.* Westminster/Knox Press, 1992.

Van Tassel, George W. *I Rode a Flying Saucer: The Mystery of the Flying Saucers Revealed.* 2nd Edition. New Age Publishing Co., 1952.
———. *The Council of Seven Lights.* DeVorss & Co., 1958.

Verner, Miroslav. *The Pyramids: The Mystery, Culture, and Science of Egypt's Great Monuments.* Grove Press, 1997 [reprint 2001].

Wall, Mike. "Rover Finds New Evidence That Ancient Mars Was Habitable." *Space.com*. June 2013. Accessed November 2013.

Wasson, R. Gordon. *Soma: Divine Mushroom of Immortality*. Harcourt Brace Jovanovich, 1972.

Watelin, L.C. *Excavations at Kish*. Vol. IV, P. Geuthner, 1934.

Watson, Rita. Wayne Horowitz. "Writing Science Before the Greeks: A Naturalistic Analysis of the Babylonian Astronomical Treatise MUL.APIN." *Culture and History of the Ancient Near East*. Vol. 48. Weippert, M. H. E. (ed.), et al. Brill, 2011.

Watson, Wilfred (ed.), Nicolas Wyatt (ed.). *Handbook of Ugaritic Studies*. Brill, 1999.

Watts, Edward J. *City and School in Late Antique Athens and Alexandria*. University of California Press, 2006.

Weadock, Penelope N. "The Giparu at Ur." *Jstor. Iraq* - Vol. 37, No. 2, 1975.

Wheelwright, C.A. (trans.), Thomas Bourne (trans.). *Pindar and Anacreon*. Harper & Brothers, 1837.

White, Michael L. "Understanding the Book of Revelation." PBS.org. (Frontline) 1999. Accessed May 9, 2013.

Wickramasinghe, Chandra, Gensuke Tokoro. "Life as Cosmic Phenomenon: The Socio-Economic Control of a Scientific Paradigm. *Astrobiology & Outreach*, Vol. 2, Iss. 2, June 27, 2014.

Wiggermann, F.A.M. *Mesopotamian Protective Spirits the Ritual Texts: Cuneiform Monographs I*, Styxx and P.P. Publications, 1992.

Wilke, Annette. Oliver Moebus. *Sound and Communication: An Aesthetic Cultural History of Sanskrit Hinduism.* Walter de Gruyter, 2011.

Wilkins, Harold T. *Flying Saucers on the Attack.* Citadel Press, 1954.

Wolfram, Herwing. *The Roman Empire and Its Germanic Peoples.* University of California Press, 1997.

Wood, Simon McGregor. "Moses Was High On Drugs, Israeli Researcher Says." *Abcnews.go.com.* March 2008. Accessed April 2013.

World Book Encyclopedia. H Volume 9. World Book, A Scott Fetzer Company. 2015.

Wright, Robert. *The Evolution of God.* Little Brown & Co., 2009.

Wyatt, N. *Religious Texts from Ugarit.* 2nd Edition. Sheffield Academic Press, 2002.

Xella, Paolo. "Resheph," *Dictionary of Deities and Demons in the Bible*: 2nd Edition. van der Toorn, Karel, (ed.) et al. Brill, W. B. Eerdman's Publishing, 1995 [reprint 1999].

Yandell, Keith E. "Shiva," The Cambridge Dictionary of Philosophy: *Third Edition.* Cambridge University Press, 1995 [reprint 2015].

Yosef, Uri. "Who is the Suffering Servant in 'Isaiah 53?': Part I—The Jewish Interpretation: Valid or Not?" The Messiah Truth Project, 2001-2011. thejewishhome.org/counter/Isa53CP.pdf

Zimmer, Heinrich. Joseph Campbell (ed.). *Myths and Symbols in Indian Art and Civilization*, Bollingen Series VI, Princeton University Press, 1972.

Zohar, Vol 5. trans. Harry Sperling, Maurice Simon, and Paul Levertoff. Soncino Press, 1934.

Zukav, Gary. *The Dancing Wu Li Masters: An Overview of the New Physics*. Harper One, 1979.

Zwiebach, Barton. *A First Course in String Theory*. Cambridge University Press, 2009.

ADDITIONAL INTERNET SOURCES:

archive.org
www.biblegateway.com
biblehub.com/strongs.htm
www.catholic.org/encyclopedia
etcsl.orinst.ox.ac.uk
gnosis.org
www.gutenberg.org
jewishencyclopedia.com
www.jstor.org
lexiconcordance.com
www.newadvent.org
plato.stanford.edu
www.sacred-texts.com
www.scripture4all.org

ABOUT THE AUTHOR

Aerik Vondenburg is a former associate member and contributing writer for The Center for Progressive Christianity (progressivechristianity.org), and former member of The Society of Biblical Literature. He studied at the Rhine Education Center (a division of the Rhine Research Center, Institute for Parapsychology at Duke University). His articles have appeared in the Consortiumnews.com, Spirituality-and-community.com, and the Esoteric Christianity E-Magazine. He is also a former member of the Church of Christ and former featured blogger at Crossleft.com. He was inspired to research this book after not only experiencing paranormal phenomenon but after personally witnessing a UFO in 1998.

www.ingramcontent.com/pod-product-compliance
Lightning Source LLC
Chambersburg PA
CBHW071732150426
43191CB00010B/1543